Plant Genetic Resources and Food Security

Issues in Agricultural Biodiversity

This series of books is published by Earthscan in association with Bioversity International. The aim of the series is to review the current state of knowledge in topical issues associated with agricultural biodiversity, to identify gaps in our knowledge base, to synthesize lessons learned and to propose future research and development actions. The overall objective is to increase the sustainable use of biodiversity in improving people's well-being and food and nutrition security. The series' scope is all aspects of agricultural biodiversity, ranging from conservation biology of genetic resources through social sciences to policy and legal aspects. It also covers the fields of research, education, communication and coordination, information management and knowledge sharing.

Published titles:

Crop Wild Relatives
A Manual of *in situ* Conservation
Edited by Danny Hunter and Vernon Heywood

The Economics of Managing Crop Diversity On-farm
Case Studies from the Genetic Resources Policy Initiative
Edited by Edilegnaw Wale, Adam Drucker and Kerstin Zander

Plant Genetic Resources and Food Security
Stakeholder Perspectives on the International Treaty on
Plant Genetic Resources for Food and Agriculture
Edited by Christine Frison, Francisco López and José T. Esquinas-Alcázar

Forthcoming:

Crop Genetic Resources as a Global Commons
Challenges in International Law and Governance
Edited by Michael Halewood, Isabel López Noriega and Selim Louafi

Farmers' Crop Varieties and Farmers' Rights
Challenges in Taxonomy and Law
Edited by Michael Halewood

Plant Genetic Resources and Food Security

Stakeholder Perspectives on the International Treaty on Plant Genetic Resources for Food and Agriculture

Edited by Christine Frison, Francisco López and José T. Esquinas-Alcázar

Published by

The Food and Agriculture Organization of the United Nations
and
Bioversity International
with
Earthscan

London • New York

First published 2011
by FAO, Bioversity International and Earthscan
2 Park Square, Milton Park, Abingdon, Oxon OX14 4RN

Simultaneously published in the USA and Canada
by FAO, Bioversity International and Earthscan
711 Third Avenue, New York, NY 10017

Earthscan is an imprint of the Taylor & Francis Group, an informa business

Earthscan publishes in association with the International Institute for Environment and Development

British Library Cataloguing in Publication Data
A catalogue record for this book is available from the British Library

Library of Congress Cataloging in Publication Data

Frison, Christine.
Plant genetic resources and food security : stakeholder perspectives on the international treaty on plant genetic resources for food and agriculture / Christine Frison, Francisco López and José T. Esquinas-Alcazar.
p. cm.
Includes bibliographical references and index.
1. Germplasm resources, Plant. 2. Crops—Germplasm resources. 3. Food crops—Germplasm resources. 4. Plant genetic engineering. 5. Food security. I. López, Francisco. II. Esquinas-Alcazar, J. T. (José T.) III. Title.
SB123.3.F75 2011
338.1'9—dc23

2011008922

ISBN: 978-1-84971-205-7 (hbk)
ISBN: 978-1-84971-206-4 (pbk)
FAO ISBN: 978-92-5-106482-5

Typeset by MapSet Ltd, Gateshead, UK
Cover design by Adam Bohannon

Printed and bound in the UK by CPI Antony Rowe.
The paper used is FSC certified.

To our children, Clara and Théodore,
who represent our Future

Contents

Part II: Perspectives on the Treaty by Stakeholders in the World Food Chain

List of Figures, Tables and Boxes

Figures

Tables

Boxes

Contributors

Editors

Christine Frison is a lawyer specializing in biodiversity, agro-biodiversity and biosafety issues. She is currently conducting a PhD research as junior affiliated researcher at the Université catholique de Louvain and at the Katholieke Universiteït Leuven (Belgium) on international law and governance of plant genetic resources for food and agriculture. Her PhD thesis is entitled "Towards Redesigning the Plant Commons: A Critical Assessment of the Multilateral System of Access and Benefit-Sharing of the International Treaty on Plant Genetic Resources for Food and Agriculture". She holds an LL.B. (Université Montpellier I, France, 2002), an LL.M. in International and Trade Law (Université Lyon I, France, 2003), and an LL.M. in Public International Law (Université Libre de Bruxelles – ULB, Belgium, 2004). Since 2004, she has been a research fellow at the Centre for International Sustainable Development Law (CISDL, based at McGill University, Montreal, Canada) and carries out consultancy contracts for various international organizations (including the United Nations Environment Programme), NGOs and governments (e.g. Belgian Public Federal Service for the Environment). She is part of the Belgian access and benefit-sharing contact group and of the Belgian contact group on PGRFA. She has been part of the Belgian delegation to the meetings of the ITPGRFA since 2006.
Email: christine.frison@uclouvain.be

Francisco López is a journalist, lawyer and political analyst and holds a master's degree in communication and education. He is currently serving at the Secretariat of the International Treaty on Plant Genetic Resources for Food and Agriculture where he has been coordinating the planning and the production of official documents for the regular sessions of the Governing Body since 2007. He joined FAO in 2002 and he has worked as information officer in the Knowledge Exchange and Communication Department and in the Right to Food Unit of the Economic and Social Development Department. He has occupied several positions in the Research and Extension Unit of FAO and coordinated numerous technology transfer projects in direct support of food security in Honduras, Costa Rica and Uganda, among other countries. Before joining FAO, he worked for the Spanish News Agency EFE in Rome and in Madrid as well as for COAG,

a farmers' organization in Spain. He has written several papers on food security, poverty alleviation and technology transfer for smallholders.
Email: francisco.lopez@fao.org

Prof. Dr José T. Esquinas-Alcázar is currently the Director of the Chair on Studies on Hunger and Poverty (CEHAP) of the University of Cordoba, and full professor at the Polytechnic University of Madrid. From 1985 until his retirement from FAO in January 2007, Prof. Esquinas has been the Secretary of the FAO Commission on Genetic Resources for Food and Agriculture, where the International Treaty was negotiated. From 1999 to 2007 he was the Chair of the FAO Committee on Ethics for Food and Agriculture. He was also the interim secretary of the treaty since its entry into force in 2004 to 2007. Prof. Esquinas holds a PhD in Genetics (University of California, Davis, USA); Doctorate in Agronomy (Polytechnic University of Madrid, Spain); MSc in Vegetable Crops (University of California, Davis, USA); and Master's Degree in Agricultural Engineering (Polytechnic University of Madrid, Spain). He has authored numerous books and publications and has received many international awards.
Email: jose.esquinas@upm.es

Authors

Jan Borring is a senior adviser in the International Department of the Ministry of Environment in Norway. He participated in the negotiations on the International Treaty on PGRFA from the outset and was also involved in the preparations for and establishment of the Svalbard Global Seed Vault. He has also been involved in policies relating to trade and environment, including in processes dealing with the relationship between intellectual property rights and genetic resources. His background is in biology and chemistry, University of Oslo.

Dr Pratibha Brahmi is a senior scientist at the National Bureau of Plant Genetic Resources (NBPGR), New Delhi. Holding a PhD from the University of Delhi in botany, she has worked at NBPGR for the past 25 years and has been involved mainly in germplasm exchange and policy issues. She has contributed towards the establishment of a plant variety protection system in India. She was a member of various national expert committees for the implementation of relevant provisions of the Biological Diversity Act 2002 (BDA) of India related to germplasm exchange. She also attended the 3rd Governing Body meeting of the ITPGRFA held in Tunisia in June 2009 as part of the Indian delegation.

Lidio Coradin works for the Brazilian Agriculture Research Corporation, linked to the Ministry of Agriculture since 1974 and was the Research Director of the Genetic Resources and Biotechnology Research Centre (Cenargen/Embrapa) between 1987 and 1989. Since 1994 he has also worked for the Ministry of the Environment where he is in charge of the Genetic Resources Division and works for the implementation of actions related to in situ, ex situ and on-farm genetic resources conservation and sustainable utilization. He is also responsible for the implementation of the National Agrobiodiversity Program and for all activities related to invasive species. He negotiated the Convention on Biological Diversity (1989–1992), the International Treaty on Plant Genetic Resources for Food and Agriculture (1995–2001) and the Cartagena Protocol on Biosafety (1996–2000) for Brazil, and follows their implementation. He received his master's degree, in 1978, from the Herbert H. Lehman College of the City University of New York.

Dr Carlos Maria Correa is Director of the Center for Interdisciplinary Studies on Industrial Property and Economics and of the postgraduate course on intellectual property at the Law Faculty, University of Buenos Aires. He has been a visiting professor in postgraduate courses of several universities and consultant to UNCTAD, UNIDO, UNDP, WHO, FAO, IDB, INTAL, the World Bank, SELA, ECLA and other regional and international organizations. He has advised several governments on intellectual property and innovation policy. He was a member of the UK Commission on Intellectual Property, of the Commission on Intellectual Property, Innovation and Public Health established by the World Health Assembly and of the FAO Panel of Eminent Experts on Ethics in Food and Agriculture. He is the author of several books and numerous articles.

Prof. Dr José Ignacio Cubero is Professor (emeritus) of the University of Córdoba (Spain) and an associated member of the Instituto de Agricultura Sostenible (Sustainable Agriculture Institute) of the National Council for Scientific Research (CSIC). He holds a PhD in Agronomy (1969) and a PhD in Biology (1973). He was Professor of Genetics and Plant Breeding, University of Córdoba (1974–2009) and has been a correspondent member of the Royal Academy of Sciences, Spain, since 2002. He was Director (Dean) of the Escuela Técnica Superior de Ingenieros Agrónomos y de Montes (ETSIAM) (Faculty of Agronomy and Forestry), University of Córdoba, from 1987 to 1991, Head of the Department of Genetics, University of Córdoba, 1974–1986 and 1995–2001, and President of the Spanish Society of Genetics, 1999–2002. He was Chairman of the Board of Trustees of ICARDA (International Center for Agricultural Research in Dry Areas, a CGIAR Center), 1986–1989, and member of that Board 1984–1989. He has been chairman and member of several international and national commissions and committees, and also Major Professor of 31 PhD theses including 2 European doctorates. He has been Supervisor of 22 national and 5 international master's theses. He is author or co-author of 374 publications (scientific, divulgation, essays and books).
Email: ge1cusaj@uco.es

Dr Tewolde Berhan Gebre Egziabher served as Dean of Science in Addis Ababa University (1974–1978) and President of Asmara University (1983–1991). Starting in 1989, he developed the conservation strategy and environmental policy of Ethiopia and established the Environmental Protection Authority, of which he became the Director General (1995 to present). He wrote the main elements of Community Rights, which become the African Model Law in 1998. Since 1991, he has led African environmental negotiators in Agenda 21, the Convention on Biological Diversity, the Cartagena Protocol on Biosafety, the Convention to Combat Desertification, and the International Treaty on Plant Genetic Resources for Food and Agriculture. He served as a member of the Interim Panel of Experts that established the Crop Diversity Trust. He is currently a member of the steering committee of the FAO High Level Panel of Experts on Food Security and Nutrition. He received the Right Livelihood Award (2000), Honorary Doctor of Science from Addis Ababa University (2004) and Champion of the Earth Award from UNEP (2006).

Dr Modesto Fernández Díaz-Silveira is an agronomist, from the University of Havana, and Doctor in Agricultural Sciences (PhD), from the Academy of Sciences of Cuba. Currently he is a senior officer for environment with the Department of Environment in the Ministry of Science, Technology and Environment in Cuba. From 1994 he has been involved, as a Cuban negotiator, in the negotiations that led to the adoption of the International Treaty on PGRFA. After the adoption of the Treaty he was elected consecutively, the Vice-chair of the 1st, 2nd and 3rd meetings of the Governing Body, and in 2008 was elected the Chair of the ITPGRFA for the 3rd meeting of the Governing Body, until 2010. He was the Vice-chair, representing GRULAC (Group of Latin America and Caribbean Countries), for the 12th regular session of the FAO Commission on Genetic Resources for Food and Agriculture and currently is the Vice-chair for the 13th session of the same commission. Formerly he was the Head of the INIFAT Gene Bank in Cuba, and has been linked to several processes of the Convention on Biological Diversity (CBD) regarding genetic resources. Senior Officer for Environment, Department of Environment, Ministry of Science, Technology and Environment, 20 and 18A, Playa, Havana, Cuba.
Email: modesto@citma.cu

Dr Cary Fowler is the Executive Director of the Global Crop Diversity Trust. Prior to that, he was Professor and Director of Research in the Department for International Environment and Development Studies at the Norwegian University of Life Sciences. He was also a senior advisor to the Director General of Bioversity International. In 1985 he was awarded the Right Livelihood Award (the 'Alternative Nobel Prize'). Amongst many other things, he headed the International Conference and Programme on Plant Genetic Resources at the Food and Agriculture Organization of the United Nations (FAO) in the 1990s, which produced the UN's first State of the World's Plant Genetic Resources. He drafted and supervised negotiations of FAO's Global Plan of Action for Plant Genetic Resources, adopted by 150 countries in 1996. Cary has been profiled by CBS *60 Minutes* and the *New Yorker*, is the author of several books on the subject of plant genetic resources and more than 75 articles on the topic in agriculture, law and development journals. Cary earned his PhD at the University of Uppsala (Sweden), and

in 2008 received an honorary doctorate from Simon Fraser University (Canada). In 2010, the Russian Academy of Agricultural Sciences awarded him the Vavilov Medal for his 'exceptional contribution' to the cause of conserving plant genetic resources for present and future generations.
Email: cary.fowler@croptrust.org

Dr Brad Fraleigh was trained in plant genetic resources at the Université d'Orsay (Paris-Sud) in France. He was Canada's chief negotiator during the negotiation of the Treaty. He represented North America as a Vice-chair of the Commission, chaired the Commission's decision to launch negotiations in 1993, and co-chaired the first extraordinary negotiating session. He continued as Vice-chair during much of the negotiations, and chaired contact groups on scope and on financial resources issues.
Email: brad.fraleigh@agr.gc.ca

Dr Emile Frison is Director General of Bioversity International (formerly IPGRI). He has spent most of his career in international agricultural research for development, starting at the International Institute for Tropical Agriculture (IITA) in Nigeria in 1979. A plant pathologist by training, Dr Frison recently led Bioversity, its stakeholders and partners in the formulation of a new strategic vision in which nutrition and agricultural biodiversity will play an important role in the overall goal of reducing hunger and poverty in a sustainable manner. He also leads the CGIAR System-wide Genetic Resources Policy Programme and is a member of the CGIAR Genetic Resources Policy Committee, and the Board of Directors of Eco-agriculture Partners. In 2006, he joined the Comité d'Orientation de l'Agence de Recherche pour le Développement, Paris. In 2007, he was appointed as Extraordinary Professor in genetic resources by the Catholic University of Leuven, Belgium. Dr Frison is a member of the International Advisory Council of the Svalbard Global Seed Vault and member of the Executive Board of the Global Crop Diversity Trust. He is author and co-author of over 150 scientific publications and is a member of several scientific societies. He obtained an MSc in plant pathology from the Catholic University of Louvain and a PhD from the University of Gembloux in Belgium.
Email: e.frison@cgiar.org

Ambassador Fernando Gerbasi is a Venezuelan diplomat. He was Vice-minister of Foreign Affairs, Ambassador to FAO (three years), Italy (two years), Colombia (two years), Brazil, German Democratic Republic, European Communities and the United Nations in Geneva. He was Chairman to the FAO Commission on Plant Genetic Resources for Food and Agriculture, 1997–2002, during which period he chaired the negotiations of the International Treaty, and was Chairman of the Interim Panel of Eminent Experts for the establishment of the Global Crop Diversity Trust, December 2002 to April 2007. He currently is Professor for International Affairs at the Universidad Metropolitana, Caracas, Venezuela.

Dr Bryan L. Harvey is Professor Emeritus of Plant Sciences, University of Saskatchewan. He served as head of the Crop Science and Plant Ecology Department, Director of the Crop Development Centre, and Vice-president Research of the University. He chaired Canada's national expert committee on plant genetic resources. Dr Harvey represented the seed sector, and chaired contact and expert groups during the negotiation of the Treaty. He chaired the First Session of the Interim Committee for the Treaty and chaired one of the two working groups at the first meeting of its Governing Body.
Email: bryan.harvey@usak.ca

Dr Geoffrey Hawtin is a British/Canadian plant breeder/geneticist. He was Director General of the International Plant Genetic Resources Institute (IPGRI, now Bioversity International, Rome) from 1991 to 2003, founding CEO of the Global Crop Diversity Trust (Rome) 2003–2005 and is currently its Senior Advisor, and was the Director General of the Centro International de Agricultura Tropical (CIAT, Colombia), 2008–2009. Geoff obtained his PhD from Cambridge University, UK, and has also worked in Uganda, Lebanon, Egypt, Syria and Canada. He currently serves on the boards of CATIE (Costa Rica) and the Royal Botanical Gardens, Kew, and was awarded the 2005 Frank N. Meyer Medal for Plant Genetic Resources.
Email: geoff.hawtin@cropturst.org

Cosima Hufler is Senior Advisor on International Environmental Affairs in the Austrian Federal Ministry of Agriculture, Forestry, Environment and Water Management, with a particular focus on matters related to access to genetic resources and the fair and equitable benefit-sharing arising out of their use, and is currently Chair of the Bureau of the 4th session of the Governing Body of the International Treaty on Plant Genetic Resources for Food and Agriculture.

Dr René Lefeber is Legal Counsel in the International Law Division of the Netherlands Ministry of Foreign Affairs, holds a chair in International Environmental Law in the Faculty of Law of the University of Amsterdam and is Visiting Professor at the United Nations University, Institute of Advanced Studies. He completed his PhD at the University of Amsterdam.

Dr Eng Siang Lim, a Malaysian national, previously held the positions of: Chairman (10th session: 2005–2006), Commission on Genetic Resources for Food and Agriculture (CGRFA), Food and Agriculture Organization of the United Nations (FAO), Rome, Italy; Chairman of the contact group for the drafting of the standard material transfer agreement under the ITPGRFA; Chairman of the intergovernmental technical working group on PGRFA, CGRFA (1st and 2nd sessions); Vice-chair of CGRFA (7th, 8th and 9th sessions); and Honorary Research Fellow with the Policy Research and Support Unit, supporting genetic resources policy in the Asia, the Pacific and Oceania region, in particular, the implementation of the ITPGRFA. He was an officer in the Ministry of Agriculture, Malaysia, from 1982 to 2004 (22 years) participating in policy formulation including the national agriculture policy, the balance of trade action plan for the food sector, the investment incentives for agriculture, the master plan for the development of the agro-based industries, the agricultural skill training plan, the plant varieties protection act and the agriculture technology development plan.

Elizabeth Matos was born in London. She obtained a BSc Biology (London University) and a MSc in PGR Conservation and Utilisation (Birmingham University). Working first at the University of Zambia (1972–1975), she is a senior lecturer in genetics, evolution and biodiversity conservation at Agostinho Neto University, Angola (1975–2011). She conducts research in sunflower breeding and PGR conservation. Director of the Angolan National Plant Genetic Resources Centre (NPGR) at Agostinho Neto University, she is also chairperson of the NPGR Committee. Besides being a board member of the SADC regional PGR network, she participated in the FAO GRFA Commission from 1994 to 2008 and in all ITPGRFA negotiating meetings from 1996 to 2005 as a representative of Angola, including as a chairperson and one of several spokespersons of the African group.
Email: fitogen@ebonet.net and liz.matos35@gmail.com

Patrick Mooney has more than four decades experience working in international civil society, first addressing aid and development issues and then focusing on food, agriculture and commodity trade. In 1977, Mooney co-founded RAFI (Rural Advancement Fund International, renamed ETC Group in 2001). Pat Mooney received The Right Livelihood Award (the 'Alternative Nobel Prize') in the Swedish Parliament in 1985. In 1998 Mooney received the Pearson Peace Prize from Canada's Governor General. He also received the American 'Giraffe Award', given to people 'who stick their necks out'. The author or co-author of several books on the politics of biotechnology and biodiversity, Pat Mooney is widely regarded as an authority on agricultural biodiversity, new technologies and global governance issues.
Email: etc@etcgroup.org

Gerald Moore is currently an honorary fellow with Bioversity International, dealing primarily with the implementation of the International Treaty on Plant Genetic Resources for Food and Agriculture. Prior to that he was Legal Counsel of FAO from 1988 to 2000 and as such was closely involved in facilitating the Treaty negotiations. Since retiring from the FAO he has been a regular member of CGIAR observer delegations to the Treaty negotiations and sessions of the Governing Body of the Treaty. He is a barrister-at-law and has law degrees from Cambridge University and the University of California at Berkeley.
Email: g.moore@cgiar.org

Dr Javad Mozafari Hashjin earned his PhD from the University of Guelph, Canada, in plant molecular genetics. He is now Associate Professor and Director of the National Plant Gene-Bank, Agricultural Research, Education and Extension Organization (AREEO) of Iran. He has served in various national and international managing and policy making bodies in the area of agricultural biodiversity, genetic resources and plant breeding for more than ten years. Dr Mozafari is presently the elected Chair of the UN-FAO Commission on Genetic Resources for Food and Agriculture and also the Vice-chair for the Near East in the Bureau of International Treaty on Plant Genetic Resources for Food and Agriculture. He has represented Iran and the Near East region throughout the Treaty negotiations since June 2000 and chaired some of the very critical committees and working groups, such as the negotiation on the Annex I Crops and Third Part Beneficiary, during the process.

Godfrey Mwila is a Zambian national. Godfrey holds a masters degree in Plant Genetic Resources Conservation and Utilisation from the University of Birmingham, UK, and a bachelors degree in General Agricultural Sciences from the University of Zambia. Godfrey has been working in the area of plant genetic resources conservation at national, regional and international levels over the past 15 years, and therefore has extensive experience in both the technical and policy aspects of plant genetic resources conservation and utilization. Prior to taking up this position, Godfrey was the Chair of the Governing Body of the International Treaty on Plant Genetic Resources for Food and Agriculture for the 2nd and 3rd sessions.
Email: godfrey.mwila@croptrust.org; godfrey.mwila@gmail.com

Wilhelmina R. Pelegrina is the Executive Director of SEARICE, a regional non-governmental organization working with farmers and other stakeholders in strengthening farmers' management of their agricultural biodiversity. She was involved in developing the Farmers' Fields School approach for on-farm conservation and participatory plant breeding. She is also involved in policy, advocacy and campaigning work around new technologies in food and agriculture, intellectual property rights and farmers' rights. She was a neophyte in the Treaty negotiations, with her first involvement only during the 1st Governing Body meeting in 2006. Her background is in agriculture and environmental sciences.
Email: ditdit_pelegrina@searice.org.ph

Renato Salazar is a senior fellow of SEARICE. He participated in the negotiations of the Treaty from the outset, and was one of the few people from civil society to be part of the Philippine delegation during the negotiations for the Undertaking. He was also active in the development of the Global Plan of Action. He is involved in organizing farmers and other stakeholders on issues around agrarian reform, democratizing agricultural science and technology through participatory plant breeding and campaigning to reform institutions and agricultural policies.
Email: searice@searice.org.ph

Dr Maria José Amstalden Sampaio has worked for the Brazilian Agriculture Research Corporation linked to the Ministry of Agriculture since 1976 and was the Research Director of the Genetic Resources and Biotechnology Research Centre (Cenargen/Embrapa) between 1984 and 1996. She is now an advisor on intellectual property rights, biotechnology and genetic resources to the directors of Embrapa. In 2007, she moved to coordinate a group of policy makers who deal with negotiations and implementation of international agreements. Her interests concern policy making at national and international levels, related with agriculture (access to genetic resources, conservation and benefit sharing, biotechnology and genetically modified organisms (GMOs), intellectual property and climate change implications).

Cinzia Scaffidi is currently Director of the Slow Food Study Center and is responsible for International Relations at the University of Gastronomic Sciences of Pollenzo and Colorno. With a background in history and philosophy, she published some research in this area in the 1990s and has maintained a historical–philosophical approach when studying scientific issues. Before joining Slow Food in 1992, she worked as a journalist, and taught and worked in the area of international cooperation. She has also been in charge of the Slow Food Award for the Defence of Biodiversity and since 2004 has been one of the coordinators of the Terra Madre meeting.

Prof. Dr Shyam Kumar Sharma is the Vice-chancellor of CSK Himachal Pradesh Agricultural University, Palampur, Himachal Pradesh, a leading hill agricultural university located in the North-Western Himalayan region of India. Formerly, he was the Director of the National Bureau of Plant Genetic Resources, New Delhi, India and had national responsibility for the management of Indian Plant Genetic Resources. A geneticist by training, he was awarded Commonwealth post-doctoral and academic fellowships, Marie Curie fellowship and Indian National Academy Royal Society International collaborative award for higher training and research in various university/institutions in the United Kingdom. He was a member of several national and international committees concerning PGR and played a key role in planning and implementation of several national and international projects.

Dr Mary Taylor, with over 20 years experience in the Pacific region, is currently the Manager of the Centre for Pacific Crops and Trees (CePaCT), and Coordinator of the Genetic Resources programme within the Secretariat of the Pacific Community, an intergovernmental organization with 22 Pacific Islands countries and territories as members. The CePaCT is the Pacific region's gene bank and holds the largest in vitro collection of taro (Colocasia esculenta). In June 2009, the Pacific region placed the ex situ collections held in the CePaCT into the multilateral system of access and benefit-sharing (MLS) of the Treaty.

Ir Anke van den Hurk is a Senior Adviser at Plantum NL, the Dutch association for breeding, tissue culture, production and trade of seeds and young plants. She participated on behalf of the seed industry in the negotiations of the standard material transfer agreement (SMTA) and in the implementation of the International Treaty on PGRFA. Ms van den Hurk is also the representative of the International Seed Federation (ISF) in the negotiations of an international regime on access and benefit sharing under the CBD. In her daily work she is also involved in various policy areas, like breeders' rights biotechnology and organic seeds that are relevant for the plant breeding sector. Her background is in plant breeding, Wageningen University.

Dr Bert Visser was born in the Netherlands in 1951. Since 1997 he is the Director of the Centre for Genetic Resources the Netherlands (CGN) which, under its own mandate, is part of Wageningen University and Research Centre. As the Director of CGN he fulfils an advisory role for the Ministry of Agriculture, Nature and Food Quality on policies regarding agrobiodiversity. His interests and activities concern genetic resources management and policy development, international collaboration in the area of genetic resources management, on-farm conservation of genetic resources, and the interface of agrobiodiversity and biotechnology.

Acknowledgements

We would like to express our gratitude to a large number of people who contributed in different ways to the publication of this book.

First, we would like to highlight that this book would not have materialized without the expertise, knowledge, patience and goodwill of every author throughout this lengthy project. They represent all of those who made the Treaty possible, by stepping back from their positions to converge in common objectives of food security and poverty alleviation. These authors include: Shakeel Bhatti, Jan Borring, Pratibha Brahmi, Lidio Coradin, Carlos María Correa, José Ignacio Cubero, Olivier De Schutter, Tewolde Berhan Gebre Egziabher, Modesto Fernández Díaz-Silveira, Cary Fowler, Brad Fraleigh, Emile Frison, Fernando Gerbasi, Bryan Harvey, Geoffrey Hawtin, Cosima Hufler, René Lefeber, Eng Siang Lim, Elizabeth Matos, Patrick Mooney, Gerald Moore, Javad Mozafari Hashjin, Godfrey Mwila, Wilhelmina R. Pelegrina, Renato Salazar, Maria José Amstalden Sampaio, Cinzia Scaffidi, Shyam Kumar Sharma, Mary Taylor, Anke van den Hurk and Bert Visser. To all of you: Thank you!

We sincerely thank Bioversity International and the Food and Agriculture Organization (FAO) of the United Nations for their trust in this book project, which has allowed us to publish this volume within the book series 'Issues in Agricultural Biodiversity' co-published by Bioversity International and Earthscan. Special thanks are addressed to Isabel Lopez Noriega and to Michael Halewood for their constructive chapter reviews and for making our publication collaboration possible.

We would specifically like to thank all of those who contributed to the practical realization of this book. We received help from many colleagues at various levels: from providing constructive advice to valuable guidance, or even reviewing chapters in-depth. These persons include in particular: Fulya Batur, Carlos Correa, Elise Denoitte, Cary Fowler, Emile Frison, Yasmine Jouhari, Gerald Moore, François Pythoud, Matthias Sant'ana, Benjamin Six, Clive Stannard, Marie Schloen, Geoff Tansey, Bert Visser, Theo van Hintum and Esther van Zimmeren.

We thank Rachel Tucker, publishing, planning and rights officers of FAO for facilitating the partnership arrangements for this publication. Special thanks go to Paola Franceschelli, who has assisted us in double checking many meeting and document references. We also extend our recognition to the staff of the Secretariat of the International Treaty. We are deeply indebted to Désirée Khoury and Gabriella Petrilli, from FAO, for their enthusiastic and helpful technical assistance in the editing, formatting and proofreading of the draft manuscript. We also pay tribute to the staff at Earthscan for their support, in particular, Tim Hardwick, Anna Rice, Claire Lamont, Nick Ascroft and Rob West.

We would also like to recognize the financial support of Spain, Italy and the FAO, which will make possible the distribution of free copies to contracting parties to the International Treaty and to international reference centres. Similarly, this publication would not have been achieved without the administrative and financial assistance of several persons and institutions. First, the Centre de Philosophie du Droit (CPDR – Centre for Philosophy of Law), Université catholique de Louvain (Belgium) has borne the financial costs for the work of the main editor of the book, Christine Frison. Additionally, we appreciate the constant support of the Cátedra de Estudios sobre Hambre y Pobreza (CEHAP) of the Universidad de Córdoba (Spain), as well as the hospitality of the Diputación de Córdoba by providing a venue for one meeting organized between the co-editors to work on the book. Finally, the book was finalized during the Fourth Session of the Governing Body of the Treaty held in Bali in March 2011. We sincerely thank the generous financial contributions from Emile Frison and Professor Jacques Lenoble (CPDR, through the Interuniversity Attraction Pole funding (IAP) – Phase VI /06, 2007–2011) for funding the participation of Christine Frison to the Fourth Session of the Governing Body.

We deeply thank Mrs Frison's PhD supervisors Professors Tom Dedeurwaerdere, Olivier De Schutter (CPDR) and Geertrui Van Overwalle (CIR) for their overall support to the project.

This project came together during the International Year for Biodiversity in 2010. A major objective of this volume is to raise further public awareness on the importance of agricultural biological diversity to human food security. We hope that this book makes a contribution to that effort.

Last but not least, we were encouraged all through the process with the love, faith and practical support of our families, friends and even neighbours, in particular, when we were under tight time pressure and constraint to work during late nights. To everyone, we sincerely thank you!

Foreword

Shakeel Bhatti and Olivier De Schutter

The International Treaty on Plant Genetic Resources for Food and Agriculture (ITPGRFA or the Treaty) is all about building bridges and connecting countries and people; it is about pooling collaborative, cooperative and common action. The Treaty provides a framework to allow the global community to work together for food security, adaptation to climate change and the sound management of agrobiodiversity – always keeping in focus the needs of farming communities, the poor and the hungry, and their right to food. States interacting with other states, people interacting with other people, with institutions (whether public or private), with civil society organizations, with research institutes and with commercial entities create multilateralism through their interactions. People are at the core of multilateralism. And it is this kind of collective and cooperative action, oriented towards the attainment of common goals, that the global crises facing the 21st century requires.

This book intends to shed light on the institutional set up that took place during the negotiation process between contracting parties and people who made this Treaty possible. By aggregating their interests, these states have established a multilateral instrument aimed at alleviating hunger and poverty in the world. They embrace farming communities, plant breeders, civil society organizations, seed industry or state's representatives.

In 2009, this book was merely an embryonic project held in the hands of a young and enthusiastic woman, driven by her will to eagerly understand how this collective action came about, and led to the birth of the Treaty. At that moment in time, the United Nations Secretary General, Ban Ki-moon and the European Commission President, José Manuel Barroso both called for 'a new multilateralism which is centred around the delivery of global public goods' to address the interrelated crises of food, energy and climate. As the Secretary General articulated at the Summit of the Americas:

> *We need a new vision, a new paradigm, a new multilateralism. A multilateralism that is organized around delivering a set of global goods. A*

> *multilateralism that harnesses both power and principle. A multilateralism that recognizes the interconnected nature of global challenges.*

Today, the International Treaty on Plant Genetic Resources for Food and Agriculture embodies this new paradigm of collaboration in an interdependent world. In that respect, its lessons reach far beyond the food and agriculture sectors. This Treaty was the first of its kind in the 21st century and it remains at the cutting edge of such a new, results-driven and output-oriented multilateralism. Together, stakeholders have established the first system to facilitate multilateral management of global public goods for the 21st century. This system covers a global gene pool of more than 1.3 million samples of plant genetic material that contracting parties govern collectively and multilaterally for the sake of the poor and the hungry. Through this gene pool, the current 127 contracting parties to the Treaty control – and are responsible for – the basis of more than 80 per cent of the world's food that is derived from plants. Moreover, it is also our most important tool for adapting to climate change in agriculture in the years to come.

The Treaty: An expression of multilateralism

The Treaty first illustrates this new multilateralism in the realms of the multilateral system of access and benefit-sharing. This mechanism is based on a wide partnership, linking the Consultative Group on International Agricultural Research (CGIAR) centres and other national, regional and international institutions and gene banks to facilitate its implementation by contracting parties and users of the system. This multilateral system still raises important questions for the various actors involved: farmers who need to be assured that the seeds, which their communities have developed over generations, will benefit humanity and that they will in return have access to the seeds they need in their farming systems; holders of gene bank collections who need to be convinced that their collections will also benefit from facilitated exchange; users who want to ship seeds but whose legal counsels notify that they first need more information on the meaning of a particular clause in the Standard Material Transfer Agreement before the shipment can take place; researchers who worry about intellectual property rights over their research results; and finally breeding companies, which are willing to share benefits in accordance with the Treaty, but wish to be assured that they will not be accused of biopiracy.

The multilateral system has been designed for all of these various actors, providing a framework under which they can cooperate. The framework must balance the needs of these different stakeholders and ensure that they will interact in ways that are both transparent and adequate for their mutual benefit. This collaboration between them is the *sine qua non* condition for addressing the challenges that the world currently faces: climate change, population growth and persistent poverty, particularly in the rural areas among the small-scale food producers.

Multilateralism as promoted by the International Treaty does not stop there, with the provision of an appropriate framework for cooperation. It also finds a concrete illustration in the funding strategy accompanying the multilateral system. A first call for proposals under the benefit-sharing fund in 2009 led to the selection of the first 11 benefit-sharing projects in the history of plant genetic resources. The successful completion of this first test-run of benefit-sharing under the Treaty has proven that international benefit-sharing within a binding legal architecture can work on a multilateral basis. Under the framework of the Treaty, international benefit-sharing is now working in practice, on the ground, for those actors who conserve and contribute to the development of the plant genetic diversity that feeds us all. These actors include, for instance, the Andean farming community that conserves in situ old varieties of potato in its centre of origin; the African genetic resource centre that is struggling to adapt its national crops to climate change and ensure food security; the Asian NGO driven by a group of local women that is developing locally adapted cultivars for small scale enterprises to ensure local livelihoods; and the Near Eastern gene bank that is conserving on-farm and in vitro its rich local citrus varieties.

While the benefit-sharing fund is still in its infancy, it shall grow rapidly in the years to come. A second call for proposals made in 2010 has led to the selection of a larger number of projects after the Fourth Session of the Governing Body of the Treaty in Bali, in March 2011. In this way, the funding strategy has begun to effectively fulfil its potential to provide tangible support for the three priorities set at the Second Session of the Governing Body, namely on-farm conservation, sustainable use of plant genetic resources and information exchange. In implementing these priorities, special attention should be given to 'farmers in developing countries … who conserve and sustainably utilize plant genetic resources for food and agriculture', as stipulated in Article 18.5 of the Treaty. Thus, the Treaty can complete the virtuous circle of facilitating exchange and practically supporting the conservation and sustainable use of agricultural plant genetic resources, particularly by and for those people who have developed and conserved these resources over the ages.

By encouraging capacity-building, the Treaty offers a third example of a new breed of multilateralism suited to an interdependent world. The capacity-building of stakeholders in the conservation and development of plant genetic resources for food and agriculture is a crucial part of this collective endeavour. At the 2nd session of the Governing Body, contracting parties created a 'Capacity Building Coordinating Mechanism' to support the national implementation of the Treaty. Enhanced collaboration between FAO, Bioversity International and the Secretariat of the International Treaty on the one hand and new partnerships on the other hand, led to the establishment of a Joint Capacity Building Programme. This programme provides assistance to developing the policies, legislation and institutional and administrative practices and arrangements necessary to implement this instrument. The International Treaty has also to be able to provide a set of information technology tools and systems that help users to find the material included and to report on their obligations. Furthermore, contracting parties have also showed interest in developing Article 17 on global information systems, taking into account existing information systems, current trends and opportunities.

Fourth, the Treaty encourages collective learning and progress through peer pressure towards the fulfilment of the goals it sets for itself. This is clear, for instance, in the area of Farmers' Rights. The International Treaty recognizes:

> *the enormous contribution that the local and indigenous communities and farmers of all regions of the world, particularly those in the centres of origin and crop diversity, have made and will continue to make for the conservation and development of plant genetic resources which constitute the basis of food and agriculture production throughout the world.* (Art. 9.1.)

It refers to the responsibility of the contracting parties to realize Farmers' Rights, by (a) protecting traditional knowledge relevant to plant genetic resources for food and agriculture; (b) ensuring that farmers can equitably participate in sharing benefits arising from the utilization of plant genetic resources for food and agriculture; and (c) protecting their right to 'participate in making decisions, at the national level, on matters related to the conservation and sustainable use of plant genetic resources for food and agriculture' (Art. 9.2). While these provisions remain vague and their implementation uneven across member states, the 3rd session of the Governing Body held in June 2009, in Tunis, agreed that contracting parties should review all measures affecting Farmers' Rights and remove any barriers preventing farmers from saving, exchanging or selling seed; and that they should fully involve farmers in national and/or regional workshops on the implementation of Farmers' Rights and report back on the implementation of Farmers' Rights at the fourth meeting, held in Bali in March 2011. This should encourage states to fully implement Article 9 of the Treaty: it illustrates that, for collective action to succeed, it may have to rely on the sharing of experiences and of information, where agreement on a detailed and binding legal framework may not be achievable at the outset.

Finally, new multilateralism can be observed within the Treaty Secretariat that developed into a lean, nimble and dynamic institution which, under its parties' guidance, ensures a transparent management of the plant genetic resources defined as a new global public good. Multilateralism also means that the Secretariat should never attempt to substitute itself for stakeholders in the conservation and sustainable use of plant genetic resources. By creating outcome-oriented partnerships, new platforms for cooperation have been provided, so that the whole can be larger than any one input. By acting so, the Treaty has become a model of a forward-looking and dynamic management for the 21st century. It is a light and flexible structure, but it is probably better suited to the task rather than larger bureaucracies whose ability to evolve in a dynamic environment is generally more limited.

The Treaty in a changing world

The Treaty is also becoming a model for other international decision-making processes: for instance, other United Nations bodies, such as the World Health Organization in its process on virus-sharing and benefit-sharing; the Convention on Biological Diversity in the elaboration of its own international regime on access and benefit-sharing; and the United Nations Convention on Law of the Sea, with regard to the genetic resources of the deep sea-bed – all are looking at the Treaty as their reference point in crafting customized multilateral systems. These new regimes for international cooperation in the maintenance and shared use of global public goods form the vanguard of public international law, combining innovative legal frameworks and practical operational systems, for the global gene pool and for the support of conservation and sustainable use through the funding strategy. In the future, similar regimes could develop, for instance to ensure the transfer of clean technologies to developing countries to support them in their efforts to mitigate climate change or to facilitate the management of freshwater resources that is based on cooperation and trust, not competition and distrust.

Therefore, the Treaty community needs to keep in mind this bigger policy picture. This international legally binding instrument is more relevant than ever in the broader policy context. It is at the crossroads where many policy-making processes converge: conservation and the sustainable use of biodiversity; recognition of traditional knowledge; trade; sustainable economic growth and development; innovation policy and intellectual property; adaptation to climate change; and above all, food security and the moral imperative to feed a still growing and often unacceptably poor world population to ensure that their human right to adequate food can be guaranteed.

May the reader of this book recall that each and every stakeholder plays an important role in reaching the Treaty's objectives of conservation, sustainable use and facilitated access to and benefit-sharing of plant genetic resources for food and agriculture. It is these same actors' interactions, and their resulting collective action, that has allowed for the creation of such an innovative multilateral system designed to safeguard food security and alleviate rural poverty in the world. Trust between the stakeholders involved, both private and public, including both providers and users, is key to the system's harmonious functioning. This book should allow each set of actors to better understand the perspective of the other actors with whom they cooperate. Finally, we deeply thank the authors and editors for their generous, and gratuitous contributions to this volume.

Dr Shakeel Bhatti
Secretary of the Treaty on Plant Genetic Resources for Food and Agriculture

Prof Dr Olivier De Schutter
United Nations Special Rapporteur on the Right to Food

List of Acronyms and Abbreviations

ABS	Access and Benefit Sharing
AG	African group
ARIS	Agricultural Research Information System
ASSINSEL	International Association of Plant Breeders for the Protection of Plant Varieties
AU	African Union
BDA	Biological Diversity Act
BSI	Botanical Survey of India
CBD	Convention on Biological Diversity
CEHAP	Cátedra de Estudios sobre Hambre y Pobreza
CePaCT	Centre for Pacific Crops and Trees
CGIAR	Consultative Group on International Agricultural Research
CGN	Centre for Genetic Resources The Netherlands
CGRFA	Commission on Genetic Resources for Food and Agriculture
CIAT	International Centre for Tropical Agriculture
CIMMYT	International Maize and Wheat Improvement Centre
COAG	Committee on Agriculture
COMESA	Common Market for Eastern and Southern Africa
CSIRO	Commonwealth Scientific and Industrial Research Organisation
COP	Conference of the Parties
CSO	civil society organization
CWANA	Central and West Asia and North Africa region
DAC	Department of Agriculture and Cooperation
DAFF	Department of Agriculture, Fisheries and Forestry (Australia)
DARE	Department of Agriculture Research and Education
DUS	distinctiveness, uniformity and stability
EAC	East African Community
ECOWAS	Economic Community of West African States
ERG	European regional group
EU	European Union
FAO	United Nations Food and Agriculture Organization
FSM	Federated States of Micronesia
GCDT	Global Crop Diversity Trust
GB	Governing Body of the ITPGRFA
GIPB	Global Partnership Initiative for Plant Breeding Capacity Building

GMO	genetically modified organism
GPA	Global Plan of Action
GRIN	Genetic Resources Information Network
GRULAC	Group of Latin America and Caribbean Countries
HOAFS	Heads of Agriculture and Forestry Services
IARC	International Agricultural Research Center
IBPGR	International Board for Plant Genetic Resources
ICAR	Indian Council of Agricultural Research
ICCAI	International Climate Change Adaptation Initiative
ICDA	International Coalition for Development Action
ICRISAT	International Crops Research Institute for the Semi-Arid-Tropics
IITA	International Institute of Tropical Agriculture
INGO	international non-governmental organization
IPCC	Intergovernmental Panel on Climate Change
IPGRI	International Plant Genetics Research Institute
IPR	intellectual property rights
ISF	International Seed Federation
ITPGRFA	International Treaty on Plant Genetic Resources for Food and Agriculture
IU	International Undertaking on Plant Genetic Resources
LAC	Latin America and the Caribbean Region
LAN	Local Area Network
MDG	Millennium Development Goal
MLS	multilateral system of access and benefit-sharing
MoEF	Ministry of Environment and Forests
MTA	material transfer agreement
NAG	National Active Germplasm Site
NBPGR	National Bureau of Plant Genetic Resources
NGO	non-governmental organization
OAU	Organization of African Unity
ODA	Overseas Development Assistance
OECD	Organisation for Economic Co-operation and Development
PAPGREN	Pacific Plant Genetic Resources Network
PGR	plant genetic resources
PGRFA	plant genetic resources for food and agriculture
PPV&FR	Protection of Plant Varieties and Farmers' Rights Act
RAFI	Rural Advancement Foundation International
SADC	Southern African Development Community
SAU	state agricultural universities
SGRP	System-wide Genetic Resources Programme
SMTA	standard material transfer agreement
SPC	Secretariat of the Pacific Community
TLB	taro leaf blight
TRIPS	Trade-related Aspects of Intellectual Property Rights (Agreement on)
UNCED	United Nations Conference on Environment and Development
UNCTAD	United Nations Conference on Trade and Development

UNDP	United Nations Development Programme
UNEP	United Nations Environment Programme
UPOV	Union for the Protection of New Varieties of Plants
USDA	United States Department of Agriculture
WANA	West Asia and North Africa region
WIPO	World Intellectual Property Organization
WTO	World Trade Organization

'Omnium autem rerum, ex quibus aliquid acquiritur, nihil est agri cultura melius, nihil uberius, nihil dulcius, nihil homine libero dignus'

Cicero, *De Officiis*, I, 42-151

'But of all the occupations by which something is built up, none is better than agriculture, none rewards more, none is more pleasant, none is more worthy for a freeman'

(Personal translation)

Chapter 1

Introduction

A Treaty to Fight Hunger – Past Negotiations, Present Situation and Future Challenges

José T. Esquinas-Alcázar, Christine Frison and Francisco López[1]

This introduction provides readers with a general overview on the content and structure of the book, the context in which the major issues related to plant genetic resources for food and agriculture (PGRFA) emerged, its relevance for humankind and some interesting details of the negotiating and implementation process of the International Treaty on Plant Genetic Resources for Food and Agriculture (ITPGRFA – the Treaty). The authors have taken this opportunity to express their personal views on some of the major challenges ahead of the Treaty, which will be further developed in the concluding chapter of this volume.

About the book

This book touches upon wide-ranging issues, such as international food policies and governance, economic and social aspects of food and seed trade, conservation and sustainable use of agricultural biodiversity, hunger alleviation, ecological concerns, consumer protection, fairness and equity between nations and among generations, plant breeding techniques and climate change adaptation. It provides for an extensive overview of the ITPGRFA negotiating and implementation process, undertaken by the stakeholders themselves. The authors identified challenges faced by the ITPGRFA and its community of stakeholders during this new and exciting phase of implementation, and explained the different interests and views of the major players in the global food chain.

Chapters have been grouped into three parts. Part I provides the views and standpoints of a number of protagonists that were part of national delegations during the negotiating and implementation process. They stand for the seven regional groups of the Food and Agriculture Organization of the United Nations (FAO): Africa, Asia, Europe, Latin America and the Caribbean, Near East, North America and South West Pacific (Chapters 2 to 9). Part II brings together the opinions of key stakeholders involved in the food chain worldwide: farming communities, plant breeders, gene banks, the Consultative Group on International Agricultural Research (CGIAR), the Global Crop Diversity Trust, the seed industry, civil society organizations (CSOs) and consumers (Chapters 10 to 17). Finally, Part III puts forward the opinions of highly recognized experts regarding key aspects of the implementation of the Treaty (Chapters 18 to 20). Five annexes complement information on the ITPGRFA and its negotiation. Annex 1 lists the meetings held at the FAO Commission on Genetic Resources for Food and Agriculture for the negotiation of the Treaty (1983–2001), as well as the meetings that took place since the signature and entry into force of the Treaty (2002–2011). Annex 2 provides the list of all contracting parties to the Treaty, by FAO regional groups. Annex 3 details the main components of the Treaty. Annex 4 gives a national perspective on the implementation of the treaty by Brazil; while Annex 5 comes back to specific anecdotes from the inception of the Treaty negotiations which express well the atmosphere in which the discussions on an international instrument for PGRFA began.

With a concern for unity, the authors were requested to focus on specific issues, following essentially the guidelines below:

- Analyse the regions' and stakeholders' positions during the negotiation process and the early implementation phase.
- Analyse the merits and drawbacks of the Treaty.
- Examine the practical legal, political, environmental and economic issues that have arisen between all involved regions and stakeholders in the negotiation and implementation, focusing on the obstacles that have been overcome.
- Identify the main challenges ahead and summarize some of the options and views on how these could be met as already expressed by regions and stakeholders.

Given the nature of the book and the heterogeneity of stakeholders, their different interests and personalities, the chapters differ in style, content and conclusions. It has been the role of the editors to harmonize them, minimize the overlaps, make the appropriate cross-references and include tables, annexes and reference material, in an attempt to ease the book's consultation and use. Every contribution bears in common the invaluable output to provide crucial information on stakeholders' positions regarding the Treaty, information that has not yet been published elsewhere. The book shows that despite the conflicting interests, which are duly highlighted, all players manage to come to an agreement to share and help conserve PGRFA for the sake of global food security and hunger alleviation. This volume also assesses the prospects for an effective and rapid implementation of the Treaty, in some cases by rescuing some old aspirations that were left behind during the negotiation process and by tabling new ideas and innovative solutions.

World food context: Plant genetic resources, food security, sustainability and equity

States have repeatedly reiterated the fundamental right of everyone to be free from hunger and the right to adequate food. In 1996, world leaders stated that: 'We consider it intolerable that more than 800 million people throughout the world, and particularly in developing countries, do not have enough food to meet their basic nutritional needs. This situation is unacceptable' (Rome Declaration on World Food Security, 1996). This assertion led to more than just the inclusion of this fundamental human right within the international legal order as such. Indeed, these states committed to implement policies aimed at eradicating poverty and inequality while improving physical and economic access by all to sufficient, nutritionally adequate and safe food. They pledged to eradicate hunger in all countries, with an immediate view to reducing the number of undernourished people to half of their present level no later than 2015.[2] A similar commitment was made at the United Nations Millennium Summit in 2000, and is included in the First Millennium Development Goal (MDGs).

Despite these pledges, the situation has worsened. Today, hunger and malnutrition reaches almost 1000 million people. As a consequence, 15 million people die every year, that is to say, more than 41,000 every day, the majority of whom are children. In addition, the world population is expected to reach 8.3 billion by 2030 and the Earth will have to feed an additional two billion people, of whom 90 per cent come from developing countries (SoW2-PGRFA, 2010).[3] It is therefore crucial to ensure not only that enough food can be produced reliably to feed this expanding population, but also that it is accessible to all.

Within this context, one should recall that food security greatly depends on the conservation, exchange and wise use of agricultural biodiversity and the genetic resources that constitute such diversity. PGRFA are essential for sustainable agriculture and food production. They provide the building blocks for farmers, breeders and biotechnologists to develop new plant varieties necessary to cope with unpredictable human needs, growing food demands and changing environmental conditions.

From a socio-economic perspective, the importance of agriculture varies by region. Only 1.9 per cent of the population in North America is dependent on agriculture whereas this number reaches 50 per cent in Africa and Asia. Agricultural production remains the major source of income for about half of the world's population (SoW2-PGRFA, 2010, p192). In spite of its vital importance for human survival, PGRFA are being lost at an alarming rate. Hundreds of thousands of farmers' heterogeneous plant varieties and landraces, which have been developed for generations in farmers' fields until the beginning of the 20th century, have been substituted by a very small number of modern and highly uniform commercial varieties. In the USA alone, more than 90 per cent of the fruit trees and vegetables that were grown in farmers' fields at the beginning of the 20th century can no longer be found. Today only a few of them are maintained in gene banks. In Mexico, only 20 per cent of the maize varieties described in 1930 are now known. In China, in 1949 nearly 10,000 weed varieties were known and used. By the 1970s, only about 1000 remained in use. A similar picture is reported for melon varieties in Spain. In 1970, one of the authors of this chapter collected and documented over 350 local varieties of melons; today no more than 5 per cent of them can still be found in the field. The picture is much the same throughout the world (SoW1-PGRFA, 1996). This loss of agricultural biological diversity has not only affected small farmers' livelihoods, but has also drastically reduced the capability of present and future generations to adapt to changing conditions.

In addition, many neglected crops and many wild relatives are expected to play a critical role in food, medicine and energy production in the near future. The FAO's first report on the State of the World on Plant Genetic Resources (SoW1-PGRFA, 1996) estimated that some 7000 species had been used by mankind to satisfy human basic needs, while today no more than 30 cultivated species provide 90 per cent of human calorific food supplied by plants. Furthermore, 12 plant species alone provide more than 70 per cent of all human calorific food and a mere 4 plant species (potatoes, rice, maize and wheat) provide more than half of all human calorific food.

Countries' reliance on foreign PGRFA is one of the oldest forms of interdependence (Frison & Halewood, 2005), which goes right back to the Neolithic when the first crops spread from their centres of origins to the rest of the world. It can be said that today no country is self-sufficient with respect to the genetic resources for food and agriculture they rely on. Indeed, the average degree of interdependence among countries with regard to the most important crops is around 70 per cent (Table 1.1). Paradoxically, many economically poor countries happen to be among the richest in terms of genetic diversity needed to ensure human survival.

Table 1.1 *Estimated range of interdependency (percentage) for regions' agricultural development on genetic resources from elsewhere*

Region	*Minimum*	*Maximum*
Africa	67.24	78.45
Asia and the Pacific region	40.84	53.30
Europe	76.78	87.86
Latin America	76.70	91.39
Near East	48.43	56.83
North America	80.68	99.74
Mean	65.46	77.28

Source: Flores Palacios (1997)

This table shows, for each region, the mean of countries' degree of dependency on crop genetic resources which have their primary centre of diversity elsewhere. The indicator used is the food energy supply in the national diet provided by individual crops. On the basis of the primary area of diversity of each crop, the estimated dependency, with maximum and minimum indices, has been calculated, showing that there is a high rate of dependency in practically all cases.

Interdependence between generations is also strong. Agricultural biodiversity is a precious inheritance from previous generations. We have the moral obligation to pass it on intact to coming generations and allow them to face unforeseen needs and problems. However, up to now, the interests of future generations who neither consume, nor have the opportunity to speak or vote for themselves have not been adequately taken into account by our political and economic systems.

Although matters related to the conservation and sustainable use of genetic resources and the management of related technologies may appear to be technical, they have, in reality, strong socio-economic, political, cultural, legal, institutional and ethical implications. Problems in these fields can put at risk the future of humanity. International cooperation in this area is therefore not a choice but a must and should focus on the fair and equitable sharing of the benefits derived from the use of genetic resources, providing an essential incentive to ensure that countries, local farmers and breeders continue developing, conserving and making their genetic diversity available to humanity. Today, the Treaty is the legal and technical instrument specifically designed for this purpose.

To accomplish this task, the United Nations, as a universal intergovernmental forum, has a fundamental role to play in the facilitation of the necessary intergovernmental negotiations. In the 1970s, worldwide systematic actions began within the FAO, resulting in the adoption the International Undertaking on Plant Genetic Resources for Food and Agriculture in 1983 and the establishment of the intergovernmental Commission on Genetic Resources for Food and Agriculture (CGRFA), the forum within which the Treaty was negotiated. Stakeholders in the field have also played, and continue to play an important role in the common commitment of alleviating poverty and promoting food security. By their continuous practices of exchanging crops, farmers and researchers have set the ground for the formal realization of the global crop commons (Esquinas-Alcázar, 1991; Halewood and Nnadozie, 2008; Byerlee, 2010). International organizations active in the field, such as the CGIAR (see Chapter 11) also contributed to pave the road for such an open approach in the management of PGRFA for research and breeding (SGRP, 2003; CGIAR, 2009). Box 1.1 illustrates the history of the development and exchange of PGRFA from the dawn of agriculture to nowadays with special details in the last decades.

The negotiations of the Treaty were not alien to, but strongly influenced by the historical and geo-political context in which they were developed. In the 1970s and 1980s, when a utopian socialism was still believed to be possible, the almost romantic concept of plant genetic resources, seen as 'heritage of mankind' to be made 'available without restriction', was defended with passion by most of the developing countries and some developed countries. This idealistic vision was

Box 1.1 History of genetic resources' development and exchange: A history of agriculture and of cooperation and dialogue among cultures

10,000 years ago: Domestication and geographic spread of crops

- Humans start their transition from nomad hunters to sedentary farmers.

In the last millennia: Development of agriculture and agricultural biodiversity

- Cultural contacts and interactions result in crop diffusion and global transfer of PGRFA.
- Sumerians and Egyptians actively collect PGRFA.
- The discovery of America boosts intercontinental exchange.

Since the 19th century: Science realizes the value and potential of genetic diversity

- Charles Darwin's and Gregor Mendel's discoveries prove the importance of genetic diversity for biological evolution and adaptation.
- In 1845, the European famine dramatically demonstrates the need for genetic diversity in agriculture.
- Between 1920s and 1930s, Nikolai Vavilov identifies the main areas of crop origin and their genetic diversity.

By the mid 20th century: Scientific and institutional developments; concerns regarding genetic erosion and vulnerability

- In the 1960s and 1970s, the Green Revolution boosts productivity but contributes to the loss of genetic diversity.
- FAO starts technical work on PGRFA collection and conservation, including through a series of international technical conferences.
- In 1972, the UN Stockholm Conference on Human Environment called for strengthening of PGRFA conservation activities. The US National Academy of Sciences raises concern over crops genetic vulnerability after a major maize epidemic.
- In 1974, what is now the International Plant Genetic Resources Institute was established to support and catalyse collection and conservation efforts.

In the last decades: First major policy developments

- In 1961, the International Union for the Protection of New Varieties of Plants was established, and revised in 1978 and 1991. National legislation restricts access to PGRFA, including through intellectual property rights.
- In 1979, FAO member countries start policy and legal discussions, leading in 1983, to the first permanent intergovernmental forum on PGRFA – the Commission on Genetic Resources for Food and Agriculture (CGRFA) – and to the adoption of the non-binding International Undertaking on PGR (IU).
- From 1989 and 1991 NGOs promote an International Dialogue on PGRFA, reaching common understandings that feed into the CGRFA's negotiations.

In the 1990s: An era of global instruments and legally binding agreements

- In 1992, the first international binding agreement on biological diversity, the CBD, is adopted. Its members recognize the special nature of agricultural biodiversity and support the negotiations in FAO.
- In 1993, the CGRFA agrees to renegotiate the IU, resulting in the adoption in 2001 of the legally binding ITPGRFA.
- In 1994, the Marrakech Agreement on Trade-related Aspects of Intellectual Property Rights (TRIPS) is adopted.
- From 1993 to 1996, the CGRFA develops the Leipzig Global Plan of Action on PGR and the 1st report on the State of the World's PGRFA.
- In 1995, the CGRFA broadens its mandate to all components of biodiversity for food and agriculture.
- In 2001, the ITPGRFA is signed (for details on the achievements of the Treaty since its inception, see Annex 1 of this book).
- In 2004, the ITPGRFA enters into force on 29 June.
- In 2006, the 1st meeting of the Governing Body of the Treaty is held in Spain. The ITPGRFA becomes operative with the adoption of the SMTA.
- In 2010, the Conference of the Parties of the CBD adopts the Nagoya Protocol on Access to Genetic Resources and the Fair and Equitable Sharing of Benefits Arising from Their Utilization to the Convention on Biological Diversity.

Source: Esquinas-Alcázar (2005), updated with the authorization of the author

reflected in the 1983 International Undertaking (IU). After the fall of the Berlin wall and the start of an era of the so called 'real politics', neoliberal economic theories prevailed. These concepts of 'heritage of mankind' to be made 'available without restriction' were consequently downgraded by those of 'global concern', 'state's sovereignty' and 'facilitated access', as reflected in the 1992 Convention on Biological Diversity (CBD) and the 2001 ITPGRFA.

A history of the Treaty's negotiating process

The negotiating process

The ITPGRFA is the end product of a long period of international debates and negotiations in the FAO (Cooper, 2002; Mekouar, 2002; Rose, 2003; Esquinas and Hilmi, 2008). Indeed, the first technical and scientific discussions in the FAO in this area started in the 1950s. Discussions focusing on the economic and social implications started in the 1970s (see Chapters 2 and 10 for more details). While formal mandate to negotiate a binding agreement did not happen until 1993, the political discussion and negotiating process had begun in the FAO Conference (the main decision-making body in the organization) in November 1979, when the Spanish delegation, later supported by numerous countries, proposed the development of an international agreement on PGRFA and a germplasm bank under the jurisdiction of the United Nations. In the 1981 FAO Conference, this proposal became a draft resolution written by Mexico and presented by the GRULAC region on behalf of the G-77. As a result, the next FAO Conference (November 1983) approved the first intergovernmental agreement on this subject – the 'International Undertaking on Plant Genetic Resources' (IU) – with the reservation of eight countries[4] (Canada, France, Germany, Japan, New Zealand, Switzerland, UK and USA). The same conference established an intergovernmental body – the FAO Commission on Plant Genetic Resources (today the Commission on Genetic Resources for Food and Agriculture, which includes 167 member countries and the European Community) to monitor its implementation. The IU is a non-binding agreement based on the principle that 'plant genetic resources are a heritage of mankind' that 'should be available without restriction'. More problematically, its definition of PGRFA included commercial varieties and other products of biotechnologies, which was considered by some countries to be incompatible with intellectual property rights (IPR). This particular issue explains why the IU was approved with eight reservations. To resolve this conflict, a number of 'agreed interpretations' of its text were negotiated in the FAO Commission between 1983 and 1991. Through these interpretations, the concepts of plant breeders' rights and farmers' rights were simultaneously recognized, while the expression 'heritage of mankind' was combined with 'subject to national sovereignty' and new concepts such as global concern and fair and equitable sharing of benefits were introduced.[5]

International non-governmental organizations (INGOs) played an essential role in this part of the process (for the civil society viewpoint, see Chapter 10). One particularly important initiative was the Keystone International Dialogue Series on Plant Genetic Resources, convened and facilitated by a neutral, non-governmental entity, between 1988 and 1991, during which several points of consensus were identified in a series of informal meetings. The process was chaired by Dr M. S. Swaminathan, who brought together key individuals from government, the private sector, research community, civil society, international organizations, and others in their individual capacity, to systematically discuss and seek consensual solutions to a range of critical issues. This initiative was very useful in paving the road for the formal intergovernmental negotiations in the Commission.

From 1988 to 1992, the CBD,[6] which aimed to become the first binding international agreement covering all biological diversity, was negotiated by the United Nations Environment Programme (UNEP) and presented for signature at the Río Earth Summit in June 1992 (Nairobi Final Act).[7] However, this agreement, which also includes agricultural biodiversity, did not sufficiently take into account the uniqueness of agricultural biodiversity and the specific needs of the agricultural sector (see Box 1.2), partly because agricultural experts were barely represented during the negotiation process. Indeed, countries' representatives related to the agricultural sector were only able to unite during the final session of the negotiations in Nairobi in May 1992. This group was able to develop and introduce a resolution at the very last minute on agricultural biodiversity that was then adopted together with the text of the CBD as Resolution 3 of the Nairobi Final Act.[7] This

Box 1.2 Uniqueness of plant genetic resources for food and agriculture and the need for multilateralism

The uniqueness of PGRFA, when compared with wild biodiversity, is based on the following:

- They are crucial to satisfying basic human needs.
- They are man-made biological diversity being developed since the origins of agriculture.
- Because of the degree of human management of PGRFA, its conservation in production systems is inherently linked to sustainable use.
- They are not randomly distributed throughout the world, but concentrated in the so-called 'centres of origin and diversity' of cultivated plants.
- There is much greater interdependence among countries for PGRFA than for any other kind of biodiversity.
- The target for conservation and use are not the species as such, but genetic diversity within each species.

The 'special nature of agricultural biodiversity, its distinctive features and problems needing distinctive solutions' was formally recognized by the Conference of the Parties of the CBD in 1995 (Decision CBD II/15), which supported negotiations within FAO for the IT.

During the FAO negotiations, the need for distinct solutions became especially apparent, particularly in relation to the application of any bilateral mechanisms for access, to and sharing of benefits derived from the use of PGRFA.

The high transaction costs (Visser, 2003) and the technical and legal difficulties (Hardon et al, 1994) in bilateral access systems such as those provided under the CBD, finally led negotiating countries to the multilateral solution: the multilateral system of access and benefit-sharing adopted in the ITPGRFA.

resolution stressed the importance of the agreements reached within FAO and called for the IU to be revised in harmony with the CBD.

The adoption of the CBD, and two years later that of the TRIPS agreement in the context of the World Trade Organization (WTO) Uruguay Round, as binding international agreements, was a wake-up call for the agricultural sector. With compliance being voluntary, the IU lacked sufficient weight to defend the specificities and interests of agriculture. Increasing pressure from other sectors, especially the commercial and environmental spheres, made possible what seemed unimaginable not so long ago. Developing and developed countries, the seed industries and non-governmental organizations (NGOs) joined together with one common political objective to transform the IU into a binding agreement that would allow (i) for equal footing cooperation with the trade and environment sectors, and (ii) guarantee conservation and access to agriculturally important plant genetic resources for research and plant breeding through a fair system for access and benefit-sharing. Consequently, the new phase of the negotiations – specifically aimed at the development of the Treaty – commenced in a highly constructive atmosphere.

These formal negotiations took place between 1994 and 2001. The FAO Commission met in three regular sessions and six extraordinary sessions. In order to speed up negotiations by reducing the number of active negotiators, the Commission appointed a regionally balanced contact group composed of 47 countries. Between 1999 and 2001, the contact group held six meetings to discuss controversial issues and to pave the road for the Commission negotiation. The 6th extraordinary session of the Committee (see Annex 1 of this publication) intended to conclude the negotiations, but its delegates could not reach agreement on several points. These pending issues were resolved during the 121st session of the FAO Council (October 2001).[8]

In a euphoric atmosphere, the negotiations were completed during the 31st Conference of FAO, on 3 November 2001, with the adoption of the Treaty (see Annex 3 of this book for a table giving an overview of the main provisions of the Treaty) by consensus with only two abstentions: Japan and the USA.[9] With an expression of disbelief and exultation after the vote, Director-General of FAO, Dr Jacques Diouf, qualified the Treaty as a milestone on North–South relationship.

The Treaty entered into force in June 2004, and became operative with the first session of its Governing Body (Madrid, June 2006). This meeting resolved important issues and resulted in the adoption of a standard material transfer agreement[10] that, through the Treaty's multilateral system of access and benefit-sharing (MLS), determines the quantity, method and terms of payment related to commercialization. During this first meeting, the Governing Body (GB) made great advances towards the resolution of other issues, such as the mechanisms to promote compliance with the Treaty and the funding strategy. An agreement between the Governing Body of the Treaty and the Global Crop Diversity Trust (GCDT) was also signed. The second (GB-2/07/REPORT, 2007) and third (GB-3/09/REPORT, 2009) sessions of the Governing Body achieved great progress on issues such as the implementation of the funding strategy, cooperation with the FAO Commission, cooperation with the CGIAR and on the sustainable use of genetic resources. It also adopted inter alia resolutions on Farmers' Rights and on the MLS. The fourth session took place in Bali, Indonesia, in March 2011. GB 4 adopted procedures and mechanisms on compliance, reached consensus on the long-standing item of the financial rules of the Governing Body, and adopted, among others, resolutions on the multilateral system, Farmers' Rights, sustainable use, cooperation with other organizations, and implementation of the Funding Strategy.

So far, the Treaty has been ratified by 127 countries and the European Union (see Annex 2 of this volume for the list of contracting parties). Significant progress has been made in the implementation of some of its provisions: countries committed to raise US$116 million to support activities for the implementation of the funding strategy during a period of five years, and during the first year US$14 million was raised. In addition, as one of the essential elements of the funding strategy, the GCDT, which focuses on activities related to ex situ conservation, had received US$136 million up to March 2010, and another US$32 million are committed. This includes contributions from public and private sources. With regard to non-financial resources, 444,824 samples of Annex I material from the CGIAR centres were transferred under the SMTA between August 2007 and July 2008, representing more than 8500 samples transferred per week.

Behind the scenes

This book is not intended to present a comprehensive history of the negotiating process. We recognize that the true story of these long and difficult negotiations took place behind the scenes and includes many interesting unpublished anecdotes and semi-clandestine contacts (see Sukhwani, 2003, Chapter 10 and Annex 4 of this book for some stories on the inception of the ITPGRFA negotiations). While it was countries that were sitting around the negotiating tables, the actual negotiators were human beings who sometimes went beyond their own mandates and occasionally in spite of them. The deep and human history which reflects the real soul of the negotiations (Sukhwani, 2003) is only partially captured in this volume.

The actual negotiations were technically complex and politically controversial. They were often based on short-term national interests that varied from country to country or within a country over a different period of time (see illustrative example in Box 1.3). However, a number of key negotiators and many observers from INGOs were moved by ideals. The dialogue between all those involved was much easier when taking into account the perspective of future generations, an issue where all interest and ideals converged.

Only some of the main protagonists of this long and fascinating process have participated as authors of chapters of this book. We therefore consider it a duty and an obligation to pay tribute in this introduction to some of those that are missing, without whose involvement, courage and perseverance the Treaty would have never been possible. Among the countries' ambassadors and representatives are: José Ramón López Portillo and Francisco Martínez Gómez from Mexico, real pioneers of the political negotiations, Carlos di Motola from Costa Rica, M. S. Swaminathan from India, Javier Gazo from Peru, Mercedes Fermín Gómez from Venezuela, Ulf Svenson from Sweden, Jaap Hardon from The Netherlands, Henry Shands from the USA, Melaku Worede from Ethiopia, Juan Noury from Cuba, Mohamed Zehni from Libya and Jan Borring from Norway. We also would like to extend our appreciation and tribute to many representatives of civil society and INGOs that often have been the real engines of the process, moved by ideals that had the privilege to call things by their name without the handicap resulting from the diplomatic language. Among them and together with Pat Mooney, pioneer and excellent thinker, were Henk Hobbelink, Patrick Mulbany, Rene Salazar, Camila Montecinos, Hope Shands and many others. We also wish to highlight the political realism and the broad vision of some of the members of the private sector such as Don Duvick and John Deusing. They all collaborated with generosity and enthusiasm in this process, facilitating a balanced result and a final consensus. Last but not least, our tribute goes to colleagues in the secretariat of FAO and its negotiating Commission on PGRFA such as Erna Bennet, Clive Stannard, Murthi Anishetty and David Cooper, as well as colleagues from IPGRI (now Bioversity Interna-

Box 1.3 Illustration of how unexpected international political events may condition the outcome of negotiations

This anecdote illustrates better than a textbook the strategic importance of genetic resources and the influence of international political developments in the negotiation of the Treaty. One of the most complex and controversial subjects in the formal process of negotiations was the selection of genera or crops to be included in the multilateral system and listed in Annex I of the Treaty. In order to provide a sound scientific and technical negotiating basis to decide which crops should be included in the multilateral system, the following two criteria were agreed: importance of the crop for global food security and countries' interdependence on the crop. After years of negotiations, countries had shortlisted 67 genera. On 1 April 2001, when negotiations on this issue were closing with the aforementioned 67 genera, a conflict over the occupation of China airspace by an aircraft of the United States[11] muddied the negotiations. China is the primary centre of diversity of soybean. The morning following this political conflict, China withdrew soy from the Treaty's list, since the United States is one of the leading soy producers and highly depends on China for this crop genetic resource. As a reaction, Latin American countries, some of which such as Brazil were among the countries most affected by this decision, withdrew peanut and tomato. Brazil and Bolivia indeed contain peanut's maximum diversity; while the Andean region is the centre of diversity for tomato. By retrieving peanuts from the list, these countries tried to force the position of China, where these products are of great importance. This explains why, instead of 67 genera, there are only 64 crops and forages included in the multilateral system of the Treaty. Although the list of crops of the multilateral system can be modified in the future, this would entail the reopening of negotiations, which would have a high economic and political cost, since any change in the text of the Treaty requires a new process of parliamentary ratification by all contracting parties.

tional) and the FAO Legal Office. All of them facilitated the negotiating process all the way through with professionalism, generosity and enthusiasm, keeping always in mind that while our duty was to serve all member countries of FAO, our heart and our ideals had to stay with the weakest. Our apologies to the many we have not cited here due to lack of space and memory. Without them the utopia of the Treaty would have never become a reality.

Challenges ahead

The Treaty is a starting point to meet new challenges posed by the 21st century to food and agriculture. Challenges ahead have technical, scientific, socio-economic, legal and institutional dimensions.

Technical and scientific challenges: The need for a Road Map with specific targets and time-table to meet the technical provisions of the Treaty

Technical provisions of the Treaty, especially those under Article 5 'Conservation, exploration, collection, characterization, evaluation and documentation for PGRFA' and Article 6 'Sustainable use of PGRFA' need to be applied at the national level. Many technical and scientific priorities and challenges for PGRFA today have largely to do with the ways in which we need to adjust our thinking on conservation and utilization methods to cope with climate change, environmental sustainability and food security. This could be facilitated by the development and adoption of a road map with specific and verifiable targets and a realistic time-table. International assistance to meet these targets should be facilitated as needed.

Various aspects should be taken into account when defining priorities and targets for a full and efficient implementation of the Treaty, including maintenance and management of genetic diversity, use of genetic resources, climate changes and food security.

Maintenance and management of genetic diversity

The following includes a number of priorities identified by countries and the FAO during the preparatory process of the 2nd report on the State of the World on PGRFA (2009):

- To carry out systematic surveys and to publish inventories to identify existing GRFA both in the field and in germplasm banks.
- To develop methods for reliably estimating plant genetic diversity and to adopt standardized definitions of genetic vulnerability and genetic erosion (FAO, 2002; Brown, 2008).
- To give greater attention to the in situ management of wild relatives; neglected crops and promising species, as well as diversity in threatened ecosystems.
- To develop a more rational global system of ex situ collections.
- To develop and implement national strategies and to strengthen national capacities to manage and use genetic resources, including a greater use of scientific methods and technologies.
- To broaden the genetic basis in crop improvement.
- To develop appropriate policies, legislation and procedures for collecting crop wild relatives, maybe by revising the 1993 FAO International Code of Conduct for Plant Germplasm Collecting (FAO, 2003).
- To carry out ethno botanical and socio-economic studies, including indigenous and local knowledge, to better understand the role of farming communities in the management of PGRFA.

Utilization challenges for food security and environmental sustainability and to face climate change

Changes in agricultural production methods, in the environment, and in consumers' demands are all likely to require a larger use of genetic resources (see Chapter 17). The utilization of a wide range of PGRFA is therefore crucial for food security, environmental sustainability and to face climate change.

Food security

The main challenge to increase food security is not just food production, but access to food. In addition, it is not simply a matter of delivering more calories to more people. It should be noted that most hungry people in the world (over 70 per cent) live in rural areas. Solutions are needed to improve stability of production at the local level, to provide increased options for small-scale farmers and rural communities and to improve quality as well as quantity of available food. Nutritional security, where dietary diversity plays an important role, is a vital component of food security.

To ensure that the benefits derived from plant genetic resources reach all those who need them, public-sector research is needed in areas in which the private sector does not invest. Most commercial crop varieties are not adapted to the needs of poor farmers, especially in many developing countries, who have limited or no access to irrigation, fertilizers and pesticides. A new environmentally friendly, socially acceptable and ethically sound agricultural model is necessary to meet their needs. This could be achieved by publicly supported programmes to breed crops that are able to withstand adverse conditions, including drought, high salinity and poor soil fertility and structure, and that provide resistance to local pests and diseases. Such programmes are likely to build on farmers' existing varieties and local crops, which often contain these traits. This is especially important at times when international prices of major crops have dramatically increased (e.g. world food crisis in 2008) and continue to be volatile and unpredictable.

Research emphasis needs to be put at the local level, often on local and underutilized crops, to support breeding and improve performance of a wide range of crops and varieties well adapted to local conditions and needs rather than just seeking uniform 'universal genotypes'. This can only be achieved by a systematic and participatory process of cooperation between breeders, farmers and consumers.

Environmental sustainability and climate change[12]

Reducing the negative impact that agriculture may have on the environment (e.g. water, energy, pesticides and herbicides) should become an absolute priority. This requires increased use of diversity in production systems through the deployment of a wider range of varieties and crops to ensure better ecosystem service provision. A good example would be the use of diversity-rich strategies to reduce damage by pests and diseases. Research is needed on how to make diversity-rich strategies more effective in terms of reaching better agriculture productivity and management.

Each predicted scenario of the Intergovernmental Panel on Climate Change (IPCC) will have major consequences for the geographic distribution of crops, individual varieties and crop wild relatives (see Chapter 7). Some recent studies have used current and projected climate data to predict the impact of climate change on areas suitable for a number of staple and cash crops (Fischer et al, 2002; Jarvis et al, 2008).

The challenges we face with PGRFA owing to climatic changes are twofold. First, climate change will accelerate genetic erosion and create a critical need to collect and conserve endangered PGRFA and wild relatives before it is too late. Second, the magnitude of change will require significant adaptation. The use of a wide range of PGRFA will thus become vital in the development of varieties able to adapt to new and unstable environmental conditions; that is to withstand conditions that are not only hotter or drier but also more variable (Hawtin et al, 2010). This will increase the need for adaptability and resilience, properties that have not been usually embedded in traditional breeding. New and innovative breeding approaches would consequently be required. Also, new genetic diversity within and between species is likely to be needed, increasing therefore the potential of underutilized crops and new promising species. All these will drastically increase countries' dependency on foreign PGRFA and therefore the need for international cooperation, in particular by facilitating access to PGRFA.

It should be emphasized that for all these areas, the question is not limited to the pursuit and discovery of specific traits from a pool of PGRFA. The research needs to be concerned with functional diversity and with diversity deployment in agricultural systems from farm fields to landscape, watershed and regional scales.

Financial and socio-economic challenges

The funding strategy of the Treaty needs to become fully operative. Indeed, it aims at developing ways and means by which adequate resources are available for the implementation of the Treaty, in accordance with Article 18. The cost of conserving plant genetic diversity is high, but the cost of not taking action is much higher. Economic resources for the conservation and sustainable use of agricultural genetic resources are well below adequate levels. This problem is particularly serious in the case of in situ conservation of traditional farmers' varieties and, increasingly, of cultivated plants' wild relatives, which are largely found in developing countries. The scarcity of economic resources in these countries is not only an obstacle to the protection of wild species, but also a major cause of genetic erosion, as people search for fuel-wood or convert virgin areas into farmland. It is estimated that conserving 1000 accessions of rice generates an annual income stream for developing countries that has a direct use value of US$ 325 million at a 10 per cent discount rate (SoW2-PGRFA, 2010).

The establishment of the GCDT (see Chapter 16), as an important element of the funding strategy of the ITPGRFA, is a step forwards in this direction. However, this fund remains specifically dedicated to ex situ conservation, maintaining the need for complementary initiatives or elements to support other aforementioned pressing priorities.

At the Third Governing Body of the Treaty in 2009, a target of US$ 116 million was agreed to be raised for the Treaty's funding strategy within the next five years. Projects have also been developed in a bottom–up, country driven process. However, most of these funds are not available yet and might be difficult to obtain. In this context, it should be recalled that only 4 per cent of Overseas Development Assistance (ODA) goes to agriculture, when more than 70 per cent of hungry people live in rural areas. The conservation and use of GRFA should, however, be seen not only as part of developmental assistance, but also as a matter of relevance to national development and food security.

The benefit-sharing fund is crucial to develop a healthy, balanced and self-financed multilateral system. The future of the Treaty may depend on it (see Chapter 18 on the importance of 'closing the circle of access-benefit sharing'). In this context and in order to ensure transparency and compliance by the users of PGRFA with the obligations established under Article 13.2(d)(ii) of the Treaty, it is important to further explore and promote the 'crop-related' royalty payment modality established by Article 6,11 of the SMTA, as adopted by the Governing Body of the Treaty. The 'crop-related' modality provides an innovative, predictable, verifiable alternative, far less bureaucratic, and much easier to administer and enforce than 'the product-related' payment scheme (see Chapter 19). There are indications that some seed industry circles are interested in investigating more deeply the potential advantages of the crop-related modality as the preferred alternative (see Chapter 12 on the seed industry). This should be taken into account by the Governing Body of the Treaty when renegotiating the level of mandatory payments established in the SMTA, in order to make the 'crop-related' modality more attractive. Other problems that could be identified with the implementation of Article 12 should be addressed by the Governing Body to ensure that there are not disincentives for its use.

From a macroeconomic perspective, PGRFA have been considered as an unlimited capital. However, PGRFA are limited resources to be used by all future generations, and their full future value continues to be ignored in market prices. In accordance with Agenda 21 of the United Nations Conference on Environment and Development (UNCED), a sustainable economic solution to the problem should be the internalization of the conservation cost of the resource into the production cost of the product. For example, when buying an apple, we could pay not only for the cost of production, but we could also contribute to the conservation cost in order to allow future generations to continue eating apples. The ITPGRFA provisions concerning benefit-sharing, including the sharing of monetary benefits that are derived from commercialization, represent a first step in that direction. Taking all the above into account, it is easy to ascertain that there is an urgent need for economic research in terms of a better understanding, description and quantification of the true value of genetic resources. Indeed, while conceptual frameworks in terms of use, future and option values exist, there is a definite lack of adequate quantification mechanisms, which would efficiently drive investment decisions and research planning.

Legal and institutional challenges

Following a country's ratification, the ITPGRFA provisions ought to be implemented at the national level, which requires the revision and development of national measures and regulations. In many cases, additional legislation is also needed to prevent genetic erosion, promote the conservation, characterization and documentation of local genetic resources, implement Farmers' Rights, facilitate access to genetic resources for research and plant breeding, and promote an equitable sharing of benefits.

Access to genetic resources and related biotechnologies is threatened by the increasing number of national laws that restrict access to and use of genetic resources, as well as by the proliferation of intellectual property rights and the expansion of their scope (Correa, 1994, 2003). In this context the adoption of the Treaty represents an important step to facilitate access to PGRFA for research and breeding. However, the Treaty cannot be seen in isolation from other relevant national and international legislation on biodiversity and related technologies. Complementarities and synergies in the implementation of existing legal instruments related to GRFA in the agricultural (ITPGRFA), environmental (CBD) and trade (WTO/TRIPS) sectors need to be ensured, possibly through the development of national *sui generis* provisions in line with the requirements of these three international agreements (see Box 1.4) (Esquinas-Alcázar, 2005). In particular, since the adoption of the Nagoya Protocol on Access to Genetic Resources and the Fair and Equitable Sharing of Benefits Arising from Their Utilization in October 2010 (COP 10, Decision X/1), coordination with this new instrument would be of utmost importance. The text of the decision adopting the Nagoya Protocol recognizes the Treaty as a complementary instrument to the international regime on Access and Benefit Sharing (ABS), as well as the special nature of PGRFA and their importance to achieve food security worldwide. It also recognizes its role for sustainable agricultural development taking into account the particular contexts of poverty alleviation and climate change.

In addition, the interests of the agricultural sector need to be well represented during the implementation processes of those instruments. The effectiveness of the Treaty in halting or reversing the tendency towards access restriction will depend on how its provisions are interpreted and implemented by individual countries and the international community.

However, there are some shortcomings: some of the provisions of the Treaty were left deliberately ambiguous in order to get consensus during the negotiating process (e.g. 'Recipients shall not claim any intellectual property or other right that limited the facilitated access to plant genetic resources for food and agricultural, or their parts or components, in the form received from the Multilateral System' (Article 12.3(d)). This ambiguity allows for different and sometimes incompatible interpretations. The development of new technologies that allows for uses of PGRFA in ways that were not foreseen when the Treaty was negotiated is an added complication in this context.

Regarding the implementation of the MLS of the Treaty, the full realization of the expected benefits might facilitate future negotiations in reaching consensus in

Box 1.4

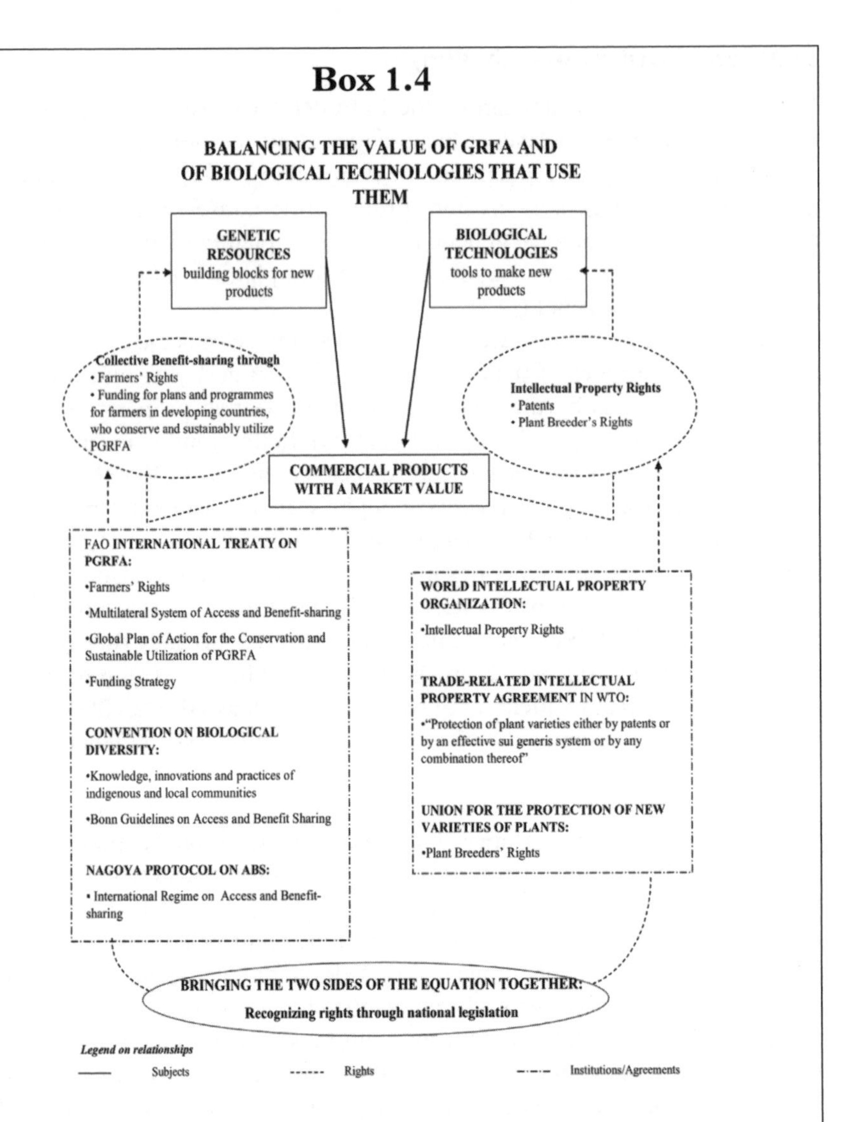

Genetic resources provide the building blocks that allow classical plant breeders and biotechnologists to develop new commercial varieties and other biological products. Although nobody can deny their importance, neither genetic resources nor the biological technologies that apply to them have an appropriate market value by themselves, while a clear market value often exists for the commercial products obtained through them. Since the 1960s, a number of international bodies and agreements (the Trade Related Intellectual Property Agreement (TRIPS/WTO), the World Intellectual Property Organization (WIPO) and the Union for the Protection of New Varieties of Plants (UPOV), have included provisions setting minimum standards for, or conferring on the developers of biological technologies, individual rights (IPRs such as plant-breeders' rights and patents) that allow the right-holders to appropriate part of the profits from any commercial products that may result from the use of those technologies. Since the 1990s, other international agreements (the CBD, the Treaty, and, more recently, the Nagoya Protocol on Access and Benefit-sharing) have conferred equivalent but collective rights (Farmers' Rights and benefit-sharing) on the providers of the genetic resources. This allows for a symmetrical and balanced system of incentives to promote, on the one hand, the developments and application of new biotechnologies and to ensure, on the other hand, the continued conservation, development and availability of genetic resources to which these technologies apply (Frison et al, 2010). It is now up to national governments to implement these provisions, including the development, as appropriate, of national legislation that takes fully into account the two 'pillars' of the system represented in the diagram, thereby allowing for harmony and synergy in the implementation of the various binding international agreements.

Source: Esquinas-Alcázar (2005), updated with the authorization of the author

other controversial and challenging issues, such as broadening the Treaty's scope by increasing the number of crops that are exchanged through the multilateral system. This is especially important at a time when climatic changes are increasing countries' interdependency on PGRFA and many so-called minor and until now neglected crops are becoming increasingly important for food security.

Therefore, there is an increasing need to ensure coherence in the implementation of the Treaty and fill in possible legal gaps. To achieve this without having to modify the Treaty's text, 'agreed interpretations' of some of its provisions may need to be developed and negotiated in due time.

The full implementation and further development of the International Treaty could be facilitated by a more active, systematic and possibly institutional participation of civil society, especially farmers and other stakeholders' organizations.

Training and public awareness

Although regulatory aspects remain crucial, legal provisions alone are not sufficient as they need to be understood, accepted and implemented. Indeed, it is of the utmost importance that provisions of the Treaty become better known by as many stakeholders and citizens as possible. Training in this area, as well as raising public awareness on the importance of genetic diversity and the dangers of its loss are very important challenges.[13]

One should recall that genetic erosion is just one consequence of mankind's exploitation of the planet's natural resources. The fundamental problem is a lack of respect for nature, and any lasting solution will have to involve establishing a new relationship with our planet and an understanding of its limitations and fragility. If mankind is to have a future, it is imperative that children learn this at school, and that adults adapt by integrating this new understanding in their everyday life.

Conclusion

The history of the exchange of PGRFA represents somewhat the history of humanity. The struggle to obtain new plants for food and agriculture has been one of the main motivations of human travel from the earliest times, and has often led to alliances and partnerships, but also to conflicts and wars between different civilizations and cultures.

The Treaty provides a universally accepted legal framework for PGRFA and an important innovative cooperating instrument in the fight against hunger. It marks a historic milestone in international cooperation. However, many things still need to be done to fully implement the Treaty, both at national and international levels. To this end, solid mechanisms to promote compliance have to be adopted.

The purpose of this book is to allow stakeholders to express their views on where we are coming from, where we are nowadays and where we should go. We are convinced that drawing this picture will help/contribute to a better understanding and implementation of the Treaty, which remains crucial to face current challenges including climate change, food security and environmental sustainability.

Notes

1 This chapter only represents the opinions of its authors. Christine Frison conducts PhD research as junior affiliated researcher at the Université catholique de Louvain and at the Katholieke Universiteït Leuven (Belgium) on international law and governance of plant genetic resources for food and agriculture. Francisco López is Treaty Support Officer for the International Treaty on Plant Genetic Resources for Food and Agriculture and is based at the FAO, Rome, Italy. José T. Esquinas-Alcázar is Director of the 'Catedra' of Studies on Hunger and Poverty at the University of Cordoba in Spain. Professor at the Politechnical University of Madrid, José Esquinas has worked as Secretary of the FAOs intergovernmental Commission on Genetic Resources for Food and Agriculture, and interim Secretary of the Treaty for 30 years.
Email: jose.esquinas@upm.es.

2 Plan of Action of the Rome Declaration on World Food Security, § 7, available at www.fao.org/docrep/003/w3613e/w3613e00.HTM

3 Agricultural production in general and crop production in particular, must increase substantially in order to meet the rising food demand of a population that is projected to expand by some 40 per cent over the period from 2005 to 2050. According to a projection by FAO, an additional billion tonnes of cereals will be needed annually by 2050 (SoW2-PGRFA, 2010).

4 The delegations from Canada, France, Germany (The Federal Republic of Germany) Japan, Switzerland, the United Kingdom and the United States of America made reservations with respect to Resolution 8/83 (the International Undertaking on Plant Genetic Resources) adopted in the 22nd Conference of FAO in Rome, November 1983. New Zealand expressed reservations regarding the IU text since it did not take into consideration breeders' rights. The same seven countries and The Netherlands expressed reservations concerning Resolution 9/83 (Establishment of a Plant Genetic Resources Commission), also adopted in the 22nd Conference of FAO.

5 For additional information on this process see Esquinas-Alcázar and Hilmi (2008), available at www.bioversityinternational.org/fileadmin/bioversity/documents/themes/policy_and_law/the_treaty/publications/Recursos_Naturales_y_Ambiente_N.53/Las_negociaciones_del_Tratado_Esquinas_y_Hilmi_RNA53_2008.pdf (last accessed November 2010).

6 See www.cbd.int/convention/text/ (last accessed December 2010).
7 Resolution 3 from the Nairobi Final Act (the relationship between the Convention on Biological Diversity and the promotion of sustainable agriculture) was adopted 22 May 1992 in Nairobi. Available at www.cbd.int/doc/handbook/cbd-hb-09-en.pdf (last accessed December 2010).
8 FAO Council, 121st session, Rome, 30 October to 1 November 2001. International Undertaking on Plant Genetic Resources, Information Pursuant to Rule XXI.1 of the General Rules of the Organization, Doc. CL 121/5-Sup.1; see also Appendix III, Doc. CL 121/5, the International Convention on Plant Genetic Resources for Food and Agriculture, as adopted at the 6th extraordinary session of the Commission on Genetic Resources for Food and Agriculture, Rome, 25–30 June 2001, and reviewed by the 72nd session of the Committee on Constitutional and Legal Matters, Rome, 8–10 October 2001. See also www.fao.org/waicent/faoinfo/agricult/cgrfa/docswg.htm (last accessed November 2010).
9 The two abstentions were Japan and the USA. See 31st session of the Conference of FAO, 2–13 November 2001, verbatim records of plenary meetings of the Conference, 4th plenary meeting, 3 November 2001, Doc. C 2001/PV, p73. See also Resolution 3/2001 (Approval of the International Treaty on Plant Genetic Resources for Food and Agriculture and provisional resolutions for its application) adopted in the 31st session of the Conference of FAO, Rome, November 2001, available at ftp://ftp.fao.org/unfao/bodies/conf/C2001/Y2650e.doc (last accessed November 2010).
10 Resolution 2/2006 (the standard material transfer agreement) adopted in the 1st session of the Governing Body of the International Treaty on Plant Genetic Resources for Food and Agriculture, Madrid, June 2006. See Doc IT/GB-1/06, report of the meeting, at ftp://ftp.fao.org/ag/agp/planttreaty/gb1/gb1repe.pdf (last accessed November 2010).
11 This incident appeared in the news such as CNN USA, available at http://articles.cnn.com/2001-04-01/us/us.china.plane.02_1_spy-plane-chinese-fighter-chinese-island?_s=PM:US (last accessed December 2010), or on the 'History Commons' journalism website at www.historycommons.org/timeline.jsp?us_military_specific_cases_and_issues=us_military_tmln_spy_plane_crash_in_china&timeline=us_military_tmln (last accessed December 2010).
12 We are thankful to Toby Hodgkin and Nicole Demers for sharing their ideas on these issues, some of which are reflected and feed the content of this paragraph.
13 Chapters 9 and 13 devote a large part to public awareness and training.

References

Brown, A. H. D. (2008) 'Indicators of genetic diversity, genetic erosion and genetic vulnerability for plant genetic resources' Report to the Food and Agriculture Organization, Rome, 26 January 2009

Byerlee, D. (2010) 'Crop improvement in the CGIAR as a global success story of open access and international collaboration', *International Journal of the Commons*, vol 4, no 1, pp452–480

CGIAR (2009) 'CGIAR Joint Declaration', Consultative Group on International Agricultural Research, Washington DC, 8th December

Cooper, H. D. (2002) 'The International Treaty on Plant Genetic Resources for Food and Agriculture', *Review of European Community and International Environmental Law*, vol 11, no 1, p2

Correa, C. M. (1994) 'Sovereign and property rights over plant genetic resources', FAO background study paper no 2, Commission on Plant Genetic Resources, 1st extraordinary session, Rome, 7–11 November 1994

Correa, C. M. (2003) 'The access regime and the implementation of the FAO International Treaty on Plant Genetic Resources in the Andean group countries', *Journal of World Intellectual Property Rights*, vol 6, pp795–806

Esquinas-Alcàzar, J. (1991) 'Un sistema globale per la difesa delle risorse genetiche vegetali', in F. Angeli (ed) *La Questione Agraria, Rivista dell'Associazione Manlio Rossi-Doria*, Rome, Italy, no 44, p53

Esquinas-Alcàzar, J. (2005) 'Protecting crop genetic diversity for food security: Political, ethical and technical challenges', *Nature*, vol 6, pp946–953

Esquinas-Alcázar, J. and Hilmi, A. (2008) 'Las negociaciones del Tratado Internacional sobre los Recursos Fitogenéticos para la Alimentación y la Agricultura', *Recursos Naturales y Ambiente*, no 53, pp20–29, available at www.bioversityinternational.org/fileadmin/bioversity/documents/themes/policy_and_law/the_treaty/publications/Recursos_Naturales_y_Ambiente_N.53/Las_negociaciones_del_Tratado_Esquinas_y_Hilmi_RNA53_2008.pdf, last accessed November 2010

FAO (2002) 'Review and development of indicators for genetic diversity, genetic erosion and genetic vulnerability (GDEV)', summary report of a joint FAO/IPGRI workshop, Rome, 11–14 September

FAO (2003) *International Code of Conduct for Plant Germplasm Collecting and Transfer*, FAO Conference, November 2003, Rome

Fischer, G., Shah, M. and van Velthuizen, H. (2002) 'Impacts of climate on agro-ecology', Chapter 3 in *Climate Change and Agricultural Vulnerability*, Report by the International Institute for Applied Systems Analysis, Contribution to the World Summit on Sustainable Development, Johannesburg, 2002

Flores Palacios, X. (1997) 'Contribution to the estimation of the interdependence of countries in the field of plant genetic resources', FAO background study paper of the CGRFA, no 7, rev.1

Frison, C. and Halewood, M. (2005) 'Annotated bibliography addressing the international pedigrees and flows of plant genetic resources for food and agriculture SGRP', Information document submitted by the System-wide Genetic Resources Programme of the CGIAR to the 8th Conference of the Parties to the Convention on Biological Diversity (COP 8) and the Ad Hoc Open-ended Working Group on Access and Benefit-sharing, International Plant Genetic Resources Institute (IPGRI), Jan. 2006, Rome, Italy, available at http://old.bioversityinternational.org/publications/publications/search.html#results

Frison, C., Dedeurwaerdere, T., and Halewood, M. (2010) 'Intellectual property and facilitated access to genetic resources under the International Treaty on Plant Genetic Resources for Food and Agriculture', Opinion Note, *European Intellectual Property Review*, vol 32, no 1, pp1–8, available at http://papers.ssrn.com/sol3/papers.cfm?abstract_id=1576823

GB-2/07/REPORT (2007) 'Report of the Second Session of the Governing Body of the International Treaty on Plant Genetic Resources for Food and Agriculture', available at www.planttreaty.org/meetings/gb2_en.htm, accessed December 2010

GB-3/09/REPORT (2009) 'Report of the Third Session of the Governing Body of the International Treaty on Plant Genetic Resources for Food and Agriculture', available at www.planttreaty.org/meetings/gb3_en.htm, accessed December 2010

Halewood, M., and Nnadozie, K. (2008) 'Giving priority to the commons: The international treaty on plant genetic resources', in G. Tansey and T. Rajotte (eds) *The Future Control of Food: A Guide to International Negotiations and Rules on Intellectual Property, Biodiversity and Food Security*, Earthscan, London, pp115–140

Hardon, J. J., Vosman, B. and Van Hintum, T. J. L. (1994) 'Identifying genetic resources and their origin: The capabilities and limitations of modern biochemical and legal systems', CGRFA background study paper no 4, FAO, Rome

Hawtin, G., Acosta Moreno, R., Swaminathan, M. S., Sekhara Pillai, B. R. and Hegwood, D. (2010) 'Expert advice on the Second Call for Proposals, including a strategy and programme for the Benefit-sharing Fund', Rome, FAO ftp://ftp.fao.org/ag/agp/planttreaty/funding/experts/bsf_exp_p01_en.pdf, accessed November 2010

Jarvis, A., Upadhyaya, H., Gowda, C. L. L., Aggerwal, P. K. and Fujisaka, S. (2008) 'Climate change and its effect on conservation and use of plant genetic resources for food and agriculture and associated biodiversity for food security', Report to the International Crops Research Institute for the Semi-Arid-Tropics (ICRISAT) and FAO

Mekouar, A. M. (2002) 'A global instrument on agrobiodiversity: The International Treaty on Plant Genetic Resources for Food and Agriculture', FAO legal papers online, 24 January 2002, pp3–5

Rome Declaration on World Food Security and Plan of Action (1996), World Food Summit, 13–17 November 1996, Rome, Italy, available at www.fao.org/DOCREP/003/W3613E/W3613E00.HTM)

Rose, G. (2003) 'International law in the 21st century: The International Treaty on Plant Genetic Resources for Food and Agriculture', *Georgetown International Environmental Law Review*, Summer 2003, pp583–632

SGRP (2003) 'Policy Instruments, Guidelines and Statements on Genetic Resources, Biotechnology and Intellectual Property Rights', Version II, produced by the System-wide Genetic Resources Programme (SGRP) with the CGIAR Genetic Resources Policy Committee, Rome, July 2003

SoW1-PGRFA (1996) 'The First Report on the State of the World's Plant Genetic Resources for Food and Agriculture', FAO Commission on Genetic Resources for Food and Agriculture, Rome

SoW2-PGRFA (2010) 'The Second Report on the State of the World's Plant Genetic Resources for Food and Agriculture', FAO Commission on Genetic Resources for Food and Agriculture, Rome, available at www.fao.org/docrep/013/i1500e/i1500e00.htm, accessed November 2010

Sukhwani, A. (2003) 'Interview with the Secretary of the FAO intergovernmental Commission on Genetic Resources for Food and Agriculture, José T. Esquinas-Alcázar', originally published in *Revista de comunicación interna: Marchamos* (magazine of the Spanish Patent and Trademark Office - OEPM)), no 16, pp13–21; English version by Helen Keefe. Both versions available at www.oepm.es. / Section Conócenos / Gabinete de Comunicación (www.oepm.es/cs/Satellite?c=Page&cid=1144260495163&classIdioma=_es_es&idPage=1144260495163&pagename=OEPMSite%2FPage%2FtplListaMarchamos&numPagActual=1)

Visser, B., Eaton, D., Louwaars, N. and Engels, J. (2003) 'Transaction costs of germplasm exchange under bilateral agreements', Proceedings of the GFAR-2000 Conference, Dresden

Part I

Regional Perspectives on the Treaty

Chapter 2

Overview of the Regional Approaches

The Negotiating Process of the International Treaty on Plant Genetic Resources for Food and Agriculture

Fernando Gerbasi

Introduction

Throughout history, humanity has suffered from famine. Its causes are multiple and stem, on a case by case basis, from certain human activities, such as war, ethnic, religious and tribal conflicts, as well as bad climate and natural disasters, like droughts, volcanic eruptions and earthquakes. Another danger is genetic uniformity.

During the last two centuries, as a consequence of the agricultural and industrial development and the progressive unification of cultural and eating habits, accentuated more recently due to the globalization and interdependence process, the number of crops and the diversity within them has been progressively reduced.

Genetic erosion is aggravated as a consequence of the disappearance of local species, wild relatives of cultivated plants, due to massive deforestation or the degradation or contamination of natural habitats: in a nutshell, due to the abusive exploitation of the planet's natural resources.

Climate change is greatly affecting the world's agricultural production. For this reason, conservation, maintenance, availability and sustainable use of the diversity of existing crop varieties is an issue of the greatest importance. These tasks are crucial to adequately satisfy the dietary needs of an ever-growing and demanding population as well as to constitute a global response to climate change.

The international community and plant genetic resources

It is well known that developing countries are the richest in plant genetic resources for food and agriculture (PGRFA). This set off a search for a reward system that covered the collective innovations carried out by farmers for centuries (for details on farmers' communities, see Chapter 13). Consequently, at the end of the 70s and the beginning of the 80s of the last century, this brought a great debate at Food and Agriculture Organization (FAO) conferences. In 1979, during the 20th Conference of FAO, the parties agreed upon the signature of an international agreement and the formation of a network of germplasm banks with international sovereignty, under the assumption that plant genetic resources are a heritage of mankind and that a legal framework was needed to ensure its unrestricted availability.

FAO Conference, in its 22nd session held in 1983, adopted Resolution 8/83 on the International Undertaking of Plant Genetic Resources (see Annex 1 of this volume for the list of all Commission and Treaty negotiating meetings). This was the first international agreement for the conservation and sustainable use of agricultural biological diversity. It is worth noting that the International Undertaking (IU), as an international instrument, was not legally binding, which was why it was adopted by several nations, especially by industrialized ones, with reserves. This was irrefutable proof of the discrepancies between North and South on such an important issue.

FAO Conference also approved Resolution 9/83, through which it established the Commission on Plant Genetic Resources, as first permanent intergovernmental body, so that countries could, among other things, monitor the implementation of the IU and advise FAO about its activities and programmes regarding plant genetic resources.

These decisions were the result of a delicate political balance among developed countries, which need access to plant genetic resources, and the wish of developing countries for a more equitable distribution of benefits, including monetary ones. The negotiation of several agreements continued, which later became part of the IU. In 1991, the national sovereignty of plant genetic resources, plant breeders' rights and farmers' rights were recognized.

When governments adopted the Convention on Biological Diversity (CBD) in 1992, they recognized the existence of two matters that required special treatment, which were not resolved by the Convention: access to ex situ collections not addressed by the CBD (as is the case of collections under the Consultative Group on International Agricultural Research (CGIAR) and the question of the Farmers' Rights (Resolution 3 of Nairobi, 1992). It was necessary that these matters be addressed within FAO's Commission. To that end, the FAO Conference of 1993 requested the Commission to negotiate the revision of the IU, in harmony with the CBD.

The regions of the FAO and PGRFA

FAO member nations are subdivided into seven geographic regions: Africa, Asia, Europe, Latin America and the Caribbean, Near East, Northern America and South West Pacific (see Annex 2 to this volume for the list of contracting parties to the International Treaty on Plant Genetic Resources for Food and Agriculture (ITPGRFA) per FAO regional groups). While the existence of these regions responds to technical needs it is also true that this has political implications in the ongoing events of the organization because it allows the rotation among countries in the bureaux of the different organs of FAO, on the basis of equitable and geographical distribution of regions and countries. However, in the Commission on Genetic Resources, during the negotiations of ITPGRFA, the regional groups were represented in the Bureau of the Commission all the time by the same persons, with the exception of Canada which replaced its vice president with another delegate, to facilitate consultations among them when necessary.

It is important to note that countries from each region consult among themselves on important decisions in order to adopt common positions. FAO regional conferences (of which Regional Conference for Asia and the Pacific; Regional Conference for Africa; Regional Conference for Europe; Regional Conference for Latin American and the Caribbean and the Regional Conference for the Near East) shall meet once in every two years in regular sessions and involve ministers for agriculture from their respective countries.

In addition, there are two major groups of countries:

- The G-77: This group was founded on 15 June 1964 by the 'Joint Declaration of the Seventy-Seven Countries' issued at the United Nations Conference on Trade and Development (UNCTAD). It integrates 131 developing countries. In the G-77 there are countries from the following regions: Latin America and Caribbean, Africa, Asia, Near East and Pacific.
- The Organisation for Economic Co-operation and Development (OECD) was created in 1960 with 31 developed countries. Country members of OECD are from North America, Europe, Latin America and Asia and the Pacific.

Therefore, in both groups there are developing and developed countries, which sometimes results in confrontation due to conflicting interests. The consultation process to arrive at joint positions in these two groups is progressive. The countries consult among themselves at regional level, and then they meet, as regions, in the G-77 or in the OECD, with the aim of reaching common positions. This process can take a long time and could lead regions to return to regional consultations before finalizing an agreement at the G-77 or OECD level. In any case, as the negotiation progresses and solves some issues and/or others arise, there are consultations, in some cases daily or even two or three times a day within and among the two groups and regions.

In respect to PGRFA, there is a great interdependence among regions (see the Introduction to this book). The agriculture of the majority of countries is greatly dependent on a supply of resources from other regions of the world. In fact, one study carried out by Kloppenburg and Kleinman (1987) shows that North America is completely dependent upon species originating from other regions of the world for its major food and industrial crops, while sub-Saharan Africa is estimated to be 87 per cent dependent on other parts of the world for the plant genetic resources it needs. The Mediterranean sub-region is dependent for 98 per cent, and Europe for 90 per cent. A large part of Asia (East and South) is dependent on species originated in other parts of the world for 62.8 per cent and Latin-America for 55.6 per cent.

The start of the negotiating process

As expected, the intergovernmental forum in charge of completing the revision of the IU was the Commission on Genetic Resources for Food and Agriculture (CGRFA) (initially the Commission on Plant Genetic Resources). The process started in 1994 through its working group (see Annex 1 of this volume for the list of all Commission and Treaty meetings). At the beginning, the IU tried to be consolidated by integrating its annexes, that is, resolutions 4/89, 5/89 and 3/91 of the FAO conferences, as well as harmonized with the applicable provisions of the CBD. It is worth noting that in that moment, maybe as a reflection of the difficulties it was facing, the working group decided to admit it had no mandate to negotiate the revised text of the IU, so it focused on making notes to the draft prepared by the FAO secretariat, which did not compromise governments but otherwise reflected the opinion of the delegates.

The Commission essentially focused, during the first two years, on three articles of the revised IU – Article 3 (Scope), Article 11 (Availability of Plant Genetic Resources) and Article 12 (Farmers' Rights). Throughout time, it prepared several drafts of the revised IU, until it reached number four halfway through 1997. These drafts, particularly the last one, contained in some cases several versions, especially in regards to the aforementioned articles. Actually, the text was not useful, since it was a mixture of concepts, without structure or guidelines to guide the negotiation. This was proof of the complexity and innovation of the matter at hand and how conflicted were positions, not only between developing and developed countries, but among the latter, particularly the United States of America, on the one hand, and the European Union, on the other.

In May 1997, I was elected for a term of two years as Chairperson of the CGRFA of FAO, which I served, after reappointment to a second term by unanimous vote, until October 2002. During that time, apart from aptly leading the normal tasks of the Commission, I took on the direction and orientation of the negotiations to harmonize the IU with the CBD. These concluded with the adoption of the IU on 3 November 2001. To effect this, I organized and convened twelve official and two unofficial meetings, as well as endless personal consultations.

Ever since assuming the Chair of CGFRA, I have tried to give negotiations for the revision of the IU a new impulse, through better organization, both to the process and to the texts to be considered by negotiators, as well as giving a more political view to the negotiating approach. Although there were ups and downs, it can be said that the process had two stages: the first, from 1994 to 1996, the period covering the first four meetings, and the second, from May 1997 to the adoption of ITPGRFA,[1] popularly known as the International Seed Treaty (Rome, 3 November 2001), adding up to a total of 12 meetings (see Annex 1 of this book for a list of these meetings).

The Montreux Consensus and the '*Chairman's Elements*'

During the 5th extraordinary meeting, held in Rome between 8 and 12 June 1998, it was confirmed that though progress had been made during the meeting, positions were still different, distant and profoundly diverging. Consequently, time was given to reflect and allow delegations to analyse the different positions, carry out pertinent consultations and identify areas of possible compromise before continuing with negotiations.

Based on the above, I carried out consultations as of August 1998, particularly with the countries that had been more active during negotiations, as well as with the other six members of the Bureau, since they represented their regions and had actively participated in the whole negotiating process. My role was to assess the situation and then take a decision about a possible extraordinary meeting that would allow negotiations to continue, so long as there was political will, a flexible attitude and a spirit of commitment among members, as well as the availability of extra-budgetary funds to perform it. I was looking for the conditions to reach an understanding and overcome the impasse under which negotiations had fallen, without generating false expectations. In the particular case of developed countries, besides the usual consultations, I also asked that they inform me whether their governments were willing to contribute financially to the preparation and performance of an extraordinary meeting of the Commission, as well as allowing the participation of delegates from developing countries in this session.

From the consultations carried out, I concluded that, although there was ample support for a swift completion of negotiations, the delegations needed more time to make more consultations. The general opinion was that a new extraordinary session of the Commission should not yet be held, and that the available time would be better used in preparing for the continuation of negotiations. As a consequence, the extra-budgetary funds to which countries committed for an extraordinary session were insufficient.

I continued consultations during the 115th session of the FAO Council, 23–28 November 1998. On that occasion, I had bilateral or plurilateral talks with the countries more committed to the negotiating process, that is, Angola, Argentina, Australia, Brazil, Canada, Colombia, Ethiopia, the European Union, France, Germany, India, Islamic Republic of Iran, Japan, Malaysia, Mexico, Norway, South Africa, Switzerland, the United Kingdom and the United States of America.

I explained that in my opinion, negotiations were completely paralysed in the absence of real commitments and due to a negotiating view where the scientific view prevailed above the political or diplomatic. I considered it necessary to resolve that impasse by calling a meeting with the head of the delegations of those countries. They were asked to act in their personal capacity, so that they could separate themselves from their instructions and negotiating postures and try to determine, jointly, the minimum elements that had to be included in what could be an agreement on plant genetic resources acceptable for all and in compliance with the conference's mandate to harmonize the IU with the CBD. I strongly pointed out that to perform this meeting I would need on the one hand, the good will of all participants, and on the other, the willingness of a country to offer the venue of said meeting and the contribution of sufficient financial funds to afford the tickets of all guests and other related costs. The idea was welcomed and to my satisfaction, the Swiss delegation informed me, in a second meeting, that it was willing to offer Montreux as a venue for the meeting and to contribute sufficient funds thereto. Germany and the United States of America also contributed, providing enough additional funds to support the participation of developing countries in this unofficial consultation meeting.

Consequently, FAO's Council decided to unanimously support my proposal to convene an unofficial meeting of experts representative of different regions and different postures, who, in their personal capacity, would deal with the following matters: a way to share benefits, Farmers' Rights, financial mechanism, legal condition of the revised IU, and other issues, such as access to PGRFA. Likewise, it decided to accept Switzerland's offer to organize and host the unofficial meeting at the beginning of 1999, under the responsibility of the Chairperson of the CGRFA. It also decided that should the Chairperson confirm that the results of the unofficial consultation provided possible progress, he would ask the Director-General to hold an extraordinary meeting of the Commission, subject to the availability of extra-budgetary funds.

In compliance with the decisions of the 115th session of the FAO Council, as Chairperson of the Commission I summoned, under my responsibility, experts from 21 countries – all of them consulted during the FAO Council plus Poland and Venezuela – and the European Union, to participate, in their personal capacity, in the unofficial meeting, held from 19 to 22 January 1999, in Montreux, Switzerland. This unofficial meeting had the support of FAO's secretariat – Mr José Esquinas-Alcázar and Mr Clive Stannard – and the International Plant Genetics Research Institute (IPGRI, now Bioversity International) Director-General, Mr Geoffrey Hawtin. These three international high officials were very useful during the whole negotiating process, due to their technical knowledge and personal expertise. A critical role was also played by Mr Gerald Moore in all legal aspects.

Without the limitations of their official orders, participants discussed the legal condition of the revised IU, the idea that it should be an internationally legally binding instrument being of greater importance, with a secretariat taken on by FAO and closely linked both to this organization and to the CBD. The structure of the IU should be such that would allow an efficient revision of all operational and administrative matters. To allow an understanding in all subjects related to the multilateral system of access and benefit-sharing (MLS), the writing of a less ambitious text with elements that would allow an ample consensus was proposed. The system would cover, at the beginning, a restricted list of crops, based on the criteria related to food security and interdependence, that would be revised and possibly widened on a periodic basis. Likewise, collections from international agricultural research centres (CGIAR; see Chapter 11 for details) would be part of the system as per conditions previously agreed with them. In regards to the Farmers' Rights, their recognition would be necessary on an international basis, understanding that the development of the Farmers' Rights would rely upon each government, who should, in due time and in compliance with national law, protect and promote said rights. Concerning the financial resources needed for the implementation of the IU, these would be obtained through a funding strategy that would use a wide range of sufficient financial resources, based on agreed upon and predictable contributions, to implement plans and programmes, particularly in developing countries.

The summons to a meeting of experts to, in their personal capacity, analyse and assess possible areas of understanding was a wise move and a crucial breakthrough, since it allowed negotiators from the main participating countries to debate amply and openly their options. These frank and open debates allowed me to write what was later known as the *Chairman's Elements* (see appendix to this chapter). These elements were simply a group of consensus proposals prepared under my total responsibility, after listening to and analysing what the Group of Experts, in their personal capacity, considered that the revised IU in harmony with the CBD needed to include to be approved by the international community. The experts did not approve the *Chairman's Elements* but they did consider the Chairperson had adequately gathered the consensus derived from the unofficial consultation.

Subsequently, I submitted the *Chairman's Elements* to the consideration of the CGRFA, which approved them, for although the elements had been introduced under the sole responsibility of the Chairman, they reflected an ample consensus and provided a solid base for the continuation and progress of negotiations.

The *Chairman's Elements* were adopted by the Commission during its 8th ordinary meeting, held 19–23 April 1999. The Commission decided to continue negotiations on the basis of the elements. Said decision was subsequently supported by the 116th session of the FAO Council and by the FAO Conference in its 30th session, held 12–23 November 1999. This political support at the heart of the Organization was of extreme importance for the continuance of negotiations, since as Chairperson I did not allow negotiators to shift their proposals, in any significant way, away from the *Chairman's Elements*.

One of the innovations I introduced in the negotiations from that moment on was the use of so-called 'contact groups', so in vogue in other negotiations at the heart of the UN. Therefore, I established a Chairperson's contact group, with 41 members (Angola, Argentina, Australia, Benin, Brazil, Burkina Faso, Canada, China, Colombia, Cuba, Ethiopia, European Union, Finland, France, Germany,

India, Iran (Islamic Republic of), Japan, Korea (Republic of), Libya (Libyan Arab Jamahiriya), Malaysia, Malta, Mexico, Morocco, The Netherlands, New Zealand, Norway, Philippines, Poland, Romania, Samoa, Senegal, South Africa, Switzerland, Tanzania (United Republic of), the United Kingdom, the United States of America, Uruguay, Venezuela, Zambia and Zimbabwe), which, in accordance with the premise of a fair and equitable geographic representation, represented the seven regions of FAO – Africa, Asia, Latin America and the Caribbean, North America, Near East, Europe and the Pacific Southwest. On countless occasions, one of the vice-chairpersons was in charge of a small contact group, to deal with a specific matter, which results were then passed on to the Chairperson's contact group, who generally accepted what was agreed upon. The Chairperson's contact group met seven times from April 1999, suggesting that the most active and positive period of negotiations was from April 1999 to November 2001. It was two years of intense negotiations and consultations, not only among the countries of the Chairperson's contact group, but also between these and the remaining members of the Commission, through the FAO regional groups.

Perhaps the most important innovation was allowing the involvement of important NGOs (see Chapter 10 for details on civil society), in representation of others, in the works of the contact group, such as the Rural Advancement Foundation International (RAFI, see Chapter 10), whose director Pat Mooney is widely regarded as an authority on agricultural biodiversity and new technology issues, and the International Association of Plant Breeders for the Protection of Plant Varieties (ASSINSEL), which gathers at its heart breeders from around the world.

Main aspects of the negotiation

In the 8th regular session of the Commission, held in April 1999, the first fundamental article, the Farmers' Rights, already established in the *Chairman's Elements*, was adopted. Farmers' Rights is a subject that was originally introduced by the FAO Conference in 1989, and has attracted much interest and controversy since that time.

The African group, the European Union and the United States of America were an integral part of this agreement. However, the African group, the region that had shown the greatest interest in this topic, was criticized by other delegations, and particularly by NGOs, as they pointed out it had made concessions too soon. The reason for this attitude was the international recognition of the national legislation as the foundation to adopt the appropriate measures to protect and promote Farmers' Rights. In any case, the adoption of said article, which was never again modified, was auspicious for the rest of the negotiating process.

For the writing of the list of crops included in the multilateral system, important research was carried out by the IPGRI and officials of FAO, with the support of the Italian government. The results of said research allowed for the negotiation of the list in 2001, particularly during the last days before the adoption of the Treaty, based on criteria of food security and interdependence. Nevertheless, it should be remembered that some regional groups had well-defined positions, as was the case of the European Union which presented a long and ample tentative list of around 270 crops, while the African group preferred a short and concise list of less than 10 crops. The Latin America and the Caribbean region preferred a list of about 40 crops while the other regions, who actively participated in the negotiating process, were not rigid in their position. In truth, last-minute negotiations on this important issue allowed Mexico and Peru to exclude certain sub-species of corn, China soybean and Brazil tomatoes. The most significant food crops missing from the final list are: soybean, cassava, groundnuts, sugar cane and tomato. To conclude, important crops from the South were unfortunately excluded from the final list, perhaps because it was never understood by developing countries themselves how important the link was between said list and the MLS. This perception should be, today, completely different in the light of the implementation of the Treaty.

Concerning benefit sharing, particularly monetary benefits, which are the true innovating concept of the Treaty, ASSINSEL (see Chapter 12 for more detail on the seed industry; see Chapter 15 for more detail on plant breeders), who always participated as an observer in the negotiating process, made a fundamental contribution when it stated, in June 1998, based on a decision of its General Assembly held in Monte Carlo on 5 June 1998, that 'in case of protection through patents, that would limit the free access to new genetic resources, the members of ASSINSEL would be ready to study a system in which patent proprietors would contribute to a fund established to collect, maintain, evaluate and strengthen genetic resources. The mechanism used to implement this system needed to be discussed.' From then on, negotiations evolved until reaching what was included in the Treaty.

Brazil always kept a conscientious posture in defence of the CBD (see Chapter 6), as it was adopted in Rio de Janeiro in June 1992. Therefore, when I proposed that the agreed upon text be named 'International Convention on Plant Genetic Resources for Food and Agriculture', the delegate from Brazil emphatically opposed the use of the word 'Convention' which led me to the word 'Treaty', that ultimately has a stronger connotation.

Conclusions

There are still many unresolved issues to make the Treaty more effective and efficient, and as it happens, the Governing Body is working on them (see Annex 3 of this book for details on the main provisions of the Treaty). However, we believe that although today 127 countries are contracting parties to the Treaty, it is more than necessary to disseminate, for both governments and the civil society, the importance of the Treaty. This can be achieved through workshops, forums and seminars, but particularly by developing and strengthening the regional and sub-regional networks of plant genetic resources in which researchers, breeders, farmers and interested members of civil society can foster the political conditions to achieve the technical exchange between them and contribute to the implementation of the Treaty. Moreover, the formation of National Focal Points in a greater number of member countries has to be promoted.

As time has passed, we have ascertained that it would be more important to verify the list of crops included in the multilateral system to include crops of great importance, such as soy and tomatoes, as well as many others from developing countries, in order for benefit sharing in the multilateral system to increase.

It is necessary that the text included in Article 12.3d be clarified: 'Recipients shall not claim any intellectual property or other right that limit the facilitated access to the plant genetic resources for food and agriculture, or the genetic parts or components, in the form received from the Multilateral System.' The European Union and several members thereof wrote the following interpretation when they ratified the International Treaty: 'The European Union interprets Article 12.3.d of the Treaty on Plant Genetic Resources as recognizing that plant genetic resources for food and agriculture or their genetic parts or components which have undergone innovation may be the subject of intellectual property rights provided that the criteria relating to such rights are met.' The Governing Body needs to determine which changes will affect intellectual property. It will not be an easy task, but undoubtedly necessary.

With the approval of the Treaty and the implementation of several of its mechanisms, such as the standard material transfer agreement and the benefit-sharing fund, the regions of the G-77, which are mostly developing countries, are acting more cohesively in order to make the Treaty a tool through which the stakeholders of their countries could obtain greatest benefit. Moreover, the stakeholders, particularly those of developing countries, are participating more actively through the creation of support networks.

The Treaty is, without a doubt, an international agreement of the greatest importance for developing and developed countries. Its provisions meet the real interests of all parties. Moreover, it appropriately takes into account the interests of other interested parties, such as autonomous communities, universities, research centres and the private sector in general. This is the first great international agreement of the new millennium.

PGRFA are *sine qua non* for the sustainable development of agriculture, which is why an agreement about the fair and equitable sharing of benefits, including those of a commercial nature, provides an incentive for farmers of every country, especially those from developing countries and countries in economic transition, to conserve and sustainably use plant genetic resources for the benefit of all.

Through the Treaty countries agreed that these plant genetic resources are vital for the survival and well-being of present and future generations, which is why conservation, maintenance and sustainable use of these resources are a transcendental cause.

Notes

1 On 3 November 2001, the 31st session of the Conference of the Food and Agriculture Organization of the United Nations (FAO) adopted, by its resolution 3/2001, the International Treaty on Plant Genetic Resources for Food and Agriculture and Interim Arrangements for its Implementation.

Reference

Kloppenburg, J. R. and Kleinman, D. L. (1987) 'Plant germplasm controversy: Analyzing empirically the distribution of the world's plant genetic resources', *Bioscience*, vol 37, pp190–198

Appendix: *Chairman's Elements*

1. Scope: Plant genetic resources for food and agriculture (PGRFA).
2. Objectives: Conservation and use of PGRFA, and the fair and equitable sharing of benefits arising from the use of PGRFA, in harmony with the CBD, for sustainable agriculture and food security.
3. National commitments towards conservation and sustainable use, national programmes integrated into agriculture and rural development policies.
4. Multilateral system, including components for facilitated access and benefit-sharing.

a) Coverage

- A list of crops, established on the criteria of food security and inter-dependence.
- The collections of the International Agricultural Research Centers (IARCs), on terms to be accepted by the IARCs.

b) Facilitated access

- To minimise transaction costs, obviate the need to track individual accessions, and ensure expeditious access, in accordance with applicable property regimes.
- Plant genetic resources in the multilateral system may be used in research, breeding and/or training, for food and agriculture only. For other uses (chemical, pharmaceutical, non-food and agricultural industrial uses, etc.), mutually agreed arrangements under the CBD will apply.
  - Access for non-parties shall be in accordance with terms to be established in the IU.

c) Equitable and fair sharing of benefits

- Fair and equitable sharing of benefits arising from the use of PGRFA, inter alia, through:
  - transfer of technology;
  - capacity-building;
  - the exchange of information;
  - funding.

Taking into account the priorities in the rolling Global Plan of Action, under the guidance of the Governing Body:

- Benefits should flow primarily, directly and indirectly, to farmers in developing countries, embodying traditional lifestyles relevant for the conservation and sustainable utilization of PGRFA.

d) Supporting components

- Information system(s).
- PGRFA networks.
- Partnership in research and technology development.

5. Farmers' rights

- Recognition of the enormous contribution that farmers of all regions of the world, particularly those in the centres of origin and crop diversity, have made and will continue to make for the conservation and development of plant genetic resources which constitute the basis of food and agriculture production throughout the world.

- The responsibility for realizing farmers' rights, as they relate to PGRFA, rests with national governments. In accordance with their needs and priorities, each party should, as appropriate, and subject to its national legislation, take measures to protect and promote farmers' rights, including:
  - the right to use, exchange, and, in the case of landraces and varieties that are no longer registered, market farm-saved seeds;
  - protection of traditional knowledge;
  - the right to equitably participate in benefit-sharing;
  - the right to participate in making decisions, at the national level, on matters related to the conservation and sustainable use of PGRFA.

6. Financial resources

Commitment to a funding strategy for the implementation of the IU, which includes:

- budget and contributions to manage the operations of the Governing Body/Secretariat etc. (some of their activities could be delegated);
- agreed and predictable contributions to implement agreed plans and programmes, in particular in developing countries, from sources such as:
  - CGIAR, GEF, plus ODA, IFAD, CFC, NGOs, etc., for project funding
  - country contributions;
  - private sector;
  - other contributions;
  - national allocations to implement national PGRFA programmes, according to national priorities.
- priority given to implementation of the rolling GPA, in particular in support of farmers' rights in developing countries.

7. Legally binding instrument

- Governing Body.
- Policy direction, and adoption of budgets, plans and programmes.
- Monitoring the implementation of the IU.
- Periodically reviewing, and, as necessary, updating and amending the elements of the IU and its annexes.
- Secretariat.

8. Provisions for amending the International Undertaking and updating and revising its annexes.

Chapter 3

The African Regional Group

Creating Fair Play between North and South

Tewolde Berhan Gebre Egziabher, Elizabeth Matos and Godfrey Mwila

Introduction

Today, Africa remains the most disadvantaged continent of the world despite having abundant natural resources. This is due to a variety of reasons, both historical and contemporary. Poverty, malnutrition and poor health, especially in sub-Saharan Africa, affect a large proportion of the people. These poor conditions are intrinsically linked with the access to food and to crops necessary for subsistence farming. For this reason, Africa has placed a lot of hope in the negotiation and implementation of the International Treaty on Plant Genetic Resources for Food and Agriculture (the Treaty or ITPGRFA).

Important data

According to 2005 estimates, 80.5 per cent of the people in this region were living on less than US$2.50 a day.[1] Africa, which is characterized by rapid population increase over the last 60 years, has now reached one billion people compared with 221 million in 1940.

In contrast to these data, Africa is the primary and secondary centre of origin of many important food crops, such as sorghum, millet, yam, oil palm, sesame, date, pea and rice (FAO, 1997). It remains nonetheless highly dependent (88 per cent) on crops (maize, cassava, plantain, banana, wheat, potato, groundnut, etc.) originating from elsewhere as Kloppenburg and Kleinman (1987) have shown in their study.

Agriculture constitutes approximately 30 per cent of Africa's gross domestic product with 70 per cent of the population depending on the agricultural sector for their livelihood. Production is mainly for subsistence and is highly dependent on rains. Because of these factors, coupled with poverty in most countries, the continent is very vulnerable to the effects of climate change. Prolonged droughts have, for instance, adversely impacted the agricultural sector in some areas of the continent.

The dependence of the so-called modern system of intensification of agricultural production on excessive amounts of agrochemicals derived from fossil fuels exacerbates climate change. The African region, though financially the poorest in the world, is perhaps still the richest in the quickly disappearing capacity of self-reliant smallholder farmers. These smallholder farmers continue feeding the bulk of Africa's population using their self-contained and decentralized agricultural systems tried out over centuries of effective performance. These time-tested and almost carbon neutral ecological systems of agricultural production by the farmers of Africa are particularly relevant for our present era of the threat of a climate chaos. These smallholder farmers and their farming systems have survived in spite of having been continuously undermined by the state and the modern international establishment since the colonial period; however, they still feed their populations in spite of all the odds. Nevertheless, owing to the imbalance of interests entrenched in the Agreement on Agriculture of the World Trade Organization (WTO), their ecologically sound produce can not compete with the heavily subsidized produce of the polluting industrial agriculture of the North. A little formal support given to these smallholder farmers, or even a mere tolerance of their existence, would thus help increase food production and improve food security in Africa.

The intimacy of African delegations with the agricultural systems of the smallholder farmers enabled the African group (AG) to have a marked impact on the negotiations of the ITPGRFA in spite of Africa's financial poverty which could have limited our chances of having preparatory meetings.

Africa as a group in regional and international forums

At the regional level, the African Union (AU) formed in 2002 as a successor to the Organization of African Unity (OAU), politically brings together all African countries, except Morocco, with the objective of accelerating political and social economic integration and promoting African common positions on issues of interest to the continent. There is no doubt that the Organization influenced, to some extent, the position of the AG during negotiations for the revision of the International Undertaking on Plant Genetic Resources (IU, adopted in 1983). This influence is tangible, in particular, with the African Model Law for the Protection of the Rights of Local Communities, Farmers and Breeders, and for the Regulation of Access to Biological Resources adopted by the OAU and recommended to African states for their domestication as national legislation.[2] Africa is also divided into sub-regional political and socio-economic groupings such as the Southern

African Development Community (SADC), the Economic Community of West African States (ECOWAS), the Common Market for Eastern and Southern Africa (COMESA) and the East African Community (EAC), which may have impacted on the negotiating positions of the AG to varying degrees.

At the international level, Africa operates as a group in all major United Nations forums. As for food and agriculture, the FAO regional group for Africa totals 48 member countries, thereby constituting the largest regional group at FAO (see Annex 2 at the end of this book for the list of countries in the African regional group, including the list of African contracting parties to the ITPGRFA). Even so, participation to the FAO Commission on Genetic Resources for Food and Agriculture (CGRFA), where the Treaty was negotiated, was quite limited.

Africa's way of thinking regarding the Treaty: Towards a just international law for plant genetic resources for food and agriculture

In the context of the negotiations of the Treaty, the industrialized North that grew in the wake of Europe still wants to treat crop genetic resources, bred over millennia by farming communities which are mostly found in the global South, as global commons. This approach would allow the Northerners to access plant genetic resources for food and agriculture (PGRFA) from the South at will, while legally protecting through national and international law the varieties that they have bred out of those very same genetic resources from the South. They use the global force of intellectual property protection, especially patenting under the WTO's Agreement on Trade-Related Aspects of Intellectual Property Rights (TRIPS), to achieve that protection.

This entrenched advantage of the industrialized North thus works by remote control through the use of skewed international law of which TRIPS is only one glaring example. Therefore, it became easy for the AG to realize that, in order to help in the evolution of a just globalizing world, the strategy should be twofold. On the one hand, it should be proactive in formulating new and just laws for Africa. On the other hand, it should grasp opportunities for fighting as hard as possible, in both making new international laws just and revising existing unjust international laws to make them more equitable. Non-governmental organizations (NGOs) became obvious allies both as sounding boards and as sources of the meagre resources needed for the battle of the AG.

An opportunity to make international law on PGRFA more just arose with the revision of the IU.[3] In the following narrative of the revision of the IU to negotiate the ITPGRFA, the sources used are the notes taken during the negotiations by the authors of this chapter and the FAO documents that were prepared for those negotiations. A history of the negotiation will not be given here, as it has extensively been presented in other contributions of this volume.

Africa's participation and strategy during the negotiations of the Treaty

The AG entered the negotiations of the ITPGRFA with confidence arising from the modest experience gained in the negotiations of the Convention on Biological Diversity (CBD). However, few African countries were able to provide representatives with the benefit of any previous experience of negotiating in other international forums. This is not to mention the outstanding exception of our leader, Dr Tewolde Berhan Gebre Egziabher, and in later years other Ethiopian representatives, including Abebe Demissie and Worku Damena. In particular, the experience and negotiating skills of Ethiopia proved useful in keeping the African spirit in the negotiations alive, especially in the initial stages. Dr Tewolde, as Ethiopian delegate and chief negotiator for Africa, was instrumental in this regard.

Negotiating strategy adopted by the AG

Even though African delegations often constituted only one or two people, a positive factor was the fairly consistent composition of the AG negotiators in terms of delegates of key countries who played a more significant role in the negotiations (at least for the last four to five years of negotiations). The major constraint was the lack of diversity in terms of expertise among the African delegates. The African region was further disadvantaged by our lack of legal experts in this field. In the early years of the negotiations in the CGRFA, Africa had the benefit of just one legal expert, namely, Worku Damena from Ethiopia. Because of our poverty, negotiators for the AG were usually one or two from each country, compared to the crowded delegations from developed countries. Therefore, at the 3rd extraordinary session of the Commission, we insisted that negotiations had to take place in plenary only and not to break into working groups. As a compromise, Africa agreed to negotiate as two working groups. At the 4th extraordinary session of the CGRFA in 1994, Africa was slightly represented at the Commission. More disturbing was the fact that while the issues under discussion were specialized, technical and political, very few of the African delegates present were primarily genetic resources specialists, purposely brought from their countries for this meeting. Indeed, at that time, the number of African plant genetic resources (PGR) specialists and national PGR programmes in the continent were still very limited. (In spite of the great wealth of PGR in farmers' fields, Africa held just 6 per cent of the world's ex situ collections.) Consequently, Africa made few contributions on the floor of the 1994 Commission session. Although we had a very strong and experienced champion in the Commission and its Bureau – Tewolde – it was clear that if concerns for the rich PGRFA heritage of our millions of African farmers were to be defended and its conservation and utilization were to be promoted, we would have to increase both the number of African states present and the technical capacity of our representatives in any further negotiations.

Lack of funds also impeded the organization of AG preparatory meetings before coming to the negotiations. The extremely slow regional coordination was mainly due to the constant need for translation as delegates often had not had previous access to documents (partly as a result of our poor communication facilities in the early years of the negotiations). Not having had sufficient opportunity to discuss the documents coming up for discussion in the Commission sessions, we found ourselves recapping on previous sessions instead of preparing for future ones. Sweden[4] was sympathetic to our plight and channelled funding through an NGO, the Gaia Foundation of London, UK, to enable us to organize at least one preparatory meeting. The AG thus met on 21–25 April 1997 and revised what colleagues in Ethiopia had written into a full draft protocol to replace the IU as suggested by the Conference of the Parties to the CBD in November 1996.

During negotiation sessions, Africa met briefly each day before the beginning or after the end of the formal negotiations. This happened because of a tremendous goodwill to work together. After each negotiation session, Tewolde analysed in writing the next session's negotiating documents to identify inconsistencies, ideas that would weaken the already weak African situation, gaps that would militate against the effective conservation and sustainable use of crop genetic resources and proposed suggestions of what ideas could thus be introduced into the negotiating documents as corrections.

In between international meetings, the AG depended on email exchanges to develop a common position on the various divisive issues that always arise in negotiations. Consultations at the sub-regional levels such as the Southern African Development Community (SADC), held immediately prior to Commission or Treaty negotiating sessions (usually during annual Board meetings of the SADC Plant Genetic Resources Centre, a coordinating centre of the SADC PGR network) came up with positions, which were shared with other delegations in the AG, contributing towards regional positions. Even so, most of us from SADC were relatively new to the PGR field. Fortunately for the region, Ethiopia came to almost all Commission and negotiating meetings with a strong delegation of two or three technical experts and with experience from other international fora. Tewolde's analyses of current stages in the negotiations were invaluable to us individually, although we usually had very little opportunity to discuss them as a group. In spite of all the odds, or perhaps because of them, the AG continued as the most united of all regional groups.[5]

Africa's participation in the negotiation meetings between 1991 and 2001

With each succeeding negotiating session, the number of African countries represented increased, as did our technical expertise and sub-regional representation. Following East Africa led by the Ethiopian team, came Southern Africa with strong voices from Zambia and Tanzania. We were joined by other consistent voices for the interests of small farmers, particularly from Burkina Faso and later from Uganda, and PGR experts from Senegal and Guinea in West Africa. On average, about 30 countries were represented during ordinary and extraordinary sessions of the CGRFA, during which time discussions on Treaty negotiations were held.

Countries that were consistent in terms of delegates and attendance throughout Treaty negotiations and that made significant contributions, included Angola, Burkina Faso, Eritrea, Ethiopia, South Africa, Senegal, Tanzania, Uganda and Zambia. Ethiopia provided leadership to the AG throughout the period of negotiations (1997–2001). Angola and Zambia complemented Ethiopia during the latter part of the negotiations from 1998 to 2001, especially in terms of facilitating regional consultations to come up with the AG positions. Angola and Zambia took up increased roles of leading the AG after 2001 during the interim period, when the focus was on the development of instruments to facilitate the implementation of the Treaty, such as the standard material transfer agreement (SMTA), rules of procedures and the funding strategy as well as the initial period of Treaty implementation, for the First and Second Sessions of the Governing Body of the Treaty. In the inter-sessional meetings of the Chairman's contact group, which focused solely on negotiations of the Treaty, the AG was represented by delegates from 11 countries (Angola, Benin, Burkina Faso, Ethiopia, Libya, Morocco, Senegal, South Africa, Tanzania, Zambia and Zimbabwe). Table 3.1 shows in broad terms the growth in African participation in Commission meetings in the principal negotiating period from 1991 to 1997, and that remained at the latter level until 2001.

By the time the contact group meetings began in 1999, the AG could call on a much stronger core of eight to ten PGR related delegates for all these meetings, including representatives from all of Africa's five geographical sub-regions. By the end of the negotiations, Africa had become the largest and one of the most united groups.

In spite of early weaknesses, when the working group on the SMTA terms and its implementation was set up, the need for African legal expertise became crucial, and by that time we were able to add a few more legal advisers, particularly from southern Africa. This included Antonieta Coelho from Angola, who played an important role in the introduction of the concept of a third party beneficiary, to oversee and ensure the fair application of the SMTA.

At every opportunity in Commission meetings, we encouraged delegates from other African countries to become contracting parties to the Treaty, and by the end of the negotiations, Africa had become not only the largest but one of the most united groups. It was an honour for the African region when Godfrey Mwila from Zambia, a consistent negotiator and champion for Farmers' Rights, was elected first as Chair of G-77 and later as Chair of the Commission at the first meeting of the Treaty's governing body in Madrid in 2006.

The main issues that the AG fought for during the negotiation

Major contributions of the AG during the Treaty negotiations were on Farmers' Rights and benefit sharing. This is not to imply that the group did not contribute in other areas. Throughout discussions on Farmers' Rights the AG pushed for the

Table 3.1 *African group representation in the ITPGRFA negotiation meetings*

Date	*Meeting*	*African Commission members*	*African IU parties*	*African countries represented at session*	*Total number of African delegates with PGR or related technical background*
June 1991	4th regular CPGR	32	29	11	• 2 PGR experts
April 1993	5th regular CPGR	33	32	15	• 1 PGR expert
Nov. 1994	1st extraordinary session	35	34	18	• 4 PGR experts
June 1995	6th regular CPGR	39	34	24	• 4 PGR experts
April 1996	2nd extraordinary session	No list available	No list available	25 6 SADC	Total 23, including: • 3 PGR experts from Ethiopia • 7 PGR experts from SADC network • 13 PGR experts from other sub-regions
Dec. 1996	3rd extraordinary session	42	34	29	Total 22, including: • 3 PGR experts from Ethiopia • 13 PGR experts from SADC • 6 PGR experts from other sub-regions
May 1997	7th CPGR	44	34	35	Total 29, including: • 3 PGR experts from Ethiopia • 12 PGR experts from SADC • 14 PGR experts from other sub-regions
Dec. 1997	4th extraordinary session	45	35	34	Total 26, including: • 10 PGR experts from SADC • 2 PGR experts + 1 legal expert from Ethiopia • 14 other technical experts
June 2001	6th extraordinary session	48	36	33	Total 20, including: • 2 PGR experts from Ethiopia • 11 PGR experts from SADC • 7 PGR experts from other sub-regions of Africa

recognition of these rights under international law. The AG was also supportive of the creation of a multilateral system, as reflected in one of the statements given on behalf of the group during one of the Treaty negotiation sessions: 'African countries would allow their sovereign rights over PGRFA to be expressed jointly with those of others' (FAO, 1998). It insisted that rights given in the CBD would have to be respected, ensuring that benefits are made communal instead of being individual.

The divisive issues in the negotiations of the Treaty became clear in 1996 during the 3rd extraordinary session of the Commission. This session began with the report of the 11th session of the negotiating working group (established by the Commission), which showed that scope, access and Farmers' Rights provisions had been discussed.

The AG agreed to push simultaneously for fair access, for Farmers' Rights and for a consistent scope, stating that there would be an agreement either on all three items or on none at all. Therefore, a refusal to agree on one of the three items would destroy the other two. Many delegations including ours[6] and groupings of delegations submitted their suggestions to improve the 'Third Negotiating Draft'. These were made on Articles 3 (Scope), 11 (Access) and 12 (Farmers' Rights).

Farmers' Rights

The major push for the acceptance of the inclusion of Farmers' Rights in the Treaty came from the AG in the working group at the 5th regular session of the CGRFA. At a critical point, the group threatened to pull out of the negotiations unless there was a clear position to accept this. The support from developed countries came from Sweden and Norway with some compromise to accommodate this coming from the EU. The AG preferred a broader text referring to 'traditional farming communities' and not 'local and indigenous communities', proposed by some delegates from the Latin and South American region. Farmers' Rights had been discussed by the working group as a mere 'concept' in spite of the long debate that had taken place in Leipzig in June 1996, in which, albeit towards the end, even the United States of America had accepted the need for its recognition. It was thus no longer a 'concept' and legislating for its implementation had been accepted as allowed at least under national law. Therefore, Tewolde objected to the use of the word 'concept' and was joined by other delegates from developing countries. On the contrary, since the beginning, the United States' delegation stated its expectation, which was perceived as unfair amongst the AG: access to all genetic resources should be free, and intellectual property rights should not be raised in this forum.

Negotiations on Farmers' Rights started with the United States delegate reiterating that such rights should be left out of the revised IU. He insisted that international law protects only individual and not group rights, and that trying instead to include group rights would destroy individual rights. In our view, his words meant that an individual should have rights, but that two or more individuals should lose those rights if they stand together; this seems somewhat odd and illogical to us. Tewolde mentioned that the rights of individuals, especially the rights of weak smallholder Southern farmers, can be protected if they are not left to fend for their weak selves individually, but rather if they are recognized as a local community, as has been done in Article 8(j) of the CBD. India also argued for community rights, and Sweden, in particular, was very eloquent in arguing for the rights of farmers as local communities. Other Scandinavian countries and all developing countries that commented supported Farmers' Rights. France gave the objection to Farmers' Rights a new twist by saying that the United Nations system recognized individuals and countries, not groups, by arguing that Farmers' Rights would run counter to the United Nations!

Ethiopia then pointed out the following: that 'if there is a will, there is a way'; that there were groups whose interests were protected; that existing law should be able to handle Farmers' Rights; that Africa will submit a written text on Farmers' Rights; that goodwill in dealing with Farmers' Rights would generate goodwill in access; and finally, that the absence of it in Farmers' Rights would remove goodwill from access also.

Many other developing countries expressed similar sentiments and, unexpectedly, so did the delegate of the United Kingdom. She indicated the following: that, in existing law, groups can have legal identities; that the international community cannot work on a top–down basis; that Farmers' Rights legislation will have to be developed nationally; that the international community can produce an enabling situation; and that examining written suggestions should start. Brazil expressed what all developing countries felt, by stating that, if there were not going to be Farmers' Rights, there would be no access. Many other developing countries and some industrialized countries (notably Sweden) called for fairness in benefit-sharing and for support to the farmers of developing countries who have given us and continue to give us the crop genetic resources, that we need to go on living. The spokesperson of the European Union then emphasized that farmers should be fairly treated and should claim their fair share of benefits, but should have no rights to be protected by law. To us, this sounded like 'double speak'.

At the end, the Chairman ruled it would be best to focus on identifying the elements and on deciding what steps are required at both the national and international levels. He also stated that identifying elements of Farmers' Rights would be useful at this juncture. After the working group's meeting ended, some representatives of the Asian countries and AGs as well as the Brazilian delegation met and discussed the specifics of merging our texts. Since the Asian text had been the first to be submitted, it was agreed that representatives from Africa and Brazil would be compared with it, and elements, not already included, would be transferred to it. Tewolde submitted the AG's draft on Farmers' Rights to the representatives of the Asian group and to the Brazilian delegation. This was accepted. Then, the Malaysian delegate announced to the Commission that the Asian group, which had submitted a draft, was going to change its submission. This was in order to enable an official submission of the draft from the developing countries (a synthesis of the Asian, African and Brazilian submissions). Norway pointed out to the Commission the need for the disclosure of the pedigree of varieties in intellectual property rights protection.[7] Ethiopia supported Norway, but the United States, Australia and Japan opposed their main argument, believing that such issues are best dealt with in the International Convention for the Protection of New Varieties of Plants (UPOV Convention). However, Ethiopia indicated that most Southern (developing) countries are not members of UPOV and therefore, they cannot use UPOV as their forum. Moreover, since the issue raised is central to the use of PGRFA, the topic should be covered in the revision of the IU. Many developing countries supported the African intervention. The EU's spokesperson (Ireland) gave lukewarm support, stating that he would, however, need to study the wording of the paragraph suggested by Norway. The United States delegate continued his objection stating that the FAO cannot administer intellectual property rights, but the Chairman intervened affirming that the issues raised were not for administration by the FAO. On 11 December in the afternoon, some representatives of the Asian and African groups as well as the Brazilians met and discussed the specifics of merging our texts on Farmers' Rights. Afterwards, developing countries submitted the first, albeit incomplete, draft of a combined document on Farmers' Rights on 12 December.

The AG realized that the compilation of the developing countries' text had left out many of Africa's important ideas. In a second exercise, we started to include them. Some representatives from Asia and Brazil were not happy to do this, but we, the AG, threatened to formally withdraw from the exercise and resubmit our own text separately. This forced the Brazilian delegation, who had become the most difficult of the developing countries' negotiators, to accept the need for a revision, and a new text was prepared accommodating all the points that we wanted. This revised text was formally submitted to the negotiation session of the Commission by China on 13 December on behalf of the developing countries.

Scope

Africa's position on the scope of the multilateral system of the Treaty was principled, pragmatic and flexible. In the interest of feeding the hungry, we proposed to place all the PGR of a short list of six or seven of the world's staple crops (rice, wheat, maize, potato, cassava and sorghum), in a worldwide common pool of facilitated access, since these are the sources of at least 60 per cent of the world's food energy needs. While recognizing that these half dozen crops in no way cover all human nutritional requirements, they clearly include the major hunger-reducing crops. This African proposal was made in good faith, not merely as a first negotiating position. It was a clear statement of generosity in providing access to the germplasm of the most important hunger-reducing crops and it was to be the first and very considerable stage in demonstrating solid commitment on our part, while giving the opportunity to show that fair and equitable benefit sharing would operate in practice. Once it could be shown that fair and equitable benefits were indeed flowing back from open access to these few crops, Africa was quite prepared to extend the list to the PGRFA of all crops. It was not so much as a result of the point blank refusal of developed country delegations to even contemplate an introductory testing period with a very short list, but eventually more in the spirit of collaboration with other developing regions, that Africa agreed to extend the list to include some other crops.

The need for scope was agreed to without undue controversy though its formulation obviously invoked all the controversies already pointed out. The working group agreed that the scope should be limited to PGRFA. The issue of whether to include animal genetic resources in future negotiations or not was raised. While all agreed that animal genetic resources should be included, the majority view was for finalizing PGR first. This view was finally accepted by all the negotiators.

Access

Possible arrangements of access, based on three lists of species used for food and agriculture, had been explored. These lists were to include species that would be accessed on (i) a multilateral basis, (ii) a bilateral basis and (iii) a combination of a multilateral and bilateral basis. All the listed species were to be accessed under mutually agreed terms consistent with the CBD. Access was, however, even more intimately linked with Farmers' Rights. Though the AG as a whole would have allowed/agreed for an access to all crops, we could only move ahead on the issue if automatic benefit-sharing was to be assured, if Farmers' Rights were to be agreed to, and if intellectual property rights protection could be prevented from withholding genetic resources, thus undoing whatever was agreed to on access.

Benefit-sharing

The AG advocated for a twofold approach to benefit sharing – monetary and non-monetary. The group strongly felt that there was need to promote information sharing and technology transfer that would contribute to enhanced capacity among developing countries for sustainable farming and crop production.

The debate around this issue came to a head during the 3rd inter-sessional meeting of the Contact Group of the Commission in Tehran, Iran, 26–31 August 2000. Norway had previously suggested that the industrialized countries consult with their respective private sector to voluntarily come up with proposals on benefit-sharing, and this had been agreed to. Therefore, the Chairman asked the industrialized countries to report on the outcomes of their discussions with their respective private sector on benefit-sharing. Nothing definite was stated as an answer by the EU. Canada stated that their private sector feels that benefits to be shared should remain voluntary, minimal and should not affect the application of intellectual property rights. The United States stated that their small firms involved in breeding feel that any sharing of benefits would throw them out of business, but that the larger companies would further consider benefit-sharing. The US delegate emphasized that the government could not pass laws to force the private sector to benefit share.

The outcome

The outcome was not a protocol to the CBD as suggested by its Conference of the Parties in November 1996 (COP 3 Decision III/11, former § 18) but a new treaty, the ITPGRFA under the FAO, as decided by the 31st session of the FAO Conference. The prime objective of the Treaty was to make a significant contribution towards sustainable world food security.

This Treaty was, like all outcomes of negotiations, achieved through many compromises and is thus far from satisfactory. Africa wanted Farmers' Rights to be recognized by international law. We managed to get an acceptance of the rights of countries to recognize Farmers' Rights through domestic law if they so wish, as Ethiopia has now done. We wanted the prevention of access to crop genetic materials through intellectual property rights regimes to be stopped. We managed to have the Treaty require those that prevent access through intellectual property rights to pay money into the multilateral system for use to help farmers. We wanted an *initial* short list of six or seven of the world's staple crops managed in a world-wide common pool and managed to exclude from the multilateral system all but the crops considered the most essential to feed humanity. This makes it possible to negotiate bilateral benefit-sharing agreements for access to the crops excluded from the multilateral system. This is not what we had wanted at the very beginning – we would have been for a totally unrestricted access to all crops if IPRs did not create so many problems.

In spite of the success of the negotiations, Africa regards some aspects as deficiencies in the Treaty: the lack of international recognition of Farmers Rights; the weak arrangements for benefit sharing; the emphasis on the multilateral system; and the restriction of access to PGR caused by IPRs. With hindsight, Africa might have made greater efforts to maintain our original position. Now, several years after the coming into force of the Treaty, we have seen very little of the fair and equitable benefit sharing that we thought was enshrined in Treaty articles and that we expected would be as binding on developed countries as the facilitated flow of germplasm from developing ones.

Challenges for Africa in the implementation of the Treaty

The low level of awareness of the Treaty and the underlying issues that underpin its key principles among major stakeholders presented some of the major challenges to the implementation of the ITPGRFA among most African countries. It would seem that stakeholder consultations, which would have helped to raise awareness by the time the Treaty was adopted and came into force, had not sufficiently taken place in most countries. The other compounding factor was that the Treaty was coming onboard during the time when country processes for the implementation of the CBD and TRIPS were underway. In a way, one would say that the Treaty became overshadowed by these and other international instruments. In Namibia, for instance, the government's intention is to wait for the international access and benefit-sharing regime to be finalized within the CBD context before legislation is drafted for the implementation of the Treaty (personal communication, Gillian Maggs Kolling, Head, national Botanical Research Institute, 30 October 2007). There are also conflicts between different institutions responsible for coordination of national level implementation of CBD and Treaty processes. In most countries, the CBD implementation is the responsibility of ministries of environment whereas the implementation of the Treaty is with the ministries of agriculture.

Challenges regarding the SMTA and the development of the funding strategy

In the contact group negotiating the development of the SMTA, 2004–2006, the African continent was represented by ten experts drawn from nine countries (Angola, Burkina Faso, Cameroon, Ethiopia, Namibia, Senegal, South Africa, Uganda and Zambia). This group combined PGRFA technical and legal experts. Among the provisions that the group pushed for was the payment method for monetary benefit sharing. The initial position of Africa on this was to create conditions where payments were mandatory to all recipients accessing PGRFA from the MLS. This implied an upfront payment and would not need to be triggered by the commercialization of a product. Having failed to get this through, the group supported inclusion of a provision that called for voluntary contributions. The other aspect on which the group pushed hard was the rate of payment. The target for the AG was perhaps the highest, being in the range of 6–10 per cent of the commercial value. The final agreed rate was of course disappointingly low for the AG. This led the group to come up with proposals on alternative methods for payment under Article 6.7, which now appears as Article 6.11h, which is a discounted rate but broadly based as it includes sales of other products that are PGRFA belonging to the same crop. All the above positions were meant to guarantee the flow and maximize monetary benefits. The prospects of getting mandatory monetary benefits, in the short term, appeared dim considering how long it could take to develop a crop variety and have it commercialized.

In the open-ended working group to develop supporting instruments for the implementation of the Treaty, an average of 27 countries represented the AG with Angola providing leadership. The AG focused its attention on the development of the funding strategy. The group was pushing for provisions that provide clear commitments on the part of developed countries to make additional funds available for the implementation of the Treaty. More specifically, the group wanted to see an indication of targets. Reference was given to failure in the implementation of the Global Plan of Action (GPA) and some of the funding targets proposed during the GPA negotiations. Again, the group was disappointed with the final outcome, in particular the absence of any commitment by developed countries to indicate any funding target. Africa shared its position on the funding strategy with most developing countries within G77+China. Norway, Spain and Italy were the few developed countries who showed sympathy to the African and developing country positions.

It would appear that the disappointments on the outcome of both the SMTA and the funding strategy removed much of the enthusiasm regarding Treaty implementation among the contracting parties of Africa and other developing countries at large.

Although disappointed with the final outcome of the Treaty, African countries showed a lot of enthusiasm and interest in supporting its implementation, at least in the initial stages. This was evidenced by the relatively large number of African countries that had ratified the Treaty and who became contracting parties by the time of the First Session of the Governing Body in June 2006. Perhaps the election of the Chair from Africa for the Second Session of the Governing Body and his re-election for the Third Session of the Governing Body is testimony of recognition of the role the AG was playing during the early stages of the Treaty implementation at the global level.

The main thrust of the AG's position with regard to their expectations in the implementation of the Treaty was an early indication that the benefit-sharing fund under the funding strategy of the Treaty would become operational, as soon as possible, so that benefits could start flowing to farmers and farming communities – especially in developing countries and countries with economies in transition as envisaged under the Treaty. It was felt that this would bring about a balance in the implementation of the main components of the MLS and other components of the Treaty. The group's perception was that there was a lot of emphasis being placed on the MLS, in particular access provisions in terms of Treaty implementation. To the AG, the apparent lack of pace in making the benefit-sharing fund under the funding strategy created a major obstacle in the Treaty implementation. The lack of clear indications on the flow of funds for this purpose created serious doubts in the minds of most African countries as to the sustainability of the Treaty, both in terms of getting contracting parties to implement the Treaty at their country level and attracting non-contracting countries to become parties. Through the voluntary contributions of a few developed-country donors, in 2008 the benefit-sharing fund had just half a million USD available for PGRFA conservation and sustainable use projects in developing countries. By 2010 this figure has risen to USD10 million a year, although we have in mind that just 0.1 per cent of commercial seed sales could provide some US200 million a year.

Challenges in the implementation of Farmers' Rights

The AG continued to push for the involvement of the Governing Body in the implementation of Farmers' Rights. As was the case during the negotiations, most developing countries were reluctant to accommodate discussions on this, insisting that the Treaty was categorical in stating the responsibility for implementing Farmers' Rights lies with national governments. The group, however, together with other developing countries and Norway, could not accept this argument and at least managed to keep this as an agenda item for the Governing Body. These efforts were made following the realization by most African countries that they were not making much headway in implementing Farmers' Rights at the national level. This was mainly owing to limited capacity in terms of legal expertise and lack of prior experience among countries of implementing such rights in African countries.

Challenges in the implementation of the multilateral system

While African countries appreciate the importance of facilitated access under the multilateral system of the Treaty, they do not seem to have prioritized this in terms of national-level implementation. It may appear that most African countries do not consider access to PGRFA as a major benefit of the MLS mainly on account of their limited financial and/or technological capacity to utilize PGRFA, both conserved in their own gene banks and those they could access from other countries. The general feeling is that access to their national germplasm has been provided and continues to be provided to other countries, especially developed countries through international gene banks, in particular, CG Centre gene banks holding their germplasm. Ethiopia, for instance, has been clear on this view in their country report to the 2nd report on the State of the World's PGRFA.

Future expectations regarding the Treaty

In terms of future expectations regarding the Treaty's implementation, the common position of African countries, which became clear during the First and Second Sessions of the Governing Body, is that the ITPGRFA will have great difficulty in generating new and additional financial resources to support programmes to conserve and sustainably utilize PGRFA at the regional, national and local farming community levels. We also hope to see concrete realization of non-monetary benefits such as information sharing, access and transfer to technologies and capacity building. Expectations regarding the slow pace in the operationalization of the funding strategy and the apparent emphasis on supporting programmes and activities relevant to access to PGRFA under the MLS constitute, to some extent, frustrations, which have been expressed by the AG in past governing body meetings. Finally, the issue of the legal protection of intellectual property rights, that prevents access to genetic resources, also contributes to the stalemate in negotiations in the WTO. Now that the need for adaptation to climate change is adding to the value of crop genetic resources, we believe that the world might have to start examining the issues all over again.

Notes

1 http//en.wikipedia.org/wiki/africa, accessed 29 September 2010.
2 This was based on ideas that Tewolde had developed in 1992 (Egziabher, 1996) augmented by Dr Vandana Shiva (1996) and better cast into legal formulation by Professor Gurdial Singh Nijar (1996).
3 Since 1995, Tewolde was involved in discussions on the Global Plan of Action on Plant Genetic Resources for Food and Agriculture and in negotiating.
4 In Southern Africa in 1989 the governments of nine SADC member states, with support from five Nordic countries, began a comprehensive 20 year project to create a PGR network in Southern Africa, including a regional PGR centre in Zambia (SPGRC), long-term capacity building programme and the establishment of national programmes in all nine SADC countries. At the 1995 SPGRC Board meeting, the chairpersons of all nine national PGR committees agreed to work towards increasing our participation in the Commission meetings and negotiations. Nordic countries, particularly Sweden and Norway, generously agreed to support the participation of all the network's chairpersons in all the ITPGRFA negotiating sessions from 1996 onwards, for seven years in all.

5 This is one of the reasons why, during the inter-sessional contact group's meeting in Tehran on 26–31 August 2000, members from developing countries asked Tewolde to be the main spokesperson of the G-77 and China.
6 Tewolde came to these negotiations with an analysis of the Third Negotiating Draft and with suggested submissions on these issues. The AG quickly adopted these suggestions and Africa as a region thus submitted drafts on the three issues and managed to stay coherent and consistent.
7 As stated in Paragraph 45 of the Secretariat's unofficial draft.

References

COP 3 Decision III/11 (Retired sections: paragraphs 1 to 12, 18, 23 and 24) Conservation and sustainable use of agricultural biological diversity, available at www.cbd.int/decision/cop/?id=7107

Egziabher, T. B. G. (1996) 'A case for community rights', in S. Tilahun and S. Edwards (eds) *The Movement for Collective Intellectual Rights*, Institute for Sustainable Development, Addis Ababa, and the Gaia Foundation London, pp1–51

FAO (1997) Background Study Paper no 7, REV.1, available at ftp://ftp.fao.org/docrep/fao/meeting/015/j0747e.pdf

FAO (1998) 'Report of the Fifth Extraordinary Session of the Commission on Genetic Resources for Food and Agriculture', FAO, Rome

Kloppenburg, J. R. and Kleinman, D. L. (1987) 'Plant germplasm controversies: Analyzing empirically the distribution of the world's plant genetic resources', *Bioscience*, vol 37, pp190–198

Nijar, G. S. (1996) 'In defence of local community knowledge and biodiversity: A conceptual framework and essential elements of a rights regime', in S. Tilahun, and S. Edwards (eds) *The Movement for Collective Intellectual Rights*, Institute for Sustainable Development, Addis Ababa, and The Gaia Foundation, London, pp71–117

Shiva, V. (1996) 'A new partnership for national sovereignty; IPRs, collective rights and biodiversity', in S. Tilahun and S. Edwards (eds) *The Movement for Collective Intellectual Rights*, Institute for Sustainable Development, Addis Ababa, and The Gaia Foundation, London, pp52–70

4

The Asian Regional Group

Eng Siang Lim

Introducing the Asian region

FAO's Asian sub-region comprises 25 members out of which 9 are not party to the Treaty (including China, Japan, Kazakhstan, Mongolia, Sri Lanka, Timor-Leste, Uzbekistan and Vietnam; Thailand having signed but not ratified the Treaty (see Annex 2 of this book for the table of ratifications per region).

Food and poverty in Asia

Despite rapid economic progress and poverty reduction, Asia and the Pacific accounts for 63 per cent of the world's undernourished (FAO, 2009a); according to the United Nations Food and Agriculture Organization:

> *In South Asia, the incidence of child malnutrition is higher than in any other region. Only a few countries are on track to meet the World Food Summit target of halving the number of undernourished by 2015. Furthermore, future progress is uncertain, especially in the wake of recent substantial gains in cereal prices that make it more difficult for the rural landless and the urban poor to afford adequate nutrition. Interest in bio-fuels as a means to achieve energy security may lead to further increases in commodity prices that will help some farmers but will have negative impacts on food security for many households.* (FAO, 2009b)

The Asian region has reported that both China and India are well on track to achieving the Millennium Development Goal of halving the prevalence of poverty and hunger, as are 17 other countries. In general terms, accelerating growth in India has put South Asia on track to meet the goal, while East Asia has experienced a sustained period of economic growth, led by China. However, a few countries in the region are continuing to face difficulties in reducing hunger sufficiently to meet the MDG and World Food Summit targets (FAO, 2009c). South Asia has the highest level of underweight prevalence in the world, with almost half (46 per cent) of all children under five being underweight. Three countries in this region drive these high levels – India, Bangladesh and Pakistan – which alone account for half the world's total underweight children. Large disparities exist for underweight prevalence among urban and rural children. On average, underweight prevalence among children in rural areas is almost double that of children in urban areas in the developing world. Malaysia has the fastest rate of improvement (FAO, 2008).

Asian countries' interdependence on plant genetic resources for food and agriculture

Next to these striking data, it is important to stress that already in the 1920s, the Russian geneticist Vavilov had identified Asia as one of the regions in the world with the highest genetic variability of cultivated food crops, through the determination of several important centres of origin including Central Asia, China, India and Indo-Malaysia. According to a background study paper of the FAO Commission on Plant Genetic Resources from 1997, Asia is indeed the primary centre of origin of many important crops such as rice, wheat, sugar, soybean, banana and plantain, grapefruit, rye, pea and onion (FAO, 1997). This study also confirms a finding from Kloppenburg and Kleinman (1987) in that the Asian and Pacific regions are the least dependent upon crop species originating in other regions of diversity for their food production (Table 4.1).

Asia is therefore a primary provider of genetic diversity to the rest of the world. This status certainly contributed to the importance given by the Asian regional group to the negotiation and implementation of the Treaty.

Table 4.1 *Percentages of regional food production dependent upon crop species originating in other regions of diversity*

Regions	Percentage of dependence
Chino-Japanese	62
Australian	100
Indochinese	34
Hindustanean	49
West Central Asiatic	31
Mediterranean	99
African	88
Euro-Siberian	91
Latin-American	56
North American	100

Source: Kloppenburg and Kleinman (1987)

The Treaty: A crucial instrument to negotiate and implement for Asia

Asia has for a long time been conscious that conserving and using plant genetic resources in a sustainable way is vital for our future. Participating in and implementing the International Treaty on Plant Genetic Resources for Food and Agriculture (ITPGRFA) is also one of the means to reach the first Millennium Development Goal (MDG). The importance given to the Treaty by Asian countries can be easily demonstrated. First, India is one of the principal Asian countries to have put the conservation biodiversity and of Farmers' Rights high on its political agenda. Already in 1981, H. E. Shrimati Indira Ghandi, Prime Minister of India, gave a Frank MacDougall Memorial Lecture on the topic, at the 21st FAO Conference. Mr Monkombu Sambasivan Swaminathan, known as the father of the green revolution in India, has also given an invaluable contribution to the promotion of the field throughout his career, first in establishing the Commission on Plant Genetic Resources as an independent chairman, FAO Council, Rome, in 1981–85. Then, he developed the concept of Farmers' Rights and the text of the International Undertaking on Plant Genetic Resources (IUPGR). He also chaired the International Congress of Genetics (1983), and between 1988 and 1991 sat as a chairman of the International Steering Committee of the Keystone International Dialogue on Plant Genetic Resources, regarding the availability, use, exchange and protection of plant germplasm. Finally, India was the first country in the world to adopt and implement legislation on Farmers' Rights, thereby recognizing the primary importance of this question.

Second, Asian countries have been very active during the negotiations of several treaties relating to genetic resources, in particular, within the Like Minded group. Asian countries have also often hosted meetings related to the conservation of biological diversity, whether specific for food and agriculture or under the scope of the Convention on Biological Diversity (CBD). In 1995 notably, Indonesia hosted the 2nd ordinary meeting of the Conference of the Parties at the Convention on Biological Diversity (COP 2), where the special nature of agricultural biodiversity, its distinctive features and problems needing distinctive solutions, were expressly recognized (CBD, 1995).

Finally, Asian non-governmental organizations (NGOs) have always been very active in the field of plant genetic resources. One of these important institutions is SEARICE, which has strongly promoted and protected farmers' communities' rights in Asia and throughout the world (for more details, see Chapter 13). The officer of SEARICE who represented the Philippines, played an important role in the negotiation on Farmers' Rights in the ITPGRFA (see Annex 3 of this book for details on the main provisions of the Treaty). Other Asian stakeholders, such as breeders and gene bank curators, have also dynamically participated in international networks such as the ones supported by the former International Board for Plant Genetic Resources (IBPGR), which later became the International Plant Genetics Research Institute (IPGRI, now Bioversity International). This significantly contributed to spread the essential need to conserve, sustainably use and share agricultural biodiversity.

Besides, I myself had the honour to chair or vice-chair several important meetings during the negotiation of the Treaty (see Annex 1 of this volume for the list of all Commission and Treaty meetings). I was therefore able to witness sensitive discussions, which I have tried honestly to articulate in the personal views expressed in this contribution. Other delegates from Asia were privileged to chair several meetings, contact groups, unofficial meetings and working groups during the negotiations of the Treaty, hereby providing and securing an Asian input in the negotiations of the Treaty. Active participation from delegates of India, the Philippines and Malaysia in the negotiations influenced the final conclusion of the provisions of the Treaty, in particular, the articles on the multilateral system (MLS) and Farmers' Rights. The positioning of the Asian region during the progress of the negotiation, reflected its social economic environment.

Today, it is recognized by most stakeholders worldwide, that agriculture and the rural economy play a crucial role in securing sustainable gains in the fight against hunger and poverty, and 'there is much greater appreciation now for the fact that agriculture has strong links with other sectors' (FAO, 2009c). Indeed, many external factors impact on the way Asia manages its plant genetic resources for food and agriculture: the tremendous growth of Asia's population and economy, the rapidly changing climate, a globalizing trade pressure, an increased recognition and implementation of democratic schemes and human rights, in particular, through the growing role played by NGOs and civil societies (see Chapter 10). These factors are taken into account at the national and regional levels, when the Asian group meets to discuss PGRFA issues. In order to facilitate the collaboration between Asian countries and allow them to take decisions on and implement the Treaty, Asian regional meetings are organized prior to each international meeting related to plant genetic resources.

This chapter will highlight some of the main issues for which the Asian region has played a role during the negotiation of the Treaty. This contribution will also spot the challenges that the region is facing in the implementation of the Treaty at the national level, as well as more global issues to be specified and agreed upon at the international level to facilitate and increase the efficient participation to the Treaty.

The principle of common but differentiated responsibilities: A key to Asia's views on the Treaty

The foundation of Asia's position during the negotiation of the Treaty was based on the Common but Differentiated Responsibilities as accepted in Agenda 21. Asia recognized that countries have common responsibilities in the conservation and sustainable use of PGRFA for food security, quality of life and environment well-being. The operational common interests to achieve the objectives of conservation and sustainable use of PGRFA for food security cover the strategic need to have access to genetic resources for research and development, technologies and information. The differentiated responsibilities lie in the strategic need of countries to provide for access to genetic resources, technology, information and financial resources in accordance with their capabilities and capacities.

The principle of common but differentiated responsibilities was pushed as the initial positioning of Asian countries. It provided the strong foundation to articulate the pillar of access to genetic resources, technology, information and financial resources. It also supported the pillar of benefit-sharing arising from the use of genetic resources. The initial positioning was necessary to support the determination of the concept of food security as a global public good. However the long and time-consuming negotiation on Farmers' Rights and intellectual property rights (IPRs) soon triggered the change towards the safeguarding of national laws in terms of access to genetic resources and IPRs.

The shift from the principle of common but differentiated responsibilities to prevailing national interests

The provisions in national laws on access to genetic resources and IPRs have influenced the negotiating position of Asia, which is to have easy access to genetic resources and to safeguard the provisions of IPRs in their national laws. The negotiation on differentiated responsibilities became more difficult as developed Asian countries needed to protect IPRs on technologies. Developed Asian countries also have national interests to protect information, in particular, technological information that gives rise to commercial/competitive advantage. Developing countries, with large rural populations engaged in small-scale agriculture, were interested in safeguarding the informal breeding and seed systems which provide the main source of rural food security and livelihoods. Some of these countries which are country of origin also have the national interest to obtain direct benefits, in particular, commercial benefits.

The 5th session of the Commission on Genetic Resources for Food and Agriculture (CGRFA) discussed the timetable for the revision of the International Undertaking (IU) (see Annex 1 of this volume for the list of all Commission and Treaty meetings). The session agreed that the revision should carefully be conducted, as a gradual pragmatic and step-by-step process, building on the consensus already achieved through the Commission's previous discussions, as embodied in the IU and its annexes. Conference Resolution 7/93 requested that the revision of the IU be negotiated (see Chapter 10 for full detail of the IU revision).

A working paper on the issues for consideration in Stage II entitled 'Access to genetic resources and Farmers' Rights', was presented to the 9th session of the working group (11–14 May 1994) and to the 1st extraordinary session of the CGRFA (7–11 November 1994). The formal negotiations of the Treaty started with the 1st extraordinary session of the Commission, in November 1994. During this session the Commission only focused on the discussion of Stage I entitled 'Integration of the annexes and harmonization with the Convention on Biological Diversity'.

At this stage, the terms of 'free access', 'availability of PGRFA' and 'conditions of access', were discussed within the framework of harmonization with the CBD. Asia pushed for the use of the term 'conditions of access'. In the end, a compromise was established with the use of the term 'facilitated access to PGRFA'.

The Annex I List negotiation: An important feature in Asia's position on the Treaty

The finalization of Article 9, Part IV on the MLS, and Article 18 also influenced the concluding negotiation of the Annex I list of crops in the MLS. Asia did not play a major role in the early phase of negotiation of the list. However, Asia had an important role in the final stage of the list negotiations, when national interests of countries of origin prevailed in excluding their genetic resources from the list. This is particularly true for major agriculture exporting countries. Their national interests were the need to safeguard/protect their competitive advantage in the export markets and to use their genetic resources for bilateral exchanges.

It was during the 6th session of the CGRFA that the proposals were submitted on a list of genera in Annex I of the proposed article on the scope of the IU, which provided an example of a list containing 231 genus and the scenario to establish a multilateral system or undertaking for those harvested species most used for food and agriculture.

Ideas to establish bilateral and/or multilateral agreements in relation to access to PGRFA, were discussed during the 10th session of the working group (3–5 May 1995), where the option for a list of crops was also proposed. This implied adding a list of mutually agreed species to which specific provisions of the undertaking would apply, particularly in relation to access to and the distribution of benefits. This option received fairly good acceptance. The idea of species or gene pool of major relevance to food security and those for which there was strong interdependency between countries was discussed.

At the 6th session of the CGRFA, 19–30 June 1995, there were proposals on the Scope of the IU and the list of 231 genera was submitted as an example under Scope. There were also proposed wordings on the Availability of PGR (access). Within the proposals on Availability, there were wordings on benefit sharing. The option submitted by EU listed 231 genus consisting of: major grain crops-grasses (12); minor grain crops (6); major grain legumes (9); minor grain legumes (12); cereals from other families (5); major starch crops (7); minor starch crops (3); oil crops (5); fruits (3); shrub fruits (6); tree fruits (30); vegetable crops (38); nuts (7); species (7); herbs (20); beverages (6); fibre (6); sugar crops (2); industrial crops (6); forage-grasses (22) and forage legumes (19). The list was incorporated into the Third Negotiating Draft.

At the 3rd extraordinary session of the CGRFA, 9–13 December 1996, the USA submitted a list of crops (genus) essential to global food security (25 crops plus forages); Brazil submitted a list of crops/genera of basic importance for human world food consumption (25 crops); Africa stated that access to and inclusion of crop species in the system could be willingly decided by members of a multilateral system; France stated that within each species, there will be different classes of genetic material: (a) First Class: designated material with unrestricted access through an international network of collections; (b) Second Class: non-designated material with negotiated access. Brazil wanted to start the multilateral system with a small window, likewise with the USA; the EU preferred it with a large window.

At the 7th regular session of the CGRFA, 15–23 May 1997, three options were provided for further negotiation:

- Option A: Designated material in the international network or PGRFA (genus) designated by national governments.
- Option B: Designated material or PGRFA (genus) listed in the Annex or Material not included.
- Option C: PGRFA (genus) listed in the Annex or Material not included.

At this stage of the negotiation, there were many possible scenarios. There were options within options and countries/regions had positions regarding access, benefit sharing and list of crops. The Fourth Negotiating Draft had 58 pages.

At the 4th extraordinary session of the CGRFA (1–5 December 1997), a major breakthrough, in terms of a proposal for a multilateral system to facilitate access to PGRFA through a list of major crops, began to take shape. From all the proposals on the list of crops, the Commission agreed to have one Tentative List of Crops for further negotiation. This list contained 37 crops (41 genus), grasses (28 genus) and legumes (33 genus).

The informal meeting of experts on PGRFA, in Montreux, Switzerland (19–22 January 1999) (see Chapter 2 for a detailed analysis of this meeting), proposed a multilateral system, including conditions for facilitated access and benefit-sharing to be applied to a specific list of crops. The Montreux meeting thus set a broad framework of agreed principles for further negotiation. The criteria used to establish the Tentative List of Crops, were (i) their importance for food security at local or global levels, and (ii) countries' interdependence with respect to PGR. The 8th session of the CGRFA (19–23 April 1999) agreed that the multilateral system shall cover PGRFA listed in Annex I to the future Treaty and established the criteria of food security and interdependence.

At the 2nd inter-sessional meeting of the contact group, 3–7 April 2000, statements were made on whether the window (list of crops) of the multilateral system, should be small or as wide as possible. Brazil wanted it small and the EU wanted it big. The USA has stated its position. Africa has stated its position. Other countries/regions remained silent. It was only at the 3rd inter-sessional meeting of the contact group (26–31 August 2000) that regions submitted a concrete list of crops. The information paper prepared by the Secretariat illustrated what the following regions proposed: Africa – 9 crops; Asia – 22 crops (24 genus); Europe – 273 crops including fruits, vegetables, nuts, herbs, species, forages, beverages and so on; Latin America and the Caribbean – 29 crops; and North America – as in the tentative list in Annex I – crops – 37 (41 genus), grasses – 28 genus, legumes – 33 genus.

At the 6th inter-sessional meeting of the contact group, Spoleto (22–28 April 2001), members of the working group on the list (Canada and Iran as co-chairs; Angola, Burkina Faso, Zimbabwe for the Africa region; China, Japan, the Philippines for Asia; France, Poland, Sweden for the European region; Argentina, Brazil, Colombia for Latin America and the Caribbean region; USA for North America; and Australia, Samoa for the Southwest Pacific) invited experts from IPGRI and the Secretariat of the CGRFA (FAO) to begin serious negotiation on the list. The working group used the criteria of food security and interdependence to select the crops. The lists submitted by the regions (FAO, 2000) were used as source material and compiled in one working document, comparing commonality of crops among regions. The working group worked on the crops most commonly suggested by the regions. The working group agreed that the working basis should be crops, with genera as indicative of crops, and species designation in cases where required. The working group achieved consensus on 30 food crops (Table I in the working group document). A further group of widely consumed food crops (Table II), where there is considerable support from a number of regions, remains under discussion. In addition, there were crops important to one or more regions that had not been discussed yet. Forage crops were highly important to all seven regions. However, requirements were diverse and highly complex. Discussions on forage crops had just begun and needed considerable further discussion, including advice from forage experts. The working group recommended that:

1 A panel of experts be asked to examine the genera in Tables I and II and make technical recommendations (including scientific sources) for further consideration and final confirmation, at the species level when required by the regions, the working group and the contact group. This study would identify and suggest the relevant genetic resources of the crop, including related genera and species that are important for breeding activities and the root stock of the crop, if relevant.
2 An opportunity be provided for discussion of the crops from the lists submitted by the regions, that have not yet been considered.
3 The working group continues to develop, with the assistance of forage experts from the regions, the list of forage crops for the next meeting of the contact group.
4 The working group finishes its work on the list of food crops before the next meeting of the contact group.

At the 6th extraordinary session of the CRGFA (25–30 June 2001), the final negotiation on the list took place, mainly among developing countries on the exclusion of such crops as soya bean, tropical forages, oil palm, sugar cane and groundnut/peanut from the list of crops. The active participation of the Asian region was focused on excluding soya bean, oil palm and sugar cane from the list in order to protect national interests in these crops. The final list consisted of 35 crops (36 genus), 15 genera of legume forages, 12 genera of grass forages and two genera of other forages. Most countries in Asia were contented with this final list. However, a few countries were not fully satisfied because rice was included in the Annex.

What are the challenges ahead for Asia?

The Treaty tried to accommodate most of the contracting parties' common interests in its MLS. However, the need for parties to safeguard their national interests will make it difficult for countries to follow a common framework of implementation at the national level. Some of the provisions in the MLS and the standard material transfer agreement (SMTA) are still very general and can be interpreted differently to suit national interests. Such provisions would require further elaboration by the Governing Body which has to agree on a common framework for implementation at the national level. Such provisions include:

Article 12.3 (e) PGRFA under development

Questions have arisen regarding what materials can be classified as PGRFA under development. The Treaty does not define PGRFA under development. However, the SMTA has a definition on PGRFA under development are defined under Article 2 of the SMTA,

> *PGRFA under development means material derived from the [original material accessed from the Multilateral System] and hence distinct from it, that is not yet ready for commercialization and which the developer intends to further develop or to transfer to another person for further development. The period of development shall be deemed to have ceased when those resources are commercialized as a Product.*

The rationale to have this definition for PGRFA under development in the SMTA is built on the idea of an unbroken chain of contractual obligations passed on from recipient to recipient until a commercial cultivar is released. It allows identification of how and when the development chain starts and how and when it ends.

Other questions that need be to resolved are:

- Article 12.3 (d) IPRs and other rights
- Article 11.2 Management and control of the Contracting Parties and in the public domain
- Article 12.3 (a) and SMTA Uses of PGRFA other than those uses provided for in the MLS
- Article 13.2 a), b) and c) Mechanism for the sharing of non-monetary benefits
- Creating legal space in national legislation on access and benefit sharing (ABS) including Article 11.3 and practical and legal implication of natural and legal persons putting material into the MLS as well as Article 12.3 (h).

The Governing Body of the Treaty has established the Ad Hoc Advisory Technical Committee on the SMTA and the MLS to consider the above issues and other issues raised by the contracting parties and other users of the SMTA and the MLS. Hopefully, the views and opinions of the committee will be useful to guide the operational efficiency and transparency as well as the legal certainty in the implementation of the SMTA and the MLS.

Conclusion

Negotiating the Treaty has been a very demanding and creative effort between all stakeholders involved in plant genetic resources and between all member countries. However, the positive outcome of the revision of the IU through the signing of the Treaty and the conception of its innovative multilateral system should not lead to a situation where states rely on what has been done, thus slowing the process down. On the contrary, more efforts should be placed in a common implementation framework to help countries efficiently apply the Treaty obligations at the national level. Particularly in Asia, integrated policy and planning, between line ministries and the private sector, and within and beyond national jurisdictions, first require that the agricultural sector becomes aware of its own environmental externalities, as well as of the impact of environmental change on its economic and societal performance. This will allow the definition of appropriate policy objectives within the agricultural sector, based on negotiated strategic actions and respecting national interests, including legal structures and resource allocation (FAO, 2009c). This will also allow for an effective application of the Treaty and will contribute to enhance and expand the recent positive outcomes of the Treaty in our region and all around the world.

References

CBD (1995) Second ordinary meeting of the Conference of the Parties at the Convention on Biological Diversity (UNEP/CBD.COP/2/19), Decision 2/15, available at www.cbd.int/doc/?meeting=cop-02

FAO (1997) Background study paper no.7, REV.1, available at ftp://ftp.fao.org/docrep/fao/meeting/015/j0747e.pdf

FAO (2000) Commission on Genetic Resources for Food and Agriculture, Fourth Intersessional Meeting of the Contact Group, Neuchâtel, Switzerland, 12–17 November 2000, document entitled 'Information provided by the regions on the list, during the 3rd inter-sessional meeting of the contact group' (Tehran, Iran, 26–31 August 2000) document 'CGRFA/CG-4/00/Inf.4'

FAO (2008) 'The state of food and agriculture in Asia and the Pacific region 2008', RAP publication 2008/03, available at ftp://ftp.fao.org/docrep/fao/010/ai411e/ai411e00.pdf

FAO (2009a) 'The State of Food Insecurity in the World. Economic crises – impacts and lessons learned' available at www.fao.org/publications/sofi/en/

FAO (2009b) FAO Regional Conference for Asia and the Pacific (APRC), 29th Session, Bangkok,26–31 March 2009, Senior Officers Meeting, document 'APRC/08/1'

FAO (2009c) FAO Regional Conference for Asia and the Pacific (APRC), 29th Session, Bangkok,26–31 March 2009, Agenda Item 7, Regional State of Food and Agriculture, document 'APRC/08/INF/5'

Kloppenburg, J. R. and Kleinman, D. L. (1987) 'Plant germplasm controversies: Analyzing empirically the distribution of the world's plant genetic resources', *Bioscience*, vol 37, pp190–198

Chapter 5

The European Regional Group

Europe's Role and Positions during the Negotiations and Early Implementation of the International Treaty on Plant Genetic Resources for Food and Agriculture

Bert Visser and Jan Borring

Introduction

European positions in the negotiations on the International Treaty on Plant Genetic Resources for Food and Agriculture (ITPGRFA) (hereafter the Treaty) were strongly influenced by developments in European agriculture during the last century. In particular since the 1960s, as a result of the creation of the European Community and its Common Agricultural Policy, the face of agriculture in Europe changed profoundly, characterized by major-scale increases in production, a strong increase in the use of external inputs at the farm and the development of a strong breeding industry making use of the latest technologies (see Chapter 12 for more detail on the seed industry). Product demands were increasingly driven by the food industry and the retail sector, resulting in a high level of product uniformity. A large proportion of European farmers would increasingly buy their seeds on the market rather than save these on the farm.

In this process, genetic erosion of plant genetic resources that had already commenced in the first half of the 20th century, continued. At first this was mainly a concern of breeders who noticed that the very basis of their work was disappearing. In the last decades of the previous century, it increasingly became a concern of segments of the general public, often in a wider context (e.g. loss of biodiversity, the need for protection of the environment and of traditional landscapes, the rise of organic agriculture and the Slow Food movement).

From the outset of the negotiations, the European regional group (ERG) attached great importance to the establishment of an international legally binding instrument for plant genetic resources for food and agriculture (PGRFA). Heavy mutual interdependence amongst regions with regard to PGRFA was recognized as a central motive for the establishment of this agreement. Underlying this was the conviction that PGRFA are of a specific nature, justifying specific regulatory measures on access and benefit-sharing (ABS), in line with the decisions laid down in the Nairobi Final Act (1992) on the Convention on Biological Diversity (CBD). Food security as such formed a less explicit but nevertheless quite widely recognized motive in the European discussions and contributions to the negotiations. In particular, a definite resolution of the status of the Consultative Group on International Agricultural Research (CGIAR) collections was regarded as a long-term guarantee for food security in poor regions. Obviously, the two motives of mutual interdependence and global food security are closely interlinked.

The European Union forms the dominant political institution in the ERG (see Annex 2 of this book for the list of European contracting parties to the ITPGRFA).[1] Although the sheer size of the European Union (EU) and its single voice made it highly influential, those two aspects did at some points also turn into a disadvantage. The EU position often resulted from lengthy internal debates, in itself a compromise and sometimes a minimum position between the views of the various member states. Such positions could only, with difficulty, be further developed in the negotiation meetings. During the early stages of the negotiations and before joining the EU, Poland played an important role in defending a special position for 'countries with economies in transition'. Moreover, two non-EU member countries[2] – Norway and Switzerland – played an important role in the process mainly because they were able to modify their positions more easily, if new developments in the negotiation process required such. For the same reason, the negotiators of these latter countries could often devote more time to exchanges with other regions. To a major extent, these distinct roles complemented each other in the negotiating process.

During the early stages of the negotiations, it appeared that not all regions had similar capacity to participate in the negotiations and to influence the outcome. Several European countries therefore contributed to capacity building as well as support for developing-country participation in the negotiations over the years. Such support was one of the factors that contributed to increased participation, involvement and influence from the African region, with significant results for the proceedings as well as the outcome of the negotiations, as acknowledged repeatedly by the African region itself.

At several stages during the negotiations, it also became obvious that the formal setting of the negotiations and the size of the meetings did not always create the best dynamics for exploring the complex issues on the table. Understanding the scientific and practical aspects of the elements under negotiation was at times at least as challenging as dealing with the more generic issues of finance, compliance and North/South perspectives. Therefore, a number of European countries at various times facilitated informal meetings where issues such as the specific nature of PGRFA, the interdependency between regions, and ways to realize benefit-sharing at the international level were explored in more detail. In addition to facilitating a better understanding of the issues at hand, such informal meetings also helped delegates from different regions in getting to know each other better, thereby contributing to better communication between negotiators in general. In a similar vein, the ERG often provided chairs or co-chairs to the contact groups in order to foster making progress in the negotiations.

Europe's positions on some key issues during the negotiations

Europe's views on the relation with the CBD

The ERG fully recognized the importance of the CBD and its objectives including the paradigm shift that underpinned the CBD – that is, a change from viewing biodiversity as the heritage of mankind and open exchange of its components to applying the concept of national sovereignty regarding the conservation and utilization of biodiversity. However, for the European region it was important that by the adoption of Resolution 3 of the Nairobi Final Act when the CBD was finalized, the specific nature of PGRFA was recognized, and that the United Nations Food and Agriculture Organization (FAO) was called to bring the FAO International Undertaking in harmony with the CBD.

In developing the new instrument, challenging factors included the legal complexities related to pre-CBD material, and the status of the CGIAR collections (for details on the CGIAR, see Chapter 11). In addition, newly enacted access and benefit-sharing legislation in some developing countries based on the CBD were perceived to ignore the needs of the agricultural sector. The challenge was to negotiate an instrument in harmony with the CBD but at the same time accommodating the needs of the agricultural sector.

Strong interdependency between regions with regard to PGRFA for important food crops was seen as a central argument for finding more effective solutions than bilateral mechanisms. Also, it was recognized that many crop varieties contained traits derived from genetic material stemming from a large number of countries meaning that frequently no clear 'country of origin' could be identified. Bilateral 'fair and equitable' benefit sharing for such materials would create extremely complex challenges of calculating and apportioning how benefits should be shared between a large number of countries.

In the negotiating process, an understanding soon developed in Europe and elsewhere that the CBD in itself did not exclude a multilateral system for ABS, provided that countries used their sovereign rights over genetic resources by agreeing to such multilateral mechanisms. In other words, 'mutually agreed terms' could be understood as multilaterally agreed rules applicable at the international level. Likewise, the scope and contents of 'prior informed consent' could be agreed on a multilateral basis. Based on this perspective, Europe strongly favoured a solution whereby facilitated access would not depend on approval on a case-by-case basis by individual countries. The ERG subsequently played a central role in developing such a model for ABS on a multilateral basis, whereby monetary benefits would flow into a financial mechanism to be managed by the parties to the Treaty for purposes of implementing the Treaty. The first text proposal for a provision linking obligatory benefit-sharing to commercialization in case of restrictive intellectual property rights (IPR) was submitted by an ERG country. Several ERG countries also facilitated informal workshops where options for such mechanisms were discussed in detail.

Until late in the negotiations (see Annex 1 of this volume for the list of all Commission and Treaty negotiating meetings), the legal and institutional status for the new Treaty remained undecided, with both a protocol under the CBD and a self-standing agreement under the FAO remaining options on the table. In the end, the ERG was content with the agreement as a self-standing instrument in the framework of FAO. However, in the subsequent negotiations on the standard material transfer agreement (SMTA) for the multilateral system, reservations on certain aspects of FAO's functioning led the EU to only agree with some hesitation to the identification of FAO as the third party beneficiary (see Annex 3 of this book for details on the main provisions of the Treaty).

The specific nature of PGRFA and the scope of the multilateral system

The presence of a strong breeding industry in Europe contributed to the recognition of one of the principle notions in the field of PGRFA: the fact that in many cases breeding strategies had resulted in crop varieties built of building blocks originating from a large number of countries and even continents, rendering the concept of countries of origin largely inappropriate. As a result of this notion, for the European breeding industry access to as many source materials as possible was important, together with the notion that wide access through international cooperation was not only essential for European breeders, but for breeding programmes in all regions alike, including for those of the CGIAR centres, for the purpose of food security.

Europe always looked at the free availability of new crop varieties and source materials as one of the most important benefits that could be realized through the development of what was to become the ITPGRFA. In promoting the concept of facilitated exchange, the ERG was willing to offer what it thought to be of high value to other regions in the world: on the one hand access to newly developed state-of-the-art crop varieties, available for further research and breeding by other parties through adhering to plant breeder's rights rather than patent rights as the IPR system of choice in plant breeding, and on the other hand large and relatively well kept gene bank collections (see Chapter 14 for an example of gene bank collections).

The European region therefore initially favoured the inclusion of all crops, and later – when this did not appear to be attainable – of a large number of crops in the list of crops that would define the scope of the multilateral system (MLS), not only because of the notion that mutual interdependence was an apparent feature for most if not all crops, but also since food security was not only a matter of access to sufficient calories but also a matter of breadth and variation in diet. During the entire course of the discussions on the scope of the MLS, it therefore defended as long a list as possible. It is a strong view in the European region that the final list is a compromise based on political interests that had little to do with food security and the recognition of mutual interdependence. In the final stage ERG proposed the addition of specific crops (e.g. temperate grasses and forage crops) on the list. They were not contested by other regions because Europe provides the major holdings of those crops.

The interpretation of Article 12.3d and IPR systems

Article 12.3d of the Treaty is the result of lengthy negotiations to reconcile opposing views on IPR systems on germplasm. This text is to some extent ambiguous. What remains a matter of legal interpretation and jurisprudence to be developed is how the phrase on the subject matter to which Article 12.3d applies – 'its parts and components' – relates to the phrase 'in the form received'. In the view of the European region the material itself, as obtained from the MLS, cannot be protected, but any product developed from that material can be protected. In fact, any other interpretation would make Article 13.2(d) meaningless. Plant breeder's rights forming the prevalent IPR system in Europe, protected varieties are freely available for further research and breeding and are in full harmony with Article 12.3(d). The remaining issue therefore regards biotechnological applications. In other words, the question is what really constitutes a 'product' in biotechnological use, and, in particular, whether the mere isolation and independent multiplication and use of a DNA sequence in its original form but in a different genetic environment is sufficient to define that as a 'product'. Such interpretation is, at least in theory, also open for claims for IPR on DNA sequences determined for DNA of germplasm obtained from the MLS. However, over the last few years obtaining patents on such very basic claims has become much more difficult in practice under most jurisdictions. This issue is strongly related to discussions in other organizations on the question of what should constitute an 'inventive step' under IPR regimes. To find generally accepted solutions remains an important challenge in order to avoid potential conflicts with other types of legislation when implementing the Treaty.

Although differences of view on some IPR-related issues occurred within the European region, witnessed in the debate on the EU Patent Directive, the ERG does regard intellectual rights systems, including patent systems, as generally beneficial for industry and for economic development and society at large. Thus, it did not accept any attempts to weaken IPR systems by the backdoor since the nature and role of these systems had been agreed upon in the framework of the World Intellectual Property Organization (WIPO) and the World Trade Organization (WTO) Trade-related Aspects of Intellectual Property Rights (TRIPS) agreement. However, on one issue that is negotiated under the CBD process towards a Protocol on Access and Benefit-sharing the region has gradually adapted its position – that is, in considering the advantages and drawbacks of modalities forcing parties requesting a patent right or plant breeder's right to disclose the origin or legal provenance of the germplasm used to develop the product. Whereas originally it questioned the need for such provision, it has accepted its potential usefulness. In the view of the region, such a modality may be integrated in IPR systems, although as an alternative the form of a self-standing requirement to market products based on biological materials was also debated for some time.

The nature of the SMTA

The negotiations on the SMTA were not part of the Treaty's negotiations but took place after its adoption. This agreement is essential for the implementation of the Treaty. Since the benefit-sharing provisions of the Treaty could not become really operational without an agreement on the SMTA, which in turn meant that the MLS could not start functioning as perceived, the region was of the view that the SMTA was the last essential component to render the negotiations on the Treaty complete.

It is the view in the European region that in many cases not the parties themselves but legal and natural entities within its jurisdictions will act as providers of germplasm for the MLS, be it under the control of governments or not. From this perspective it is easier to understand why the European region regards the SMTA first and foremost as a contract between the two signatories of the SMTA, and therefore as a contract under civil law. The only peculiarity of the SMTA is formed by the fact that from its very nature as standard (not as model) it followed that its contents could not be negotiated by its parties/signatories, clearly placing the instrument in a special category. This position also explained the region's pleas during the negotiations of the SMTA to accept recourse to the International Chamber of Commerce as an appropriate component of the SMTA. In line with this position, the region holds the view that any dispute about the adherence to any specific signed SMTA is only a responsibility of its signatories and not necessarily of the parties to the Treaty, in so far as they are not themselves a signatory to that specific agreement. Instead, the role of the parties is to oversee whether the SMTA fulfils its tasks and is appropriate and functional. Only an eventual review of the SMTA is seen as a major responsibility of the parties to the Treaty.

Since its adoption in 2006, some countries within the ERG have actively promoted the acceptance and use of the SMTA by its collection holders and by its breeding sector, both the public and the private sector. The European Seed Association has advised its members to accept the SMTA for access to genetic resources under the MLS.

The concept of the third party beneficiary was favourably evaluated by European parties. It was realized that in many cases providers of germplasm under the MLS would not have the means to follow up the utilization of that germplasm, including in cases where serious doubts on the adherence to the obligations might arise. For the European region, defining the roles and functions of the third party beneficiary formed a major issue that was only partly addressed in the negotiations of the SMTA. In particular, a development leading to a large number of tasks, an added bureaucracy and inspection of individually signed SMTAs was seen as highly undesirable by the ERG. In particular, some concrete proposals were interpreted as being against the principle of the MLS that no need of follow-up of individual transfers would be needed. At times, some of these proposals were seen as representing a dangerous slide back into bilateral approaches and an emphasis on the countries of origin concept. The interpretation of the relevant articles of the SMTA by the European region is that the third party beneficiary should only act in cases of serious doubts on the adherence to the provisions of the SMTA, and by no means act as an agency controlling the issuing and implementation of the SMTAs.

Farmers' Rights

In general, the European region did not show a strong interest in the Farmers' Rights concept, (see Chapter 13 for positions of Farmers' Communities) although some individual member countries were active on this issue. The ERG felt that it would be difficult to develop the concept of Farmers' Rights into a legal mechanism due to a number of inherent complications, such as identification of the rights holder, the absence of novelty in case of traditional farmers' varieties and the challenge to sufficiently define a variety in the absence of uniformity. In addition, it was pointed out that a legal interpretation of Farmers' Rights might clash with the existing IPR systems. The region also noted that some stakeholders in the global NGO community (see Chapter 10 for details on civil society) argued that the legalization of the concept of Farmers' Rights would only introduce private property thinking into the sector of small-scale agriculture that until then still operated under the concept of heritage of mankind.

The European region was quite flexible during the negotiations with regard to accepting specific language on the rights of farmers to traditional seed management as long as the language remained within the limits of the TRIPS agreement. However, this possible compromise language became irrelevant when the present wording for this chapter was proposed and adopted, leaving the interpretation of such rights largely to national legislation. Internal differences of views on the balance between IPR protection and Farmers' Rights became visible in the form of country declarations when the Treaty was adopted by the FAO Conference in 2001.

Financing the implementation of the Treaty

During the negotiations the ERG argued for utilizing existing funding channels and for avoiding major additional implementation costs. It was, however, accepted that some specific provisions would be needed for the proper implementation of the Treaty. The funding strategy came to be seen as consisting of three components: obligatory benefit sharing under the MLS, voluntary contributions to the account under the control of the Governing Body, and other self-standing financial mechanisms allowing the implementation of the Treaty. With regard to other funding agencies and mechanisms, at the outset different perceptions within the

ERG existed, but gradually a consensus evolved that governments would have a responsibility to act coherently across funding agencies to ensure resources for prioritized activities including the Treaty.

In addition, it was recognized that the Treaty's secretariat needed an operational budget, but the European region argued that these needs were to be covered from the core budget of FAO. Furthermore, it pointed to a number of international and bilateral mechanisms in place to implement the funding strategy, and although acknowledging that funds from recipients bound to the obligatory benefit-sharing under the MLS would be slow to come for many years, it did not wish to commit itself a priori to the provision of extra, additional funds for the benefit-sharing fund under the funding strategy. It probably misjudged how developing countries interpreted this as a lack of commitment to the operationalization of the Treaty.

Summing up: Merits and drawbacks of the Treaty from a European viewpoint

In the view of many European member countries the mere existence of the Treaty represents a major merit in itself in that it challenges governments and the community at large to recognize the importance of plant genetic resources.

In particular, the reconciliation of different perspectives on the functioning of the Treaty, namely as an international agreement between states, as well as an agreement that should bind legal and natural persons exchanging and using plant genetic resources, is regarded by the ERG as a major accomplishment.

For the European group, a major feature and merit of the MLS is that it recognizes that the availability of new plant varieties as such is a major benefit of the MLS. The continued need for a distinction between products that are or are not freely available for research and breeding may be stressed again in future European positions.

For the European region, the MLS is the core of the Treaty, although it also places much weight on Articles 5 and 6 on conservation and sustainable use, emphasizing that these articles apply to all plant genetic resources, not just the crops listed in Annex I.

A notion widely shared within the ERG was that monetary benefits stemming from the application of Article 13.2 (d) will remain limited at least for a number of years to come,[3] and the volume of such benefits will remain limited given the fact that obtaining plant breeder's rights on a product developed from germplasm accessed from the MLS will not lead to obligatory benefit-sharing.

The alternative payment option, proposed by the African region in a late phase of the negotiations on the SMTA, was gradually being perceived as an interesting option (see Chapter 19 for more detail). However, it can be expected that the alternative payment option will only be preferred by users if the ratio of the payment levels between the default payment arrangement (by individual product) and the alternative option (by access to a crop listed in Annex I) would have been substantially larger. Possibly, the alternative payment option may still evolve into a major merit, if the ratio between these two payment levels can be revisited in the future.

The Global Crop Diversity Trust (see Chapter 16 for detail on the GCDT) is seen as a major instrument for benefit-sharing, and ERG governments have made substantial donations to the Trust.[4] The ERG regards the GCDT as an essential component of the funding strategy and indeed as a major building block to the Treaty. Furthermore, the grant conditions of the GCDT request that the germplasm that is regenerated or characterized with support from the Trust will be available under the conditions of the SMTA. The Trust currently supports the regeneration and characterization of both international collections of the CGIAR and national collections in a large number of countries. The Global Environment Facility has also been identified as an important instrument to facilitate benefit-sharing at the multilateral level. In addition, various bilateral programmes explicitly include the strengthening of genetic resources conservation and utilization. In situ conservation, management and use remains a major challenge, however, although the strategic plan for the implementation of the funding strategy adopted by the Governing Body offers options to specifically address such needs.

Challenges ahead

Although significant progress has been made, the authors of this chapter recognize a number of outstanding issues. In our view, solving these challenges will increase mutual trust between the stakeholders in the Treaty: governments of developing countries and developed countries, the private sector and farmers' organizations, breeders and conservationists alike.

Funding strategy

Full and proper implementation of the Treaty depends on the funding strategy. Ex situ conservation of Annex I crops is to a large extent taken care of by the contributions of the GCDT. However, the ITPGRFA also refers to the need for complementary in situ measures. In spite of the potential offered by the Global Environment Facility and bilateral funding, a case can be made that the Treaty needs funds under its own control, to develop a coherent portfolio of projects for proper in situ management. An adequate funding strategy is also needed in order to address neglected and underutilized crops, as well as the capacity building needs for the implementation of the Treaty and following from the Global Plan of Action on Plant Genetic Resources for Food and Agriculture.

It was a major achievement to reach consensus on the budget during the early stages of the Treaty implementation. Lack of an agreement on the financial rules still hampers the Treaty to fully function at this crucial stage.

Various stakeholders in the European region feel that success in fundraising will depend on a clear focus and on the ability to render the benefit-sharing fund attractive to additional non-state donors. In addition to the ground-laying contributions of Spain, Italy and Australia, the innovative Norwegian pledge for benefit-sharing as a function of seed sales in the country (0.1 per cent of all sales), as well as the ensuing strategic partnership between the Treaty and United Nations Development Programme (UNDP) can be seen as highly interesting initiatives.

Building the MLS, including introducing the SMTA

Currently, the major holdings brought into the MLS are those of the CGIAR centres. Large collections maintained by national gene banks and other public sector institutions should also become part of the MLS. To the extent that such collections come under the management and control of the contracting parties and are in the public domain, these collections automatically form part of the MLS upon ratification of the Treaty by the corresponding country. Where such collections are held by institutions outside the government, collection holders themselves should decide to bring their collections into the MLS.[5] The extent and pace by which this can be realized will strongly influence the success of the Treaty.

An important challenge will be to aim for the broadest possible participation in the Treaty. Some countries that are important players in the field of PGRFA have not ratified the Treaty yet. Universal membership, expansion of Annex I, filling the MLS and realizing facilitated access for all PGRFA important for food security, will form important goals for the European region in the future, and Europe will have to consider how best to foster such development.

The Treaty will also be judged by the progress made with the introduction of the SMTA. A shift to the SMTA takes time, both for technical and for policy reasons. Governments should identify all material that automatically falls under the MLS and promote inclusion of all other germplasm listed in Annex I held in their jurisdictions. Regulatory measures might be necessary to arrange for contributions to the MLS. Providers should develop an administrative system to archive signed SMTAs and to report on such transactions to the secretariat of the Treaty. Clear progress in the introduction of the SMTA, across regions and sectors, should be demonstrable in order to boost the profile and recognition for the Treaty.

Transparency, mutual trust and risks of misappropriation

The issue of misappropriation is clearly linked to the question of how Article 12.3 (d) of the Treaty should be interpreted. The ERG shares the view that nothing in this article should prevent the granting of IPR on products developed from materials from the MLS. However, obtaining such rights should not limit access for others to the same materials in the MLS. It is highly likely that this issue can only be gradually resolved by discussing case studies as they will develop over time in the Governing Body, and resorting to arbitration in those cases where no agreement between the contract partners can be reached.

A related challenge is the need to provide sufficient transparency in the transactions that take place with germplasm in the MLS and materials under development derived from that germplasm. The level of detail in the reporting of transactions and the means by which parties can gain access to that information still needs to be elucidated and agreed upon. The ERG acknowledges that reporting is essential, but it also holds that in accordance with Article 12.3 (b) policing of individual transactions in the system should not be an objective of the reporting. In the European view, the provider – whether a provider of germplasm in the MLS or of materials incorporating germplasm from the MLS – remains responsible for documenting and respecting the details of each transaction.

Access to non-Annex I crops

Various crop collections that do not fall under the MLS will still play a role in reaching global or regional food security. In the view of the European region, taking into account that it regarded Annex I as too limited, it would be preferable to make germplasm of all PGRFA available under the terms and conditions of the MLS. Some European parties (such as The Netherlands and Germany) have indeed already adopted this approach.

Access to germplasm maintained in situ

Some stakeholders hold the view that the MLS should only effectively deal with access to ex situ collections. However, in the European view this is not in line with the text of the Treaty, and, in particular, Article 12.3 (e) which explicitly refers to access to plant genetic resources being developed by farmers, and Article 12.3 (h) stating that access to plant genetic resources found in in situ conditions may be ruled by national legislation or in accordance with standards as may be set by the Governing Body. Thus, the Governing Body may wish to discuss the need for such standards. In that process, it might be considered to what extent the Treaty can provide any basic rules for access to germplasm held under in situ conditions, and to what extent such access to Annex I germplasm may depend on national policy and legislation. Paying due attention to access and benefit-sharing on genetic resources held on farm or occurring in in situ conditions will greatly increase the impact of the Treaty.

Raising the political profile of the Treaty

In order for the funding strategy to attract sufficient funding, the importance of the Treaty will have to be 'mainstreamed'. The various stakeholder groups should contribute to an increased awareness on the importance of conserving crop genetic diversity and to raise its profile. In addition, for proper implementation of the Treaty more interdepartmental cooperation at the national level will be needed. In the process, the notion that agriculture is simply a threat to biodiversity will have to be replaced by the realization that, as part of agriculture, genetic resources for food and agriculture form a major component of our total biodiversity and should thus be conserved and cherished. In addition, there could be great potential in referring to the cause of food security in order to raise the awareness on biodiversity in general.

Raising the profile is not only a challenge at the national level but also at the international level. Whereas FAO has fully recognized the importance of genetic resources, this issue has not been given the proper attention in other multilateral agreements and organizations. In a future in which climate change will increasingly affect agriculture in all regions, conserving our agrobiodiversity should be recognized as an important insurance policy providing us with an essential tool for adaptation to changed circumstances. Some early developments already point to a strong need for more systematic exploration of crop genetic resources in order to adapt to climate change.

Conclusions

The ERG member countries regard the ITPGRFA as a very important agreement, that will allow breeding efforts to continue and to develop further in order to meet the challenges of food security, and that – to that purpose – will contribute to an enhanced conservation of our genetic resources in international cooperation. The European region is also aware of the fact that the Treaty will complement and contribute to a future International Regime on Access and Benefit Sharing, that is currently negotiated under the CBD and that will have regard to all genetic resources. It is of the opinion that major first steps towards implementation of the Treaty have already been made, but that continued efforts will be needed to complete the process of implementation.

Notes

1 Europe's role, positions and perspectives should be understood as those of the European regional group, representing the countries of the region. The European regional group encompasses all European countries including the countries of Central and Eastern Europe. Russia also belongs to the European regional group in FAO.
2 At the start of the negotiating process in 1995, the number of EU member states was still 15, whereas at the conclusion of the Treaty in 2001 the number had increased to 25, and at the 2nd Governing Body meeting it had reached 27 member states.
3 If the breeding cycle is taken as a reference, substantial income can be expected only 7–15 years after distribution of germplasm for the purpose of breeding has occurred.
4 At April 2008, the total payments of European governments to the Global Crop Diversity Trust had reached an amount of US$75 million.
5 A growing number of national collections are now placed in the MLS, as reported on the website of the Treaty. So far, ERG member states have placed more than 200,000 accessions in the MLS or made these accessions available under the terms and conditions of the SMTA.

Chapter 6

The Latin American and Caribbean Regional Group

A Long and Successful Process for the Protection, Conservation and Enhancing of PGRFA

Modesto Fernández Díaz-Silveira

A necessary introduction

The process leading to the adoption of an International Treaty on Plant Genetic Resources for Food and Agriculture (ITPGRFA) has been developed in other chapters of this book. This allows starting from the perspective of the Latin American and Caribbean region (LAC), instead of repeating in detail what the scenario was during the last part of the second half of the 20th century, and the complex and at the same time interesting process leading to the adoption of the first environmental multilateral agreement of the 21st century.

A key characteristic of plant genetic resources for food and agriculture (PGRFA) is based on what breeders – formal ones in research centres or enterprises as well as farmers and local and indigenous communities – in all countries need: that is, to get access to genes in the form of PGRFA to keep their breeding programmes running, or to maintain and enhance local or traditional varieties for sustaining local communities and cultures. The problem is that all countries are dependant on genes coming from the entire world, as nowadays, there is no single country known to be self-sufficient on PGRFA (Crucible Group, 1994; FAO, 1997; Correa, 2000; Gerbasi 2004; Moore and Tymowski, 2005). Countries, institutions, researchers, farmers and people in general, need this flow of genetic resources to support their breeding programmes and to ensure food security.

As a reminder, the most important issue was that PGRFA were considered until the 1980s as a common heritage of mankind. In a way, this concept did not recognize the enormous contributions of people for centuries: mostly farmers and local and indigenous communities who conserved and enhanced the wealth that crop plants can represent to current generations for food, feeding, fibres, housing and so many other uses that man needs for everyday life.

In this interesting and unique process, mostly after the 1980s and, of course, after the decisive adoption of the Convention on Biological Diversity (CBD) in Rio de Janeiro, 1992, the international community developed the ITPGRFA on the basis of ensuring the conservation, the sustainable use and the sharing of benefits derived from the use of PGRFA. It is something interesting that the Treaty was able to achieve primarily for PGRFA, what the CBD is still negotiating without a concrete outcome yet, for access and benefit sharing for biodiversity in general.

A good support for the efforts developed by FAO, its country members and farmers' communities in the entire world, was the decision adopted during the 2nd meeting of the Conference of the Parties to the CBD in Jakarta, Indonesia, in November 1995. This decision on 'The Global System for the Conservation and Utilization of PGRFA' recognized 'the special nature of agricultural biodiversity, its distinctive features and problems needing distinctive solutions', and at the same time, recalled Resolution 3 of the Nairobi Final Act of the Conference for the Adoption of the Agreed Text of the CBD, which recognized 'the need to seek solutions to outstanding matters concerning plant genetic resources within the Global System for the Conservation and Use of Plant Genetic Resources for Food and Sustainable Agriculture, in particular (a) access to ex-situ collections not acquired in accordance with this Convention; and (b) the question of farmers' rights' (CBD, 1995). This decision gave an inestimable impulse to the process.

In the statement delivered during the First Session of the Governing Body of the Treaty in Madrid (ITPGRFA, 2006), it was recalled that the 6th meeting of the Conference of the Parties to the CBD in 2002 'recognized that the International Treaty on Plant Genetic Resources for Food and Agriculture will have an important role for the conservation and sustainable utilization of agricultural biological diversity, for facilitating access to plant genetic resources for food and agriculture, and for the fair and equitable sharing of the benefits arising out of their utilization' (CBD, 2002). It thus recognized that the Treaty will make a significant contribution to the achievement of the three objectives of the Convention in the strategic area of agricultural biodiversity. For this reason, the Conference of the Parties (COP), at the same meeting, stressed the need for the expeditious entry into force of the Treaty and called on the 188 parties to the CBD and other governments to give priority consideration to its signature and ratification. The CBD Secretariat in its message recalled also that the 7th meeting of the COP of the CBD, held in Kuala Lumpur in 2004, again urged parties of the CBD and other governments to ratify the Treaty as an important instrument for the conservation and sustainable use of genetic resources, leading to hunger reduction and poverty alleviation.

Complexity of negotiations

The complexity of negotiations regarding PGRFA can be explained quickly recalling the close link of agriculture with the fundamental issue of food security, with trade – including the role of the World Trade Organization (WTO) and Trade-related Aspects of Intellectual Property Rights (TRIPS) agreements – without forgetting that plants are also a major source of pharmaceutical products and products for industrial use (several of them with a very high value) that surpass the value of those plants as food or any other common use. Tobin (1997) made an analysis on this issue and mentioned data from the Rural Advancement Foundation International (RAFI) that estimated losses of Southern nations from forgone royalties for the use of genetic material for the pharmaceutical industry around US$5079 million, and compared the use of genetic resources for agriculture with losses of US$302 million, highlighting that the use of genetic resources for agricultural purposes obviously offers less potential benefits than its use for pharmaceuticals.

The fact that the scope of the Treaty includes all 'plant genetic resources for food and agriculture' makes it a very important instrument for the LAC region, which recognized from the very beginning of the process the relevance of having a unique instrument for dealing with the diversity and the uniqueness of PGRFA.

The characteristics of the Latin American and the Caribbean region

LAC is a wide region comprising the southern part of North America (Mexico), whole Central America, the Caribbean and South America, which show different ecosystems, climates, cultures and people. Although mostly tropical climates predominate, also sub-tropical and even temperate climates are present. At the same time, vegetation in LAC goes from sea-level territories to very high mountains, reaching for some of them more than 6000m in height, as the examples of Peru (Nevado Huascaran, 6768m in the Andes) and Bolivia (Nevado Illimani, 6462m, also in the Andean region) considered being some of the highest mountains in the world.

All those different conditions contribute to facilitate the existence of a large diversity of genetic resources, which constitute a real wealth for humankind. It is common to have a very wide range of crops, forms and varieties, many of which are still not found, described nor used by people other than some indigenous or local communities that know and use those plants, as they are living in the wilds in close contact with nature.

Different cultures living in LAC were developed, in a great extent, having plants and crops closely linked to people. This fact made it easier for those indigenous and local communities to learn the characteristics, properties and usefulness of different plants and allowed them not only to conserve, but also to improve, in a very primary way, those plants leading to improved varieties and crops, that at the same time became closely linked to those cultures. This is a clear example of traditional knowledge incorporated and interlinked with PGRFA.

Together with that, LAC is the centre of origin of some of the most valuable food crops for humankind, like potatoes, maize and some very valuable tubers and roots, making an even heavier responsibility for the region, as unique genes have to be preserved to ensure the existence of the necessary and valuable variability on those PGRFA for the future.

From a total amount of 126 countries that are parties to the Treaty as of September 2010, 16 of them are countries from LAC (see Annex 3 of the book for the tables of participation to the Treaty by regional groups), and there are still five countries more that signed the Treaty and are in the process of becoming parties in the future.

The role of LAC in the process previous to, during the adoption of, and after the adoption of the Treaty

The active role of LAC represented in international negotiations by GRULAC[1] can be pointed out by the fact that Mexico, as part of GRULAC and supported by the G-77, started a decisive debate during the 21st Conference of FAO in 1981. This debate led, two years later, to the adoption of important steps for the international process on PGRFA.

As Esquinas-Alcázar and Hilmi (2008) rightly pointed out in an article on the negotiations of the ITPGRFA, at the end of the 1970s and the beginning of the 1980s, the problem of PGRFA shifted clearly from a scientific problem to a political problem. The sensitiveness of developing countries raised valid questions as difficult to answer as the following:

- If PGRFA are distributed through the whole world, but the majority of biodiversity is present in tropical and subtropical countries where we find most of developing countries, then, when seeds are collected and placed in gene banks, frequently belonging to developed countries, to whom do the seed samples stored there belong?
- If the new varieties obtained are the result of applying the technology to the raw material or genetic resources, then why are the rights of donors of technology recognized in the form of breeder's rights, patents, and so on, whereas the rights of donors of germplasm are not?

These questions were key for starting a negotiation process. The fact that LAC is a region with a very rich biodiversity, that the region contributed for decades to the collection of gene banks of several international institutions and developed countries, and that some of the mega-diverse countries of the world belong to our region, conditioned an active role of almost all of our countries in the pre-negotiation and negotiation process of the Treaty, leading to its adoption in 2001. The active role of LAC countries, as members of FAO, contributed to set up the scenario for the consecutive steps that allowed: first, to decide during the

22nd Conference of FAO in 1983 to establish a Commission on Plant Genetic Resources, and second, to adopt an International Undertaking (IU) on PGRFA that, although not legally binding, became the first example of the desire of LAC, as part of the international community, to address the loss of plant biodiversity in the world.

At that point, nobody could challenge the idea that the process of establishing an international regime for PGRFA was clearly a political issue, requiring negotiations and the establishment of national, regional and international policies to allow all countries to be able to use those fundamental resources to ensure food security for all.

The adoption of the CBD in 1992 marked a shift in the process of an international regime for PGRFA. The 1st extraordinary meeting of the Commission on Plant Genetic Resources in 1994 initiated the harmonization of the IU with the CBD; this could be considered a milestone for the process. Participation of GRULAC countries was fundamental in clearly highlighting issues that must guide the negotiations, according to the needs and requirements of developing countries.

A significant fact was the decision of the members of the Commission on Genetic Resources to elect Venezuelan Ambassador Fernando Gerbasi as its Chair in 1997. I consider it obligatory, to recognize the role played by Mr Gerbasi from this moment onwards guiding the Commission to succeed in the adoption of the Treaty four years later, making an extremely wise use of his diplomatic skills and unbelievable conciliatory power (see Chapter 2 of this book). This was a clear example of the capacity and commitment of GRULAC with the process on PGRFA. It is an interesting exercise to consult the book of Ambassador Gerbasi (Gerbasi, 2004) that summarizes very well some of the important moments of the whole process of negotiations of the Treaty.

We have to mention that, together with the whole community of Latin America and the Caribbean Countries, Cuba, as a member of the Commission on Genetic Resources, played a relevant role, not only presenting the national positions for the negotiations, but also representing GRULAC and the G-77, when this last group of developing countries placed the responsibility on Cuba to be the Chair of the G-77 in Rome during the last part of the negotiating process and the adoption of the Treaty in 2001. This was a responsibility and an honour, that the Permanent Representative of Cuba to FAO in Rome, Ambassador Juan Nuiry, performed with skilfulness and an inimitable ability to bring everybody together, and to allow having a balanced outcome of those difficult and complex negotiations.

Developing countries, including GRULAC members, gave the G-77 the responsibility of negotiating on their behalf, with all the power given by the membership of more than 120 countries at that time.[2] We should remember that some of the very skilful negotiators based in the United Nations headquarters in New York were sent to FAO by their respective countries to reinforce the negotiations at the final stage of adoption of the Treaty. Some of the very important and well informed negotiators belonging to the LAC countries that participated in this group.

After the adoption of the Treaty, GRULAC continued placing great importance on implementing the instrument that had been adopted. The active role played by GRULAC during the meetings of the Commission acting as the Interim Committee of the Treaty, was fundamental to prepare the arena for the first meeting of the Treaty, once it entered into force in 2004. The financial rules for the Treaty, the rules of procedure, the draft budget for the Treaty, the financial strategy, as well as the first steps for designing a standard material transfer agreement (SMTA) for the multilateral system (MLS), and the draft procedures to promote compliance were the most important negotiations after the adoption of the Treaty. Without those proposals, prepared for the consideration of the Governing Body, the operation of the Treaty would not have been possible (Earth Negotiation Bulletin, 2002).

After that, different groups were created under the Commission acting as the Interim Committee for addressing different relevant issues. One of the groups focused on developing the SMTA for the MLS. The countries of the LAC region contributed to a great extent to the design of this valuable instrument.

Another group was the 'Open-ended Working Group on the Rules of Procedure and the Financial Rules of the Governing Body, Compliance, and the Funding Strategy', which convened at the FAO headquarters in December 2005 (CGRFA, 2005). This working group prepared draft resolutions for the consideration of the Governing Body at its first meeting. The G-77 proposed the nomination of Cuba as the Chair of the meeting, which was elected and acted as such during the meeting. This again was an expression of the recognition of the active role performed by Cuba and GRULAC.

Without being exhaustive, and recognizing the key role played by all LAC countries, I want to mention the active role of some countries during the whole process of the Treaty: Brazil, with substantive contributions in a wide list of issues, placing its well known negotiation capacities and expertise for the benefit of GRULAC; Argentina with a relevant role in the negotiations for the adoption of the Treaty and in the Commission acting as the Interim Committee for the Treaty, mainly through the important contribution of their legal experts, when developing important proposals for the draft SMTA, the rules of procedure, the third party beneficiary mechanism of the MLS, and others; Colombia with the relevant technical expertise on PGRFA and the very important considerations on biodiversity, as a well known mega-diverse country; Ecuador, with the technical expertise as well as diplomatic and negotiating skills of their experts that contributed notably to sustain GRULAC positions; and Uruguay, with substantive technical contributions that allowed progress with a solid text on several occasions, as with the very important contribution for the development and refinement of the financial strategy. Those are only examples, and I want to emphasize my personal recognition to all countries in the region for the wonderful work developed.

The need to harmonize existing legislations at national level within countries is a prerequisite for allowing the LAC countries that are still not parties to the Treaty, to complete the process for becoming a party. It is perhaps the case of some very important and active LAC countries in the FAO Commission on

Genetic Resources. Ruiz (2008) made a good analysis of the implications of some regional decisions and national legislation that consider the issue of access and benefit sharing, like the Decision 391 of the former Andean Pact. As Ruiz rightly commented in his article, some very restrictive steps adopted by some countries are now very difficult to implement in a practical way. An exercise for the harmonization of those national and regional legislation with the Treaty could perhaps allow implementing the latter at national levels without any confrontation with national legislation or with sovereign decisions made by each country.

The position of the LAC region regarding PGRFA and an international regime for them

Negotiations for the ITPGRFA were, as in any intergovernmental multilateral agreement, made among government delegations, with civil society representatives included in the delegations.

A common theme for almost all LAC delegations was the consideration that the region contributed for centuries to PGRFA and it was not adequately compensated for making this wealth available to countries of all regions of the world. At the same time, during the last decades, some PGRFA from the LAC region were collected by research institutes, international organizations and different groups of collectors, and used in different countries for developing pharmaceutical or industrial products that represented several times the value known for those plants, without sharing a fair part of this value with the LAC countries. LAC countries, like stewards for those materials, had preserved the valuable materials that contained active substances or compounds that were used for saving lives or making human life more comfortable. Correa (1998) mentions concrete examples of a legal dispute, where the The Andean Court of Justice had to rule against the intentions of the pharmaceutical industry to obtain protection on pharmaceutical products already patented, that were produced from plants prospected in LAC.

For success in the negotiations towards the adoption of the Treaty, a more political approach was necessary when considering PGRFA and this vision was part of the position of the LAC region for the negotiations. A clear view on what was missing was fundamental for LAC at that time. As a result, LAC contributed in a substantial way to it, with concrete proposals, firm positions, while at the same time showing flexibility and understanding towards the needs and positions of other countries and regions.

This situation of inequity marked the negotiations of the Treaty for our region and explains why it was so difficult to reach a final consensus for the agreement in 2001, why there was such a strong position of almost all our countries, regarding the role of indigenous and local communities in the conservation of our resources for centuries, and why in the end we only had a small list of crops in Annex I of the Treaty, constituting the core of the MLS.

At the same time, LAC realized that access to PGRFA was also necessary for developing new and important plant forms and varieties, and recognized that all our countries are in need of PGRFA for ensuring food security. Mainly research institutions and gene banks in LAC wanted to have access to the necessary genes for their plant breeding programmes, and negotiators managed the situation in a very smart way, contributing a lot of ideas and proposals that allowed negotiations, although slowly and step by step, to go forward until succeeding.

Although the MLS of the Treaty is limited in its scope because of the still limited number of crops included in Annex I, it constitutes an innovative mechanism serving as a model, facilitating the access to genetic resources, and at the same time ensuring that any benefits derived from their use are shared in a fair way with the people that maintained and enhanced those PGRFA for centuries. This is the most novel characteristic of the Treaty, which allowed progress in this very sensitive issue thanks to its objectivity and fairness. To allow the sharing of benefits derived from the use of PGRFA, although in a collective way, as it is designed for being distributed through the benefit-sharing fund to all developing countries, constitutes the first and still unique approach to this aspiration of humankind.

The special challenge imposed by climate change

At a time when climate change is a real concern for all humankind, the LAC region should focus on putting in place concrete measures, strategies and programmes to address the challenges that climate change will impose on our countries in the future. Highly dependant on agriculture, some recent examples of natural disasters are a call of alert that our agricultural systems must be improved for the imperative of subsistence. Indeed, flooding due to continuous and heavy rains in Central America, heavy storms and severe hydro-meteorological events in the Caribbean, and extreme temperatures and displacement in time of the occurrence of the seasons during the year have affected the planting and harvesting of food crops in all LAC.

To cope with climate change adaptation challenges, it is absolutely necessary to have PGRFA available to researchers, farmers and all people involved in plant breeding and in food security issues. Genes that would allow plants to adapt to those changing climate patterns should remain available. Because of the different climates and conditions prevailing in LAC, we will need genes adapted to raising or lowering temperatures; to resist droughts or heavy rains; to be able to grow in altitudes different varieties for crops like potatoes, in places as high as the mountains in the Andean region; or even to adapt crops to migrate to lower or higher latitudes.

Because the adaptation of plants to changing climate conditions is not a process that could be achieved in a short time, we give the Treaty a relevant role to guide this process in a way that food security could be achieved in the medium term, in spite of challenges imposed by climate change. Adaptation of agriculture to climate change challenges will require the availability of adequate PGRFA, with enough time to allow its incorporation to new improved and adapted varieties, and a good quantity of work by farmers and researchers.

Walking towards the future

To ensure having an operational Treaty will require sufficient available funding for its implementation as a whole, including the implementation of all and every article of the Treaty, as well as a working funding strategy and an enhanced MLS for access and benefit-sharing, and to ensure that the Governing Body remains sovereign and able to implement the Treaty in the way it decides, without any constraints or limitations. This is crucial if we want to continue supporting developing countries to conserve and use, in a sustainable way, all PGRFA to achieve food security for all their inhabitants and to widely develop further, the innovative concept of Farmers' Rights.

Regarding the issue of widening the scope of Annex I of the Treaty, my opinion is that it remains a very sensitive one. As Annex I is part of the letter of the Treaty, its modification would imply opening the letter of the Treaty, and I think that there are some risks that might not be faced, even today. It is true that all countries, developed and developing, are in need of the whole rainbow of plant species and varieties for research and breeding. For example, Brazil has been asking to include crops like garlic and onion, peanuts, tomato, soybeans and sugarcane in the Annex I list, but perhaps at this time we should find some alternative ways to overcome this difficulty. I am sure that the continuous implementation of the Treaty, in the successful way it is occurring, will allow, after a number of years, to reconsider the list of crops included in Annex I. Construction of trust and a successful implementation and operationalization of the MLS, the effective control of the SMTA, the growing and ensured sharing of benefits derived from the use of PGRFA from the MLS with an effective functioning of the third party beneficiary mechanism, will all constitute the basis for this important and necessary step.

I see in the future all countries of the LAC region becoming parties to the ITPGRFA. The strength of a whole region, negotiating and constructing an operative and strong Treaty, will be the best contribution we could make to the international efforts to allow all people to have access to food. I am sure that some remaining issues, mainly legal issues on national legislations, could be overcome and the whole LAC region will work together for this aim.

The enormous challenge of having an international instrument for PGRFA was faced and overcome by the international community. This is something that seemed impossible 15 years ago. The quick rise in the number of parties to the Treaty, in a very short time for an international multilateral agreement,[3] is a clear response from countries that the Treaty is well designed and responds to the expectations of them all. It is a good reason and a good incentive to continue working very hard to ensure our contribution to eliminating hunger from all parts of the world, and ensuring food security for all.

A final note

A book addressing the process of the Treaty cannot forget to pay tribute to a very special person, who played a key role from FAO and the Commission on Genetic Resources for Food and Agriculture: Don José Esquinas-Alcázar. A convinced herald for the need to conserve, protect, enhance and use in a sustainable way all PGRFA, our good friend 'Pepe Esquinas' dedicated part of his life to the development of the Treaty. I am sure that Latin America and the Caribbean region will join me in this well-deserved homage in recognition of the work Mr Esquinas-Alcázar and his continuous support to the LAC region for so many years.

Notes

1 GRULAC is the Latin American and Caribbean group. It comprises the following countries: Antigua and Barbuda, Argentina, Bahamas, Barbados, Belize, Bolivia, Brazil, Chile, Colombia, Costa Rica, Cuba, Dominica, Dominican Republic, Ecuador, El Salvador, Grenada, Guatemala, Guyana, Haiti, Honduras, Jamaica, Mexico, Nicaragua, Panama, Paraguay, Peru, Saint Kitts and Nevis, Saint Lucia, Saint Vincent and the Grenadines, Suriname, Trinidad and Tobago, Uruguay and Venezuela.
2 The G-77 now comprises 131 countries from all over the world.
3 Only 9 years after its adoption and 6 years after its entry into force, the Treaty reached 127 contracting parties.

References

CBD (1995) Second ordinary meeting of the Conference of the Parties at the Convention on Biological Diversity (UNEP/CBD.COP/2/19), Decision 2/15, available at www.cbd.int/doc/?meeting=cop-02

CBD (2002) Sixth ordinary meeting of the Conference of the Parties at the Convention on Biological Diversity (UNEP/CBD.COP/6/20), Decision VI/6, available at www.cbd.int/decisions/cop/?m=cop-06

CGRFA (2005) *Report of the Open-ended Working Group on the Rules of Procedure and the Financial Rules of the Governing Body, Compliance, and the Funding Strategy*, Rome, Italy, 14–17 December 2005, FAO document CGRFA/IC/OWG-1/05/REP

Correa, C. M. (1998) 'Aplicación de los TRIP en los países en desarrollo', *Boletín Tercer Mundo Económico*, no 112, Agosto de 1998

Correa, C. M. (2000) 'Implications of national access legislation for germplasm flows', Global Forum on Agricultural Research (GFAR), Document No: GFAR/00/17-04-01

Crucible Group (1994) *People, Plants and Patents: The Impact of Intellectual Property on Biodiversity Conservation, Trade, and Rural Society*, International Development Research Centre, Ottawa, ON

Earth Negotiation Bulletin (2002) 'Summary of the First Meeting of the Commission on Genetic Resources for Food and Agriculture Acting as the Interim Committee for the International Treaty on Plant Genetic Resources for Food and Agriculture', 9–11 October 2002, vol 9, no 245, Monday, 14 October 2002

Esquinas-Alcázar, J. and Hilmi, A. (2008) 'Las negociaciones del Tratado Internacional sobre los Recursos Fitogenéticos para la Alimentación y la Agricultura' (English summary), *Recursos Naturales y Ambiente*, no 53, pp20–29

FAO (1997) *Background Study Paper n°7*, REV.1, available at ftp://ftp.fao.org/docrep/fao/meeting/015/j0747e.pdf

Gerbasi, F. (2004) '*Tratado Internacional sobre los Recursos Fitogenéticos para la Alimentación y la Agricultura*, Imprenta de Miguel Angel García e Hijos, Caracas, Venezuela, 136pp

ITPGRFA (2006) 'First Session of the Governing Body of the International Treaty on Plant Genetic Resources for Food and Agriculture', Madrid, Spain, 12–16 June 2006, FAO document IT/GB-1/06/Report

Moore, G. and Tymowski, W. (2005) *Explanatory guide to the International Treaty of Plant Genetic Resources for Food and Agriculture*, Environmental Policy and Law Paper No. 57, International Union for Conservation of Nature, Gland, Switzerland and Cambridge, UK,

Ruiz, M. (2008) 'Una lectura crítica de la Decisión 391 de la Comunidad Andina y su puesta en práctica en relación con el Tratado Internacional' (English summary), Recursos *Naturales y Ambiente*, no 53, pp136–147

Tobin, B. (1997) 'Certificates of origin: A role for IPR regimes in securing prior informed consent', in J. Mugabe, C. V. Barber, G. Henne, L. Glowka and A. La Viña (eds) *Access to Genetic Resources: Strategies for Sharing Benefits*, Nairobi, ACTS Press, pp329–340

Chapter 7

The Near East Regional Group

Centring the Diversity for Unlocking the Genetic Potential

Javad Mozafari Hashjin

Introduction

The Near East region, as defined in the Food and Agriculture Organization (FAO) or some of the other United Nations bodies, covers countries in West and Central Asia and North Africa regions, which encompass four ecological and socio-economical sub-regions:

- the Mediterranean region: Egypt, Jordan, Lebanon, Libya, Palestine and Syria;
- the West and Central Asia region: Iraq, Iran, Afghanistan;
- the sub-Saharan Africa region: Sudan and Somalia;
- the Arabian Peninsula region: Bahrain, Kuwait, Oman, Qatar, Saudi Arabia, United Arab Emirates and Yemen.

The Near East region also known, by the Consultative Group on International Agricultural Research (CGIAR) circle, as the Central and/or West Asia and North Africa Region (CWANA or WANA), spreads over two large continents – Africa and Asia – and includes some of the largest deserts, highest mountains and deepest land depressions in the world. Although the region is predominantly dry, it is very rich in agricultural plant diversity. It represents one of the three nucleus centres of origin of agricultural crops in the world and encompasses three or four Vavilovian centres of crop diversity (Figure 7.1). Two of these centres (the Mediterranean region and the Near Eastern region) are considered the centres of origin of more

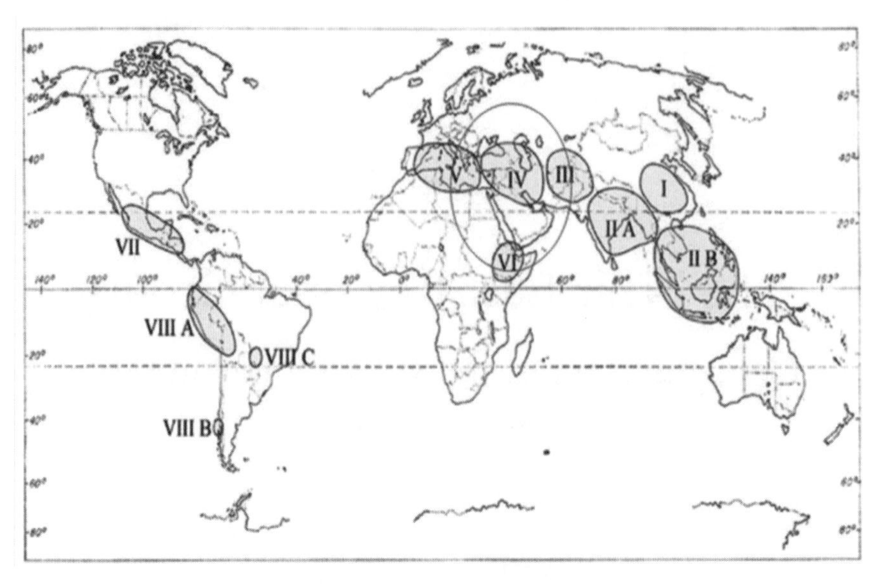

Source: Janick (2002), adapted from Zeven and Zhukovsky (1975)

Figure 7.1 *The geographical status of the Near East region among eight Vavilovian centres of origin for crop plants*

than 150 grown plant species. About ten thousand years ago, the Near East region was the centre of domestication (origin of agriculture) for wheat, barley, lentil, forage species and many fruit trees that still support today's agriculture (Zeven and Zhukowski, 1975). It is estimated that the species that originated from this area are feeding over 38 per cent of the world's population. Wheat alone accounts for about one-third of the global food production. The composition of the Near East region may vary because of practical or mandate-related reasons which lead to inclusion or exclusion of certain countries within the regional group (Zehni, 2006; Amri et al, 2008).

In the Near East region, which is considered the cradle of agricultural civilization, agricultural systems are seriously threatened with drastic instability in climatic conditions and consequently with the erosion of basic natural resources, supporting agricultural productivity, including soil, water and genetic resources. The Near East is a food insecure area. Member countries spend a significant part of their foreign trade earnings on importing food and feed materials. Productivity levels in the region, both for crops and animals are low. Therefore, the Near East region continues to be a net importer of wheat and other commodities with the exception of Turkey, Syria and Pakistan. Iran has achieved a fragile self-sufficiency in wheat since 2004 (Amri et al, 2008). Agricultural exports of vegetables and fruit trees have increased in the last decade. Conservation and sustainable utilization of the immense agro-biodiversity of the region are key to increasing the sustainability and productivity of agriculture, which will lead to enhanced food security and the livelihood of people in the region. However, this cannot be achieved unless the plant biodiversity is fully and effectively employed to improve high-yielding plant cultivars adaptive for stressed environments. Fortunately, substantial research capacity for plant improvement work exists in the region.

Contribution of the Near East region to world food security

Contribution of the Near East region to world food security has been enormous throughout the history of humanity. This region is considered the centre of origin and/or diversity for 22 out of 35 food crops and 19 out of 30 forage crops listed under Annex I of the International Treaty on Plant Genetic Resources for Food and Agriculture (ITPGRFA or the Treaty). The Near East region has also been recognized as the hotspot for genetic diversity of 10 out of 16 crops, whose global conservation strategies have been developed by the Global Crop Diversity Trust (Zehni, 2006).

Considering the fact that agriculture is at the heart of sustainable development and that plant genetic resources for food and agriculture (PGRFA) are at the heart of agriculture, PGRFA continue to be an important asset for humanity facing the challenges of global climate changes in sustaining agriculture and the environment. Therefore, unlocking the genetic potential of these resources will play a very effective role in the mitigation of an adaptation to climate change. Also with other new challenges of globalization such as soaring food and energy prices, market demands for 'diversity rich' food, growing environmental concerns of consumers on food safety and debate over genetically modified crops, the need for the contribution of the Near East's plant genetic resources is even greater in the 21st century.

Despite the global importance of plant genetic diversity of the region for food security, it is unfortunate that the Near East region witnesses one of the highest levels of genetic erosion in the world due to unfavourable conditions discussed above. This requires immediate attention (FAO, 1995).

Climate change

The Near East region is among those food-insecure regions of the world that are particularly vulnerable to the effects of climate change on crop production. As a result there will be significant risks to wild biodiversity, including crop wild relatives. However, together with rapidly growing and changing demands for greater production, these changes are likely to result in increased pressure to cultivate marginal lands.

The range and migration patterns of pests and pathogens are likely to change. Switching to new cultivars and crops adapting to new conditions will require a greatly increased use of genetic diversity and a substantial strengthening of plant breeding efforts. Breeding must take into account the environmental conditions predicted for the crop's target area 10 to 20 years later. Certain underutilized crops are likely to assume greater importance as some of the current staples become displaced. It will be very important to characterize and evaluate a range of germplasm as wide as possible for avoidance, resistance or tolerance to major stresses such as drought, heat, waterlogging and soil salinity. Research is also needed to gain a better understanding of the physiological mechanisms, biochemical pathways and genetic systems involved in such traits. Global partnership for generating, exchange and use of genes and molecular information is crucial for global success in adaptating to climate change.

However, having the basic programmes with adequate human and financial resources to screen germplasm and to run variety trials in key agro-ecologies is of paramount importance. This will require global collective efforts and closer collaboration among the breeding programmes of different regions.

Broadening the genetic base of world agricultural crops

The use of plant genetic resources as building blocks of agriculture has become even more prominent, facing newly emerged challenges such as global climate change, production of biofuel crops as new sources of energy, and elevated standards of food security, safety and diversity. Progress made in research and development of genomics, proteomics and bioinformatics has drastically expanded the horizon of utilizing these resources. More attention is now being paid to increase the levels of genetic diversity within production systems as a mean of reducing risk, particularly in light of the predicted effects of climate change. Thus, a global strategy should be developed for broadening the genetic base of world agricultural crops through the use of genetic resources of crop wild relatives (Cooper et al, 2001; Mozafari, 2008).

Special attention needs to be given to the conservation of crop wild relatives in their centres of origin, major centres of diversity and biodiversity hotspots. The involvement of local communities is essential in any in situ or on-farm conservation strategy for plant genetic resources. Development of an early warning system for monitoring genetic erosion of crop wild relatives, in all countries in the centres of origin and major centres of diversity, should be an important component of such strategy (FAO, 2009).

Development of a knowledge base and research capacity for in situ conservation and utilization of crop wild relatives has been tabled as an essential element of global efforts in this direction (FAO, 2008). Studies on the mechanisms, extent, nature and consequences of gene-flow between wild and cultivated populations will provide information needed for development of appropriate strategies or technologies for the conservation and use of crop wild relatives. Lack of skilled staff is considered a major constraint for conservation and the use of poorly researched species.

Near East commitment to dialogue and collective efforts in meeting global challenges

Genetic resources in general and PGRFA in particular, are considered humanity's most important assets for meeting major global challenges, including alleviation of hunger and poverty, climate change and sustainability of agriculture, soaring food prices, demand for bio-energy crops, and growing food safety and environmental concerns. These challenges have put plant genetic resources into the centre stage of agriculture, health, trade and industry sectors.

Global common concerns on conservation of and access to these resources spurred international dialogue and coordination during the last two decades for developing a fair and equitable system of access and benefit sharing (ABS) for sustainable use of these resources.

The Near East has played a major role in the development of two important international instruments targeting genetic resources: the Convention on Biological Diversity (CBD) and the ITPGRFA. The Near East countries were among the first nations who embraced the CBD (Table 7.1 below) and continue to support this international agreement as an umbrella framework for ABS on genetic resources in general. The recognition of countries' sovereign rights over their genetic resources under the CBD caused a paradigm shift in the legal perception of genetic resources including PGRFA and triggered very significant changes in the international legal and policy frameworks of these resources. However, Near East countries, conceived as the cradle of agricultural civilization, along with the agricultural sector and farmers around the world, were perceived to be interdependent on each other for the plant genetic resources they needed, to produce enough food and to meet the increasing challenges of food security. Therefore, easy access to these resources was practised among farmers of the region for centuries and was still considered essential for sustainable development of agriculture and achieving world food security. Due to the special features and needs of the agriculture sector, the establishment of more specialized ABS systems for PGRFA was favoured in the Near East region.

Near East countries unanimously supported the start of the negotiation for revising the non-binding International Undertaking on Plant Genetic Resources (IU) towards reaching a legally binding international agreement on conservation, access, utilization and benefit sharing of PGRFA, in harmony with CBD, in 1993. The Near East representatives were trying hard, throughout the negotiation, to specifically strike a balance among issues such as: sovereign rights of the countries on their genetic resources, intellectual property rights (IPR), facilitating the exchange of both germplasm and technical capacity, fair and equitable sharing of benefits, and taking the fair share of responsibility towards the conservation of PGRFA for future generations. The negotiation was difficult in nature and experienced considerable obstacles along the way (for details on this negotiation, see Chapters 2 and 10 of this book). An important breakthrough in the negotiation was made on the commercial benefit sharing provisions in the Tehran meeting of July 2000, in the Near East region. The Near East was the only region that hosted a round table of negotiation of the Treaty outside Europe (Cooper, 2002).

The Near East region, together with the North American region, had also co-chaired one of the most difficult parts of the negotiation and the core of the Treaty – the list of crops in Annex I to the Treaty. The Near East, generally, played a very constructive role in bridging the gaps among extreme views of parties for achieving consensus, before and after the entry into force of the Treaty. One of the major contributions made by the Near East region to the Treaty was the consensus reached on the rate of monetary benefit accrued from the seed sale of a PGRFA accessed under the standard material transfer agreement (SMTA).

The International Treaty, securing the access and promoting the use

One of the most important international agreements in the plant genetic resources sector developed in harmony with the principles of the CBD has been the adoption and entry into force of the ITPGRFA. The Treaty draws together the threads of the non-binding IU and those of the CBD based on the principle of national sovereignty over genetic resources, their conservation and sustainable use for a global advancement on food security and sustainable agriculture (Esquinas-Alcázar, 2005).

Table 7.1 *Status of membership to major PGRFA-related international agreements and fora in the Near East region*

Countries	*Date of accession*		*Membership in*	
	CBD	*Cartagena Protocol*	*ITPGRFA*	*FAO-CGRFA*
Egypt	1994-06-02	2004-03-21	2004-03-31	Yes
Sudan	1995-10-30	2005-09-11	2002-10-06	Yes
Somalia	No	No	No	No
Jordan	1993-11-12	2004-02-09	2002-05-30	Yes
Lebanon	1994-12-15	No	2004-05-06	Yes
Libya	2001-07-12	2005-09-12	2005-04-12	Yes
Syria	1996-01-04	2004-06-30	2003-08-26	Yes
Afghanistan	2002-09-19	No	2006-09-11	Yes
Bahrain	1996-08-30	No	No	No
Iran	1996-08-06	2004-02-18	2006-04-28	Yes
Iraq	No	No	No	Yes
Kuwait	2002-02-08	No	2003-09-02	Yes
Oman	1995-02-08	2003-09-11	2004-07-14	Yes
Qatar	1996-08-21	2007-06-12	2008-07-01	Yes
Saudi Arabia	2001-10-03	2007-11-07	2005-10-17	Yes
UAE	2000-02-10	No	2004-02-16	Yes
Yemen	1996-02-21	2005-03-01	2006-03-01	Yes

Source: Amri et al (2008)

The multilateral system: A unique feature of the Treaty

The Treaty establishes a unique multilateral system (MLS) of facilitated access and benefit sharing for those plant genetic resources that are most important for food security and on which countries are mostly interdependent. For such genetic resources, which are listed in Annex I of the Treaty, the contracting parties have agreed on a SMTA that governs the terms and conditions of use of PGRFA accessed from the MLS and the sharing of benefits arising from such use. It is important to note that the Treaty is much more than its MLS and its Annex I crops. It relates to any PGRFA. The list of the Annex I crops of the MLS may be expanded by the decision of the Governing Body of the Treaty. Such decision will depend upon the future developments in the successful implementation of the Treaty and conclusion of the new international ABS regime being negotiated under the CBD.

Actions for implementation of the Treaty from the Near East perspective

National policies

The Treaty creates an enormous potential and paves the ground for significant progress in sustainable crop production. This potential can be materialized through developing and implementing national legislations and policies in line with the Treaty. Such policies should facilitate:

- strengthening linkages between policy makers of agricultural and environmental sectors, scientists, educators, farmers and all other key stakeholders of plant genetic resources;
- establishing a strong national programme that has a clear national status and mandate to develop and implement policies for management of plant genetic resources;
- enhancing capacity in all facets of plant genetic resources management including in situ conservation, application of emerging technologies for ex situ conservation and use, farmer participatory approaches and public awareness methodologies;
- streamlining and improving coordination among all gene banks in a given country within the context of one harmonized national PGRFA conservation and utilization programme.

Synergistic effects of regional collaboration

Despite the importance of developing a national capacity on policies and legislation related to PGRFA conservation and utilization, harmonizing PGRFA related views, policies, regulations and action plans based on the Treaty at the regional level could be very effective for the implementation of the Treaty. The concerted efforts among countries will synergistically promote the use of PGRFA through the facilitated flow of germplasm, information and technologies, and enhanced seed trade. The international and regional collaboration on effective use of PGRFA in meeting the regional and global challenges of agriculture will produce speedy results on sustainable agricultural development and enhance food security. For this purpose, national and international centres of excellence in the region should be identified and their research and capacity building activities should be increased towards assisting national programmes in unlocking the potential of PGRFA for achieving sustainable development and food security.

The establishment of a regional institutional mechanism officially recognized by national governments in the region, such as an intergovernmental working group on genetic resources, can foster such harmony in views, policies and legislation and spur a more effective contribution of the region to harmonized implementation of the relevant international agreements including the Treaty. Similar harmony and integrated cooperation should also be developed among international instruments, international research centres and funding bodies for supporting PGRFA conservation and utilization initiatives at regional and global levels.

Practical action in conservation and utilization

The agro-ecosystems of the Near East region are facing some of the highest erosion risks in the world. Due to the importance of agricultural biodiversity of the region for world agriculture, this should not be merely considered a Near East problem but a serious world problem (FAO, 1995, 2008). Therefore, it should be considered a global responsibility of all contracting parties to sustain these threatened agro-ecosystems and conserve endangered plant genetic resources. Identifying the agro-ecosystems at risk and closely monitoring the status of crop-associated agro-biodiversity in these hotspots can provide a good basis for developing sustainable PGRFA conservation and utilization options, including among others:

- enhancing national capacities in all areas related to conservation, research and management of PGRFA;
- promoting a multidisciplinary and integrated approach to in situ conservation activities, involving farmers and farming communities, local governing bodies, scientists and policy makers;
- providing greater support for conservation and utilization of wild relatives of crops of global importance within the ecosystem;
- establishing a strong research programme in the region which is known as the cradle of agricultural civilization by Near East countries which believe in building a solid knowledge base and developing appropriate approaches for in situ conservation of crop wild relatives. In situ conservation is not only considered the most appropriate strategy for conserving crop wild relatives but can also guarantee the continued evolution of plant genetic resources in their agro-ecosystems and possible development of new genetic resources;

- assessing and identifying constraints to seed and field gene banks and taking measures to raise the standards and enhancing practices of gene banks;
- applying best practices for securing long-term conservation of collections which are kept under unreliable conditions;
- making arrangements and mobilizing resources for the safety backups of nationally, regionally and globally important PGRFA collections;
- considering climate change, threatened agro-ecosystems and gaps existing in the regional collections need to be identified by PGRFA to be then collected and secured in ex situ collections;
- improving documentation systems and developing strong information sharing mechanisms;
- promoting the use of new conservation technologies;
- strengthening national PGRFA programmes in the region by repatriation of material collected from the region and held in national and international gene banks;
- building capacity and developing well trained human resources in all aspects of PGRFA conservation.

The effective continuum between conservation and utilization of PGRFA and of the seed system is crucial for commercialization, for generating benefits and for making an impact from the use of these resources. In order to make significant progress in using PGRFA to meet present and emerging challenges, the following points should be seriously considered (FAO, 2008, 2009):

- developing national strategies and programmes for the utilization of PGRFA involving all key stakeholders;
- developing trained human resources, and enhancing research and development capacity in all facets of PGRFA utilization, characterization, evaluation, breeding and seed production;
- promoting the application of new tools and technologies such as genomics and biotechnologies;
- broadening the genetic base of commercial crop cultivars by utilization of PGRFA adapted to the region and its emerging biotic and abiotic constraints;
- enhancing research to improve the productivity and added-value options of landraces and underutilized species for better access to markets;
- strengthening seed production and supply programmes;
- diversifying farming systems through the use of new and adapted PGRFA and promoting underutilized species to sustain agricultural development;
- improving market access and opportunities for farmers of developing countries.

New directions for the rolling Global Plan of Action

In Article 14 of the Treaty, the Global Plan of Action for Conservation and Sustainable Utilization of Plant Genetic Resources for Food and Agriculture (GPA) is seen as an important component and supporting pillar of the Treaty as a whole. The GPA is indeed an essential scientific and technical framework or manifesto for taking action in implementation of the Treaty at both the national and international levels, particularly for the benefit-sharing provisions in Article 13 of the Treaty (Moore and Tymowski, 2005). Thus, the priority activity areas of the GPA have already been integrated in the set of priorities of the funding strategy of the Treaty. Generally, the priority areas are still very effective and applicable; however, in light of the entry into force of the Treaty, the priority areas of activities prescribed in the GPA need to be updated accordingly. The Near East countries

Table 7.2 *Recommended changes to priority activity areas for updating the rolling Global Plan of Action*

Priority activity	*Recommended changes*
Activity #4	Elaborating on strategies and policies as well as conducting research for developing methodologies on in situ conservation with the involvement of farmers and local communities benefit.
Activity #10	A greater emphasis and an explicit reference should be given to plant breeding and the use of biotechnology tools in plant breeding. The sustainable utilization of PGRFA should be promoted through appropriate policy making, capacity building, research and advanced technologies to meet climate change and other new challenges.
Activity #11:	Involvement of extension services for raising farmers' awareness and facilitating transfer of technologies.
Activity #13	A greater support should be given to the development of sound and harmonized seed production and distribution systems in the region.
Activity #15	Extending capacity building for the following priority areas with respect to climatic changes and new challenges: • in situ conservation; • utilization and breeding activities; • biotechnology; • seed technology; • policies and legislation.
Activity #16	Establishment of regional committees for PGRFA.
Activity #19	Expanding education and training to cover advanced technologies and methodologies. Promoting public awareness of the role and value of PGRFA in sustainable development and food security for all categories of people, from farmers to decision makers.

Source: FAO (2008)

have recommended several changes to be considered in updating the GPA (Table 7.2). Most of the proposed changes in activities require a strong regional mechanism of cooperation and coordination. Despite the rich genetic resources, the Near East region is presently lacking a sustainable institutional mechanism for an efficient networking and partnership among centres of excellence and regional centres conserving, managing and utilizing these valuable resources.

Enhanced sustainability of agriculture with sustainable use of PGRFA

Sustainable agriculture has been defined as agriculture that meets the needs of today without compromising the ability of future generations to meet their needs. In a sustainable agriculture system, special attention is paid to the conservation of natural ecosystems and resources (biodiversity, soils, water, energy, etc.) and social equity. Promoting the healthy functioning of ecosystems helps ensure the resilience of agriculture as it intensifies in meeting growing demands. Biodiversity plays a central role in the sustainability of productivity and other services provided by agro-ecosystems (e.g. nutrient cycling and carbon sequestration, pest regulation and pollination). This is particularly important in the face of increasing global challenges, such as feeding expanding populations and climate change. The role of farmers is very critical to sustaining such ecosystem services and with appropriate support they can enhance and/or manage these ecosystem services.

The Near East region has stressed throughout the Treaty negotiation, the importance of breeding for resistance or tolerance to pests and diseases, salt, drought, cold and heat, as a means to reduce pollution and biodiversity loss. Crops that are genetically improved for such resistances can contribute to sustainable agriculture by helping reduce requirements for agrochemicals.

Regional importance and global impact of underutilized species

Compared to major crops, there is relatively little research or breeding programme on less-utilized species, even though they can be very important locally or regionally. Such crops often have important and unique nutritional qualities or can grow in environments where other crops fail. Production of locally adapted crop species will diversify overall cropping systems and reduce the risk of food insecurity. Therefore, global and/or regional strategies or initiatives should be developed to promote research on, and improvement of, underutilized crops (FAO, 2008, 2009).

The livelihood of farmers in the Near East region (West and Central Asia and North Africa) strongly depends upon regionally important crops such as: date palm, pistachio and other nut crops, pomegranate, stone fruits, saffron, safflower and many other local crops. Food security in the region cannot be achieved without enhancing the production of such crops. This requires development of research based on new technologies for the production of such crops.

There is also a growing recognition of health problems associated with inadequate food quality and lack of specific nutrients in diets. Different plants are rich in different dietary constituents, the combination of which underlies the health-promoting effects of a diverse diet. Therefore, both problems can be addressed through the increasing diversity of food crops in diets and breeding crops, especially in the major staples for improved nutritional quality (Genc et al, 2009). However, little is known on genetics and breeding of biofortification of specific nutrients in food crops. As far as breeding crops are concerned, varieties that are richer in such compounds, characterization and evaluation of both cultivated and wild germplasm for nutritionally related traits are important steps. The application of biochemistry, genetics and molecular biology used to manipulate the synthesis of specific plant compounds, has been promising for the increased nutritional value of crops (FAO, 2009; Genc et al, 2009;). An example of this application is HarvestPlus, a programme of the CGIAR that targets the nutritional improvement of a wide variety of crop plants through breeding and focuses on the enhancement of betacarotene, iron and zinc.

Establishing a knowledge base for conservation and utilization

Enhancing global knowledge and developing technical capacity in conservation and utilization of the diversity of crop wild relatives under ever increasing environmental pressure, is crucial for meeting global agricultural challenges and food security, as also highlighted in all global crop diversity conservation strategies. Towards that goal, the establishment of an international research site on conservation and utilization of crop wild relatives in the Near East region will be a practical step forward in enhancing the conservation (both in situ and ex situ) and the use of crop wild relatives genetic resources, regionally and globally. Iran, on several occasions, has volunteered to host such a site (FAO, 2008).

Most of the ex situ collections in the centres of origin such as the Near East, are cross-sections of national (local) or regional diversity, which are assumed to be very unique and have not been completely duplicated anywhere else in the world (FAO, 2009). In addition, more than 95 per cent of the accessions in these collections are heterogeneous with a considerable diversity comprising many genotypes within the sample.

Regenerating, phenotyping and genotyping these materials remain challenging for most national programmes due to the extent and diversity of species in these countries, and due to a lack of funding, facilities or technical capacity. Lack of facilities and technical know-how, in particular, jeopardize the genetic integrity of the germplasm accessions and lead to their erosions in the gene banks. The problem is even more serious with cross-pollinated species. Such gene banks, in developing countries, holding important collections of cross-pollinated crops that are threatened by the loss of viability or genetic integrity, need to be urgently identified for sufficient financial and technical assistance towards the meeting of conservation standards (FAO, 2009).

Benefit sharing for a full implementation of all components

The success of any international agreement is measured by the fulfilment of its objectives. Achieving food security through sustainable agriculture is the driving force for the Treaty, which has been developed in harmony with the CBD. The conservation and sustainable use of PGRFA and the fair and equitable sharing of benefit arising out of their use are identified as objectives of the Treaty. To fulfill these objectives, the MLS and its unique mechanism for access and benefit sharing are particularly important (Esquinas-Alcázar, 2005). The full implementation of the Treaty with its supporting components is a key element to the success of this important legally binding international agreement.

Therefore, from the Near-Eastern perspective, obligations of contracting parties as regards to access and benefit sharing should be taken as equally important. Among these obligations, sharing of non-monetary benefits, exchange of technical information, transfer of technology and building the capacity of contracting parties of developing countries are among the main components of the Treaty, which should not be overlooked. Mobilization of the required financial resources for the full implementation of all components of the Treaty is fundamental and therefore should be the prime concern of all contracting parties, but the key responsibility of the contracting parties of the developed countries. It is particularly crucial for developing countries to get due financial and technical support in building their capacities for fulfilling their obligations towards the implementation of the Treaty. Small-scale farmers in centres of crop origin and diversity have contributed enormously to the development and conservation of plant genetic resources. Encouraging governments to address the rights of these farmers, particularly in sharing the benefits, based on model laws, already enacted in some countries such as India, will be very useful for the success of the Treaty.

Farmers' Rights

The issue of Farmers' Rights has been a topic of PGRFA discussions for a long time, particularly around the time of the final negotiations of the Treaty. The importance of farmers as custodians and developers of genetic diversity for food and agriculture was recognized in the Treaty through the provisions of Article 9 on Farmers' Rights. Such rights include: the protection of PGRFA associated traditional knowledge; the participation in decision-making mechanisms related to the conservation and sustainable use of PGRFA; equitable sharing of benefits accruing from the use of PGRFA; and to save, use, exchange and sell farm-saved seed/propagating material, subject to national law.

In the Near-Eastern countries, no legislation has specifically been developed on Farmers' Rights. Countries that have enacted legislation promoting such rights have done so within their seed acts and plant breeders' rights laws, as, for example, in Iran. Some other countries such as Turkey and Pakistan are currently developing legislation on access to biological resources and community rights.

Adoption of specific legislations on Farmers' Rights in India has provided a good example for developing countries. In industrialized countries, where farmers' organizations are well connected to policy processes, there was no need to push for Farmers' Rights and the debate on the use of farm-saved seed is held in the framework of IPR and seed legislation. In Europe, only Italy and Spain have adopted regulations on Farmers' Rights, and a number of countries are considering how they might support the implementation of Farmers' Rights in developing countries (Anderson, 2009).

The key to success of the International Treaty

In the Near-Eastern view, taking a sincere, fair and equitable responsibility towards both access and benefit sharing obligations is the key for a successful implementation of the Treaty. Although sharing of PGRFA under the MLS, itself, is recognized as a major benefit, this is true when benefits arising from the use of PGRFA are shared on a 'fair and equitable' basis. The fairness and effectiveness of the benefit sharing arrangement will be reflected on the achievement of food security, enhanced sustainability of agriculture and improved status of PGRFA conservation and use. The exchange of information and results of technical, scientific, and socio-economic research on PGRFA, and access to and transfer of technology are among the most important benefits to be shared. The Treaty lists various means by which the transfer of technology is to be carried out, including participation in crop-based or thematic partnerships, commercial joint ventures, human resource development and making research facilities available. Access to technology, including new PGRFA developed using the MLS should be provided and/or facilitated under fair and most-favourable terms, while respecting applicable property rights and access laws.

Capacity building in developing countries through facilitating scientific education and training, development of technical infrastructure for the conservation and use of PGRFA and carrying out joint scientific research, has been envisaged as an important prerequisite for fair and equitable sharing of benefits (FAO, 2008). The financial benefits arising from commercialization form part of the funding strategy under Article 18 of the Treaty. This strategy also includes the mobilization of funding from other sources. It is crucial for the success of the Treaty that all elements of the funding strategy, including the Global Crop Diversity Trust, work in coherence as part of one strategy.

As the monetary benefits flow not to the individual country providing the resources, but to the MLS, the provider of the resources from the developing countries has limited interest and financial resources to enforce the terms of the agreement when they are breached. The role of FAO as the third party beneficiary appointed by the Governing Body to represent its interests and initiate action where necessary to resolve disputes is very important.

Future perspectives

The Treaty has been perceived in the Near East region as a great achievement at the global level. Its impact on this region can be a good indicator of its success to be seen in the future. That is why there is still a great deal of work to be done by all contracting parties, developed countries, in particular, to successfully implement the Treaty and materialize their obligations. In addition, in defining a comprehensive international ABS regime, the specific needs of the agriculture sector need to be taken into account. Mutual supportiveness between the Treaty and the international ABS regime should also be developed. There is also a need for stronger coordination and synergy in the development of policies, legislation and regulations among the international instruments, various ministries, governments and other institutions having responsibility for different aspects of PGRFA. Countries need to adopt appropriate and effective strategies, policies and legal frameworks and regulations that promote the use of PGRFA, including appropriate seed legislation. Greater efforts are needed in order to materialize the real benefit of the Treaty to increase plant breeding capacity worldwide, especially in developing countries by mainstreaming new biotechnological and other tools in unlocking the potential of plant genetic resources.

The Treaty as a whole and its MLS as a unique feature of this Treaty should be put into context of all other related issues and communicated properly to other international agencies beyond the agriculture sector. Enhancing coordination, collaboration and synergy among concerned international agencies is vital for the success of the Treaty. Collaboration of UNDP is particularly important for materializing the funding strategy of the Treaty and its benefit sharing mechanism.

References

Amri, A., Mozfari, J. and Rukhkyan, N. (2008) 'Near East and North Africa regional analysis of PGRFA', a contribution to the Second State of the World's Plant Genetic Resources for Food and Agriculture, ICARDA, Aleppo, Syria

Andersen, R. (2009) 'Information paper on Farmers' Rights', input paper submitted to the Secretariat of the Plant Treaty, 19 May 2009 (IT/GB-3/09/Inf. 6 Add. 3)

Cooper, H. D., Spillane, C. and Hodgkin, T. (2001) 'Broadening the genetic base of crops: An overview, in H. D. Cooper, C. Spillane and T. Hodgkin (eds) *Broadening the Genetic Base of Crop Production*, CABI, Wallingford, pp1–24

Cooper, H. D. (2002) 'The International Treaty on Plant Genetic Resources for Food and Agriculture', *Review of European Community & International Environmental Law*, vol 11, no 1, pp1–16

Esquinas-Alcázar, J. (2005) 'Protecting crop genetic diversity for food security: Political and technical challenges', *Nature Reviews Genetics*, vol 6, pp946–953

FAO (1995) Report of the sub-regional preparatory meeting for West and Central Asia, Tehran, Iran, 9–12 October 1995 available at http://typo3.fao.org/fileadmin/templates/agphome/documents/PGR/GPA/prepWCAS.pdf

FAO (2008) Report of the Regional Analysis of PGRFA Conservation and Utilization in NENA Region, 29–30 November 2008, Aleppo, Syria

FAO (2010) *Second Report on the State of the World's Plant Genetic Resources for Food and Agriculture*, FAO, Rome, Italy

Genc, Y., Humphries, J. M., Lyons, G. H. and Graham, R. B. (2009) 'Breeding for quantitative variables. Part 4: Breeding for nutritional quality traits', in, S. Ceccarelli, E. P. Guimaraes, and E. Weltzien (eds) *Plant Breeding and Farmer Participation*, FAO, Rome, Italy, pp419–449

Janick, J. (2002) *History of Horticulture*, Purdue University, West Lafayette, IN

Moore, G. and Tymowski, W. (2005) 'Explanatory guide to the International Treaty on Plant Genetic Resources for Food and Agriculture', Environmental Policy and Law Paper No. 57, International Union for Conservation of Nature, Gland, Switzerland and Cambridge, UK

Mozafari, J. (2007) 'Effective use of plant genetic resources: A key global strategy and a national necessity for improving stability in crop production', Key papers of the 10th Iranian Crop Sciences Congress, 18–20 August, Karaj, Iran

Zehni, M. (2006) *Towards a Regional Strategy for the Conservation of Plant Genetic Resources in West Asia and North Africa* (WANA), published by the Association of Agricultural Research Institutes of the Near East and North Africa (AARINENA) available at www.aarinena.org/aarinena/documents/Strategy.pdf

Zeven, A. C. and Zhukovsky, P. M. (1975) *Dictionary of Cultivated Plants and their Centres of Diversity*, Pudoc, Wageningen, The Netherlands

Chapter 8

The North American Group

Globalization That Works

Brad Fraleigh[1] *and Bryan L. Harvey*

Scientists and policy makers in North America share the view that genetic improvement of crop plants is a great benefit to humanity. It is one of the least costly and most effective ways to increase production of food, fibre and plant-based products, to resist pests and diseases, to meet new market opportunities, and to address the challenges of abiotic stresses such as drought, temperature and climate change.

Three conditions appear necessary for these benefits to be realized: plant genetic diversity for food and agriculture must exist; the plant genetic resources must be available; and the capacity must be present to use them – that is, human, scientific and financial resources. These three conditions gave rise to important issues in the negotiation of the International Treaty on Plant Genetic Resources for Food and Agriculture (ITPGRFA or the Treaty), namely: conservation, access to genetic resources, and sharing the benefits arising from their use, in order to build capacity to generate more benefits.

It is well understood that all countries are interdependent when it comes to seeking plant genetic resources for food and agriculture (PGRFA) (Palacios, 1997). There is no country that considers itself to be self-sufficient for all the genetic diversity in all of its crop plants for all time. Each country benefits from having access to plant genetic resources kept in other countries. On the other hand, quite a few countries need better capacity to optimize the use of the genetic resources they might possess or acquire. In fact, the more a country can benefit from the use of plant genetic resources, the more they should be willing to share genetic resources with other countries.

The North American region and its perspectives regarding the Treaty's negotiation

The United States and Canada did not endorse the non-binding 1983 International Undertaking on Plant Genetic Resources, which neither country had negotiated. In February 1988, Canada wrote to the Director General of the United Nations Food and Agriculture Organization (FAO) detailing four concerns with the International Undertaking (IU). In particular, its concept of 'common heritage of mankind' appeared to conflict with existing property rights, including real property or intellectual property owned by individuals or governments. The statement eventually adopted by the Commission to the effect that 'common heritage of mankind' was not intended to conflict with national sovereignty or property rights, was helpful in that respect but not conclusive. That is one reason both countries were prepared to negotiate a better instrument when the opportunity arose during the 5th regular session of the Commission on Plant Genetic Resources[2] ('the Commission') in 1993, to adapt the IU in light of the newly adopted Convention on Biological Diversity (CBD).

North America was concerned with the growing trend of isolationism and 'access chill' towards the end of the last century, fuelled by unrealistic expectations of enormous profits to be made by selling genetic resources. A number of countries were increasingly reluctant to grant permission to collect crop germplasm or to give access to samples of genetic resources kept in their national collections. At the same time there was continued pressure to increase production to feed the growing world population and reduced ability of developing countries to support germplasm conservation. Thus the time was ripe for the establishment of the Treaty.

The first negotiating session, which one of us (Brad Fraleigh) had the honour to co-chair with Dr R. S. Rana (India), took place in November 1994. All told, there were 17 negotiating sessions, including 4 regular sessions of the Commission and 12 extraordinary sessions or meetings of contact groups or working groups, before finalization of the Treaty at the 2001 session of the FAO Conference (see Annex 1 of this volume for the list of all the Commission and Treaty meetings). There were also at least three informal consultations. If everyone realized how long it would take, and how hard it would prove, many people would have thought twice about the whole thing!

The North American region at FAO consists of only two countries: the United States and Canada (see Annex 2 of this book for the current list of contracting parties to the Treaty in each FAO regional group). The region is the centre of origin for certain crops such as sunflower, tobacco and Jerusalem artichoke. Wild relatives of cereal crops (e.g. wheat, barley and oats) and small fruits (such as raspberries, blueberries and strawberries) are also found here. Native North Americans developed adapted landraces of crops such as beans, maize and squash.

Significantly, both countries are major agricultural producers and have invested in sophisticated national crop research systems. Like every country in the world, we have benefited from the use of germplasm accessed from many locations around the globe (see e.g. Shands and Wiesner, 1991, 1992). In return for such inputs to their own agriculture, American and Canadian scientists and researchers have generated an enormous amount of crop genetic diversity and made improved germplasm and associated knowledge widely available for use by plant breeders and researchers around the world. This includes crop varieties, elite germplasm, genetic stocks and breeders' lines. North American institutions have provided funding for capacity building and infrastructure development in numerous developing countries. Thousands of graduate students and postdoctoral fellows have been trained in their universities.

Both the United States and Canada have long-standing commitments to conservation and the sustainable use of agricultural plant biodiversity. Each has extensive gene bank collections, which are well characterized. The United States has the largest single national genetic resources system in the world. Canada also has extensive collections of crop species found in temperate climates. According to the second report on the State of the World's Plant Genetic Resources for Food and Agriculture (FAO, 2010), North America preserves almost as many genetic resources samples in its ex situ collections as the entire Consultative Group on International Agricultural Research (CGIAR). Access to collections in both countries has been unrestricted for research and further development.

Canada and the United States have a long-standing tradition of cooperation in PGRFA. For example, they have maintained reciprocal membership on their respective national plant germplasm committees, and both use the Genetic Resources Information Network (GRIN) database management system, originally developed in the United States. Given this history of cooperation, it is not surprising that the two countries in the North American region have taken similar perspectives on issues related to conservation and utilization of plant genetic resources. It was therefore relatively easy to develop common regional positions during the Treaty negotiations. Frequent regional consultations were held throughout the negotiation process. It was clear to Canada that it would be essential to reach an agreement that the United States could ratify.

Some practical legal, political, environmental and economic issues that arose in the negotiations

It is well known that genetic resources are situated at the intersection of many domains: scientific disciplines like genetics, conservation biology, plant breeding and plant health, and social dimensions related to trade, economics, law and culture. Many issues in the negotiation of the Treaty were closely related to each other, and it was necessary to make incremental progress on all of them simultaneously, or in rapid alternation. Negotiators had to make serious efforts to listen carefully in order to understand different points of view and seek common ground.

The scope of the multilateral system

North America initially proposed that the multilateral system for access and benefit-sharing (MLS) should cover the full scope of the Treaty – that is, all PGRFA. It became clear early in the negotiations, however, that for many delegates the extent of coverage of the MLS was closely related to benefit-sharing. Many developing countries wanted proof that the Treaty would generate benefits for their countries to build their capacities to conserve genetic resources and use them sustainably, and would only agree to address a list of crops as a starting point. In the spirit of compromise, North America agreed; one of us (Bryan L. Harvey) chaired the first committee that discussed this topic, and Canada's John Dueck later co-chaired the committee which negotiated the list that eventually became the Treaty's Annex I.

The next logical question was which genetic resources would be covered for a given crop. Until quite late in the negotiations, the North American region affirmed their willingness to entertain two options. In one scenario, the entire gene pool of a crop could be included, provided property rights were respected and monetary benefit-sharing was not directly connected with individual transactions.

Maintaining and defending real and intellectual property owned by individuals, including farmers, by legal entities such as plant breeding companies, and by governments, was an important consideration for North American negotiators. The offer made by some African delegates to provide access to 'all our farmers' landraces' if other regions guaranteed access to 'all of the private collections' proved unacceptable to North America, because of the implied expropriation of the private property rights of farmers and other owners of genetic resources.

Under the other option, if monetary benefit-sharing was to be directly connected to individual transactions, only genetic resources under the management and control of national governments could be included in the MLS, because governments could not legally ensure that benefit-sharing arrangements would apply to genetic resources owned by other entities. Of course, this option does not exclude coverage, on a voluntary basis, of other genetic resources owned by anyone else. Developing countries eventually preferred this second option, which became the basis of the terms of access and benefit-sharing in the Treaty.

Farmers' Rights

The discussions on Farmers' Rights were often quite bewildering for developed countries. Canada, for example, has publicly stated that it takes numerous measures to ensure the contribution of farmers to the conservation and sustainable use of PGRFA, without the need for a specific law on 'Farmers' Rights'. Canadian farmers share in benefits arising from the utilization of PGRFA, in particular, the availability of new crop cultivars that are better suited to the challenges they face and to new market opportunities. Crop research is directed toward developing and evaluating new crop varieties that will enable producers to access new markets, diversify production, improve the quality of their products and enhance resistance to pests and pathogens. Increased crop diversity enables farmers to use new crops in rotation, aiding pest management strategies and ensuring more balanced soil nutrient distribution both spatially and temporally. Crop breeding programmes emphasize crops that represent large acreage, strong production potential in northern latitudes, or have the capability to act as effective components of a system for diversification or sustainable cropping practices. The production of new crop varieties with pest resistance is a key component. Major pest threats to crop production have been documented and potential new threats are monitored so that all significant pests are considered. Breeding programmes include pest resistance screening as a routine feature, and may use biotechnological tools to introduce genetic resistance into new varieties.

Many research centres are studying new crops and varieties for rotation, intercropping, replacement, niche markets and market opportunity. A wider diversity of alternative crop options supports the use of effective crop rotation as pest and resource management tools. Programmes that support crop diversification are carried out in Canada in conjunction with provincial initiatives and in cooperation with the private sector. Better management of inputs is an important aspect of crop diversification initiatives and the thrust is to reduce inputs through the use of new crop varieties, improved management practices and better timing and selection of inputs.

Farmers' associations in Canada participate in making decisions at the national level, on matters related to the conservation and sustainable use of PGRFA. For example, many producers' organizations were consulted in the development of Canada's agriculture and agri-food policy framework known as 'Growing Forward' (Agriculture and Agri-Food Canada, 2008).

Canada is of the view that all communities create culture and some of these cultural expressions may be considered traditional knowledge. Knowledge, traditional or otherwise, evolves over time. Community-level procedures for accessing traditional knowledge differ from one community to the next, and for many reasons. To a large extent, decisions regarding what is 'protected' are taken by key individuals and/or the community as a whole. In many cases, how an indigenous community achieves informed consent for access to traditional knowledge within its community is privileged information and therefore not for disclosure to users, the public, governments or the parties to the Treaty.

The preservation of traditional knowledge may take many forms, including (but not limited to): maintenance and transmission of traditional practices; preservation of aboriginal languages; preservation in national collections (e.g. artefacts and records); support for cultural organizations and activities; and preservation and distribution through print and broadcast media.

Under Canada's national intellectual property system, there is no specific protection for traditional knowledge. Nevertheless, a creator or inventor who meets the specific requirements of a particular piece of intellectual property legislation will receive intellectual property protection. Examples of such protection can be found in relation to copyright law, patent law, industrial design law and trademark law. Additionally, trade secrets law may be of use to holders of traditional knowledge if such knowledge is susceptible to commercial application.

North America was therefore not initially inclined towards subscribing to the concept of Farmers' Rights in a legally binding international agreement. Moreover the demand for Farmers' Rights was often inconsistent and contradictory. At one point, no less than 13 different themes were proposed by various delegations under this heading! This situation led to many misunderstandings. At one point our Head of Delegation, the late John Dueck, read out a statement explaining that it was a problem for Canada to consider Farmers' Rights as a new 'human right' because in our country human rights were for everyone, and cannot apply to a single occupational group, even one as important as farmers. A representative of a Canadian-based civil society organization (CSO) became upset, stating that if Canada awarded plant breeders' rights to plant breeders we could award Farmers' Rights to farmers. This is inaccurate, because anyone can obtain a plant breeders' right if their crop variety or line meets the criteria in the legislation, and they don't have to be designated as a 'plant breeder'. The CSO representative then announced he would mobilize Canada's farmers against this position. The department of agriculture did in fact receive a letter from the association that represents the vast majority of Canada's farmers, which supported the official Canadian position. Eventually common ground was found among the negotiators, and Article 9 of the Treaty entitled 'Farmers' Rights' was one of the first major issues to be agreed. One essential element was the provision that the responsibility for realizing Farmers' Rights rests with national governments. This remains a challenge for the Governing Body of the Treaty, where certain delegations are tempted to propose resolutions which try to tell national governments what to do.

Financial issues

Some delegations presented an 'entitlement' or 'compensation' approach to financial resources under the Treaty. On the other hand, many North American stakeholders pointed out that since everyone would benefit from improved access to crop genetic diversity, there should be no need for any dedicated funding at all for the Treaty. Other North American policy-makers, including the authors (see Fraleigh, 1987), argued that many countries lack capacity, or require additional capacity, to optimize the use of the genetic resources they might possess or acquire. Why, if they are not assisted to build their capacities, would such countries be motivated to share plant genetic resources with others? In definitive, the North American approach to financial resources and benefit-sharing was strongly linked to capacity-building in the conservation and sustainable use of PGRFA.

Interactions with stakeholders

In Canada, there is a strong tendency to consult. Many stakeholders contributed to the development and consideration of the Canadian approach to the Treaty. The country's official positions were determined by a federal government interdepartmental committee on genetic resources at the FAO. This committee was chaired by the department of Agriculture and Agri-Food Canada and included representatives from federal departments and agencies dealing with foreign affairs, international trade, international development, industry (mostly regarding patent policy), food inspection, environment and forestry. A series of legal advisors working for the Ministry of Foreign Affairs provided invaluable counsel about the conformity of various proposals with Canadian law and international obligations. The Canadian International Development Agency had great influence in determining Canada's positions on financial resources.

Advice was requested regularly during the negotiations from other national stakeholders by way of Canada's national expert committee on plant genetic resources. These stakeholders included representatives of provincial departments of agriculture, academia, scientific societies, CSOs and industry associations. The industry associations, especially the Canadian Seed Trade Association, which represents private sector plant breeders, followed the negotiations closely, recognizing that a global agreement on terms of access to genetic resources and benefit-sharing might contribute significantly to legal certainty in their work.

Canada had one of the few delegations that regularly included non-government members. Agriculture and Agri-Food Canada offered to fund half the cost of participation of one industry representative and one civil society representative in the Canadian delegation for each negotiating session. These representatives were nominated by the national expert committee, not by government. Industry took up this offer more often, and named one of us (Bryan L. Harvey) as their representative. He later served as Chair of the second meeting of the Interim Committee for the Treaty in 2004 during the period between the Treaty's adoption and the first meeting of its Governing Body. CSOs sent Ms Sharon Rempel, at the time a member of Seeds of Diversity Canada to one negotiating session. She served as a fully fledged member of the delegation and attempted to act as a link between the delegation and international CSOs. As officials representing the Canadian government, negotiators were always cognizant of the need to be aware of the views of all national stakeholders and to understand these as clearly as possible, in order to provide advice on Canadian positions from the perspective of national interest and good public policy.

The stakeholders remain involved during the implementation of the Treaty. Canada initiated use of the standard material transfer agreement (SMTA) on 1 July 2008, which leads to monetary benefit-sharing when the recipient commercializes a product under its terms. In 2009, Canada announced its first voluntary contribution to the Treaty's Benefit-Sharing Fund, relative to commercialization of a superior line of triticale developed by Canadian researchers working with the CGIAR and in particular the International Maize and Wheat Improvement Centre (CIMMYT). It will be an annual contribution for the duration of the commercial life of the variety.

Strengths and weaknesses of the Treaty

North America perceives a number of strengths in the Treaty (see Annex 3 of this volume for information on the main components of the Treaty). It recognizes the special status of PGRFA, and was tailored for this sector. Its scope covers all PGRFA, not just Annex I species. Establishment of a multilateral system of facilitated access and benefit sharing for crops important to global food security is a strong positive step. The MLS defines and codifies the rights and obligations of contracting parties to conserve and provide access to their germplasm and to ensure that appropriate benefits flow from its utilization, with a good balance of provisions contributing to both these objectives. It is a benefit that a large number of countries have ratified the Treaty; Canada has been a contracting party since 2002. The United States is in the final stages of its complex international treaty ratification process and is expected to ratify in the not too distant future. This would further strengthen the Treaty. North America views the inclusion of the collections in the international agriculture research centres supported by the CGIAR and other international organizations to be an important positive inclusion.

North America has also identified some weaknesses in the Treaty. In the view of its spokespersons, the species list in Annex I is far too short and should be expanded; soybean is a clear example of a crop that should logically be included, because it so obviously fulfills both the criteria of food security and interdependence stated in the Treaty's Article 11.1. It is also a weakness that uses in agro-forestry, industrial agriculture and ornamentals are not currently envisaged, bearing in mind that food security requires that farmers have access to a range of cash crops which can generate revenue for them to purchase inputs and necessities to improve their lives.

Benefit-sharing provisions specified in the SMTA may be weaker than they needed to be, and may not optimize the generation of revenue in the short term. There may be more willingness to pay for benefits at a reasonable rate, bearing in mind the very low margins in the international seed industry. Thus, for instance, if an acquired accession contributed more than 25 per cent by pedigree, or contributed a significant trait, such as disease resistance, to a resultant commercial cultivar, then a requirement for payment could be triggered regardless of whether the resulting product was freely available for further research and breeding. The current provision, that a payment is only triggered if the product is not freely available, means that a smaller percentage of varieties will generate revenue for the Treaty. Thus revenue may rely in the short term, for the most part, on voluntary contributions such as Canada has already made.

The practical fact that the effective operation of the Treaty will rely on the good sense and good will of the participants is both a strength and a weakness. Implementation of the Treaty will be largely self policing. Fortunately, the overwhelming majority of people involved in crop plant germplasm conservation and utilization are committed to doing the right thing for the betterment of the human condition.

Challenges ahead and how these could be met

The major challenges for the Treaty are to articulate and enhance its role to accompany and assist member countries in addressing the interrelated global problems of food security, climate change and habitat destruction, as well as the increasing urgency to address these issues. PGRFA are threatened by these problems but can also contribute to solving them.

When the text of the Treaty was adopted by FAO Conference in November 2001, the *Earth Negotiations Bulletin* wrote:

> *... major hurdles still remain. First is the issue of ratification, which raises the need to educate national policy-makers and those actually using plant genetic resources for food and agriculture on what the system is and how it will work. Several delegates also mentioned that negotiations on the standard MTA could easily occupy them for another seven years. As countries turn to the future they will have to identify the necessary capacity for national implementation, a process well evidenced in delayed ratifications of the Cartagena Protocol on Biosafety and in related discussions on access and benefit-sharing under the Convention on Biological Diversity. Negotiators will also remain busy with discussions on how other* ex situ *collections of genetic resources ... should be handled.* (*Earth Negotiations Bulletin*, 2001)

The situation is quite different today in many respects. The Treaty entered into force in 2004 and there are 127 contracting parties at the time this chapter was written – ratification by all eligible countries would be even better. The Treaty's Governing Body adopted the SMTA at its first meeting in 2006, and it is being used in many member countries. Standards for ex situ gene banks are being updated by the Governing Body and the Commission working together. In many areas, science and technology, especially the enhanced use of molecular technologies, have the potential to increase the contributions of plant genetic resources to solving food security problems.

Significant financial support has been provided under the Treaty's benefit-sharing fund and by the Global Crop Diversity Trust to build capacity in the conservation and sustainable use of PGRFA. However, strengthened research capacity is required in many areas, for example, to address gaps in characterization and evaluation data. Human resource capacity and needs should be assessed and prioritized by countries requiring international assistance, as the basis for drawing up education and training strategies. Human capacity, funds or facilities, are not adequate in some parts of the world to manage ex situ collections at the required standards.

The use of plant genetic resources has been stimulated by the creation of FAO's Global Partnership Initiative for Plant Breeding Capacity Building (GIPB), but its investigations have so far demonstrated that overall, global plant breeding capacity has not significantly changed.

In Canada, following the entry into force of the Treaty, one of us (Bryan L. Harvey) presented the provisions and impact of the Treaty to several meetings of plant breeders and scientists in academic and government institutions across the country. This was helpful to educate users of plant genetic resources and decision-makers. However, many feel the Treaty needs to generate more information about continuing genetic erosion, in other words the loss of crop genetic diversity. Genetic vulnerability, which is strongly correlated with the diversity of crops grown in the field, also needs to be accorded more attention. More information is needed for policy- and decision-makers about the contributions of plant genetic resources to solutions for the many challenges faced by agriculture, including the need for increased production, threats from pests and diseases, climate change, and so on.

These gaps and needs are detailed in the second report on the *State of the World's Plant Genetic Resources for Food and Agriculture* (FAO, 2010). Many are susceptible to enhance national implementation. The priority activity areas to address these issues should emerge from the process of updating the first Global Plan of Action on Plant Genetic Resources for Food and Agriculture (FAO, 1996), which will be considered for adoption by the Commission on Genetic Resources in 2011. The updated Global Plan of Action should also establish the funding priorities for the Treaty.

North America has evidently maintained its long-term vision of a MLS expanded to cover the full scope of the Treaty, in terms of crops as well as agricultural uses. Views tending in the same direction are being expressed by people in other parts of the globe, and such expansion may well take place in due time, as the generation of benefits thanks to the MLS becomes progressively more evident.

In closing, one challenge for the Treaty, at least during the next few years, will be to determine its interaction with and relative field of activity relative to the new 'Nagoya Protocol on Access to Genetic Resources and the Fair and Equitable Sharing of Benefits Arising from their Utilization to the Convention on Biological Diversity'. The Nagoya Protocol was adopted quite recently, in October 2010. Many of its provisions will require interpretation, and will no doubt be discussed for many years to come. The article entitled 'Relationship with International Agreements and Instruments' is particularly interesting for parties to the Treaty. It states in particular that:

> *... the provisions of this Protocol shall not affect the rights and obligations of any Party deriving from any existing international agreement, except where the exercise of those rights and obligations would cause a serious damage or threat to biological diversity.*

and later in the same article states that:

> *... where a specialised international access and benefit-sharing instrument applies that is consistent with, and does not run counter to the objectives of the Convention and this Protocol, this Protocol does not*

> *apply for the Party or Parties to the specialised instrument in respect of the specific genetic resource covered by and for the purpose of the specialised instrument.*

These provisions may present a useful basis for mutually advantageous interaction between these two legally binding international instruments.

Notes

1 Dr Brad Fraleigh of the Department of Agriculture and Agri-Food, Government of Canada. © Her Majesty the Queen in Right of Canada, as represented by the Minister of Agriculture and Agri-Food Canada, 2011.
2 Renamed 'Commission on Genetic Resources for Food and Agriculture' in 1995.

References

Agriculture and Agri-Food Canada (2008) *Growing Forward Agricultural Policy Framework*, www4.agr.gc.ca/AAFC-AAC/display-afficher.do?id=1200339470715&lang=eng, accessed 3 December 2010

Earth Negotiations Bulletin (2001) 'Negotiations on the International Treaty on Plant Genetic Resources for Food and Agriculture: 30 October – 3 November 2001', vol 9, no 213, International Institute for Sustainable Development (IISD)

FAO (1996) *Global Plan of Action for the Conservation and Sustainable Utilization of Plant Genetic Resources for Food and Agriculture*, UN Food and Agriculture Organization, Rome

FAO (2010) *The State of the World's Plant Genetic Resources for Food and Agriculture*, ISBN 978-92-5-106534-1, UN Food and Agriculture Organization, Rome

Fraleigh, B. (1987) 'Plant genetic resources of Canada: Four years of international germplasm exchanges', *Plant Genetic Resources Newsletter*, no 69, pp48–49, FAO and IBPGR

Palacios, X. F. (1997) 'Contribution to the estimation of countries' interdependence in the area of plant genetic resources', Background Study Paper No.7, Rev.1, Commission on Genetic Resources for Food and Agriculture, FAO

Shands, H. L. and Wiesner, L. E. (eds) (1991) *Use of Plant Introductions in Cultivar Development Part 1*, CSSA Special Publication no 17, Crop Science Society of America, Madison, WI

Shands, H. L. and Wiesner, L. E. (eds) (1992) *Use of Plant Introductions in Cultivar Development Part 2*, CSSA Special Publication no 20, Crop Science Society of America, Madison, WI

Chapter 9

The Southwest Pacific Regional Group

A View from the Pacific Island Countries and Territories

Mary Taylor

Putting the Pacific region in context

Visitors to the Pacific region are often amazed by the diversity that exists. The region is geographically, ecologically, sociologically and economically diverse. The Pacific region, with a land area of 550,000km^2 surrounded by the largest ocean in the world, is home to 9.5 million people. Five islands (Fiji, New Caledonia, Papua New Guinea, Solomon Islands and Vanuatu) account for 90 per cent of this land area, and more than 85 per cent of the population. In contrast to these relatively large landmasses, the world's smallest island states and territories, for example, Nauru, Tuvalu and Tokelau, can be found in the Pacific (see Annex 2 of this volume for the list of contracting parties by regions). The importance of agriculture in sustaining livelihoods varies across the region. In the larger islands, such as Papua New Guinea, Solomon Islands and Vanuatu, agriculture, and forestry also, remain the mainstay of the economy and employment, contributing significantly to household income and, increasingly, export earnings, whereas subsistence dominates in some of the smaller islands (SPC-LRD, 2008b).

Islands, by their very nature, have unique diversity and the Pacific is no exception. The region is the centre of diversity for coconut (*Cocos nucifera*) and breadfruit (*Artocarpus altilis*). Secondary centres of diversity have arisen for crops such as sweet potato and yam, moving with people as they migrated from different regions. Banana (*Musa* spp.), yam (*Dioscorea* spp.) and taro (*Colocasia esculenta*) emerged from Southeast Asia, but are now very important staple crops in the Pacific, reflecting the interdependence among regions. This interdependence continues to this day, with pest and disease outbreaks and climate change, highlighting the vulnerability of the majority of Pacific Island countries. The importance of crop diversity to food and nutritional security in the Pacific is further discussed in the following heading.

The Pacific region faces numerous social and physical challenges in maintaining and improving the productivity of their agriculture sectors and protecting their biological diversity. The geographical isolation of the region and the small size of many of the islands have resulted in a narrow genetic and production base with limited opportunities for scaling up production. These constraints do little to support recovery from natural disasters which are an increasing occurrence. Movement of goods and people, through trade and tourism, have heightened the risk of introducing unwanted plant and animal pests, weeds, diseases and other alien invasive species, threatening the fragile ecosystems and resource base of the region.

Significant social challenges exist which affect the agriculture sector. Populations are projected to grow at an annual rate of 2 per cent in Melanesia, 1.84 per cent in Micronesia and 0.7 per cent in Polynesia.[1] Urban populations are growing at a faster rate, and are expected to double in 25 years in Melanesia. Rural to urban migration has the potential to reduce agricultural production and increase reliance on imports. Diets that include an increasingly higher proportion of imported food with little nutritive value are causing or contributing to escalating rates of non-communicable diseases, malnutrition and micronutrient deficiencies (SPC-CRGA, 2008a).

Climate change will exacerbate many of these challenges – the fragile ecosystems and in the majority of cases the fragile infrastructure will be tested to the limits. The region is used to disasters but it is foreseen that these disasters will increase in intensity and become more unpredictable with climate changes. This impact has been demonstrated very clearly in 2009–2010 in Fiji with the occurrence of severe flooding and two cyclones, Cyclone Mick and Cyclone Tomas. These disasters impose serious constraints on development in the islands, so much so that some of the islands seem to be in a constant 'recovery-mode'. With urbanization and an increase in imported food consumption comes also a loss of traditional knowledge and practices of local farmers – this knowledge and these practices are likely to be critical in finding solutions to future challenges, such as climate change.

Food security in the Pacific falls under the mandate of the Secretariat of the Pacific Community (SPC), an intergovernmental organization providing technical and policy advice and assistance to its Pacific Island members. SPC was established as an international organization in 1947 and has 26 member countries and territories, 4 of which are founding members (Australia, New Zealand, France and the United States of America). SPC services are provided primarily in the form of technical assistance, training and research. The organization has six divisions, one of which is the Land Resources Division which covers sustainable forestry and agriculture, genetic resources, plant health, crop production, animal health and production, and biosecurity and trade.

Australia as a founding member of SPC, provides significant support to the organization, through support for ongoing priority core programmes, and also additional funding for the implementation of specific initiatives, such as climate change adaptation through the International Climate Change Adaptation Initiative (ICCAI). It is also the leading donor of aid to the independent countries of the Pacific, and has significant trade and commercial interests in the region – the Pacific is Australia's closest market. Within the context of the International Treaty on Plant Genetic Resources for Food and Agriculture (ITPGRFA or the Treaty), Australia is a contracting party and a member of the Southwest Pacific group (see Annex 2 of this book listing the SWP contracting parties to the ITPGRFA). With its significant capacity advantage in conventions and legal matters, Australia has represented the Pacific region at Treaty negotiations within the Southwest Pacific group.

The importance of crop diversity in the Pacific region

Crop (PGRFA) diversity is an essential tool to assist the region in responding to the many challenges it faces, providing the means to manage climate change, to meet market needs and, importantly, ensure food and nutritional security. A wide range of PGRFA diversity will be required to satisfy this basket of needs. The Pacific is a centre of diversity and/or origin for a number of crops, but in general, due to its history of human colonization, genetic diversity in the mostly vegetatively propagated crops of the region declines markedly from west to east. In 1998, the 'Taro Genetic Resources: Conservation and Utilization' (TaroGen) project, funded by AusAID,[2] was established. Over 2000 taro accessions were collected from within the region. Morphological and molecular approaches were utilized to determine what diversity existed in the collection and to identify accessions for the core collection, representative of the diversity in the whole collection, reducing the size of the collection from 2000 to 200 (Mace et al, 2006). The subsequent molecular comparison between Asian and Pacific taro germplasm confirmed the limited genetic diversity that exists in the Pacific, compared to Asia (Lebot et al, 2004) and set the direction for taro breeding programmes in Samoa and elsewhere.

The vulnerability of a limited genetic base was clearly demonstrated in Samoa in 1993 when taro production was brought to a halt by taro leaf blight (TLB), a disease caused by the fungus, *Phythphthora colocasiae*. Taro was the main staple food in Samoa as well as a lucrative cash crop, with exports worth US$7 million annually. TLB wiped out the entire taro industry in a matter of months, raising food security concerns, and significantly reduced export revenues, which impacted on the nation's foreign reserves. Across the food sector, taro was soon replaced by less nutritious starchy staples in the form of instant noodles and rice. There was also fear that the disease might spread to the Cook Islands, Fiji, French Polynesia, New Caledonia, Niue and Vanuatu, with equally devastating results. At the time of infection taro production was based on just one variety, that of taro Niue, which was highly favoured by the overseas market. Niue was highly susceptible to the disease, and combined with ideal weather conditions and the movement of planting material, these factors enabled the disease to reach epidemic proportions.

Chemical and cultural control methods were evaluated but were neither effective nor realistic. At the same time as chemical and cultural control methods were being tested, local varieties were also being evaluated for their resistance to TLB, but no resistance was found; they were all, in fact, highly susceptible. Consequently the call went out to other countries both within and outside the region for taro varieties with known resistance/tolerance to TLB. From outside the region the Philippines was the first to respond and provided a variety known as PSB-G2. Varieties from the Federated States of Micronesia and Palau were also considered, and a variety called Ngerruuch from Palau was particularly successful, both in its response to TLB and also its acceptability by the Samoans. It was therefore PSB-G2 and Ngerruuch that supported the revival of taro production in Samoa, a case of 'crop diversity to the rescue'.

Sharing crop diversity

The TLB outbreak in Samoa highlighted the importance of diversity, of which there was limited awareness. It had such major consequences for the country, there was no ignoring the fact that diversity was important and should be an important component of any crop production chain. The need to be able to access diversity from elsewhere, demonstrating that no country is self-sufficient in PGRFA diversity, had also been highlighted. Countries can only have access to diversity outside their borders if others are willing to share. This was a key message for the Pacific region where many of the major staple crops, including taro, have very strong cultural associations. This cultural connection strengthened the belief that all crops and varieties required for food security could be found at least within national borders and at most within the region.

After 1993 there were a number of developments that acknowledged the importance and renewed interest in PGRFA. Of significance was a meeting in 1996 of the Pacific ministers of agriculture, where they pledged to put in place at the national and regional levels, policies to conserve, protect and utilize plant genetic resources. SPC's response to this recommendation was twofold: the establishment of the then Regional Germplasm Centre (RGC), now the Centre for Pacific Crops and Trees (CePaCT) in 1998 and the Pacific Plant Genetic Resources Network (PAPGREN) in 2001. These two components of the Genetic Resources programme within the Land Resources Division of the SPC ensure an effective regional hub, but equally important, an active and wide-reaching network, which supports both national and regional activities.

The basic aim of the CePaCT is to provide the region with the means to safely and effectively conserve their PGRFA, and to facilitate access to useful diversity both within and outside the region. In vitro methodology is used, and collections exist for the aroids, yam, sweet potato, banana, breadfruit, cassava, and other more minor crops. Since its establishment in 1998, the Centre has significantly expanded its operations, both with regards to collections conserved, crops/accessions distributed and research conducted. CePaCT now holds a globally unique collection of 878 accessions of taro, and is building up its collections of other edible aroids and yam species. CePaCT has generated interest in diversity through its distribution programme; countries are keen to evaluate new varieties, increasingly so with the concern over climate change. The offer of crops and varieties in the 'climate ready collection' further that interest for 'new' diversity.

In 2001, PAPGREN was launched, with funding from NZAID[3] and ACIAR.[4] Technical support was made available by Bioversity International. At the time of inception, the membership of the network included 11 Pacific Island countries; membership now stands at 17, and includes 2 French territories. PAPGREN was the perfect framework within which to nurture the importance of diversity and the need to share that diversity. In 2003 a publication commissioned by PAPGREN, 'Policy Issues Relating to Plant Genetic Resources in the Pacific' was released (SPC, 2003). At the First Regional Conference of the Ministers of Agriculture and Forestry Services, held in Fiji (2004), a paper was presented that emphasized the importance of the Treaty and urged countries to ratify. At the end of the meeting, the ministers acknowledged:

> *... that access to genetic resources (crop, tree and animal) is necessary to ensure food security in the long-term. Broadening the genetic base of crop, trees and livestock, and genetic improvement and diversification are crucial in coping with climate change. Access to and utilization of genetic resources will be enhanced through active participation in PGR networks, both at the regional (PAPGREN) and international level (COGENT [International Coconut Genetic Resources Network] and BAPNET [Banana Asia and Pacific Network]). To ensure continued access to genetic resources, the countries of the region should consider endorsing the RGC MTA, ratifying the International Treaty, and signing the Establishment Agreement for the Global Crop Diversity Trust Fund* (SPC, 2005). (See Chapter 16 for details on the GCDT.)

This was basically the first exposure the Pacific region had to the Treaty.

In May 2006, a Plant Genetic Resources for Food and Agriculture Workshop was held in Fiji by SPC in collaboration with the Australian Government DAFF[5] and CSIRO.[6] At this meeting DAFF clarified the elements of the Treaty, and participants formed working groups to address the standard material transfer agreement (SMTA); regional issues; Farmers' Rights and implementation issues. The workshop showed there was a general willingness throughout the Pacific to participate in the Treaty and the multilateral system (MLS), as exemplified in the 2004 resolution of the Heads of Agriculture and Forestry Services (HOAFS) meeting. There was consensus that the main challenge for people working in the area of plant genetic resources is the ability to influence their governments on the costs and benefits of the Treaty. The workshop agreed that SPC had, to date, performed a key role in brokering initial information sharing on the Treaty (workshops, policy advice, draft Cabinet submissions). The outcomes from the Fiji workshop indicated that the Pacific region was strongly committed to the Treaty and wished to be fully engaged in the process.

The first meeting of the Governing Body of the ITPGRFA was held in Madrid in June 2006 (see Annex 1 of this volume for the list of all Commission and Treaty meetings). Contracting parties attending from the Southwest Pacific region were Australia, the Cook Islands, Kiribati and Samoa. New Zealand and Fiji sent observers. The Southwest Pacific region held discussions prior to the Governing Body meeting, and daily discussions prior to each day's proceedings. At this meeting the Southwest Pacific group agreed that DAFF would represent the region on the Bureau of the Governing Body.

During their country presentations at the 2006 annual PAPGREN meeting, the Cook Islands, Kiribati and Samoa encouraged other countries to ratify the Treaty thereby having a voice in negotiations. It was evident at this meeting that PAPGREN representatives fully supported the aims and objectives of the Treaty and would endeavour to progress ratification in their countries. However, the number of contracting parties to the Treaty did not increase until 2008 when Fiji and Palau acceded to the Treaty after PAPGREN conducted national consultations in both countries. One impediment to ratification of the Treaty has been the relatively frequent changes in government in many countries, especially the larger ones. The Treaty ratification process is a lengthy one, and in several cases ratification has not been achieved after significant work has been carried out by the PGR focal point due to a change in government or key people within the government.

Although a significant number of countries in the Pacific are yet to accede to the Treaty, the progress made in the overall understanding and acceptance of the importance of sharing diversity has to be recognized. It has been achieved through both a top–down and bottom–up approach, acknowledging the key contributions of the ministers in both 1996 and 2004, and the PAPGREN national focal points. An independent survey recently conducted showed that, since the establishment of PAPGREN, the understanding of PGRFA issues and their contribution to food and nutritional security has significantly increased. Annual meetings mean that PGRFA researchers and workers meet, exchange ideas and skills on how to use and enhance PGRFA. National priorities and problems are highlighted and the network strives to find solutions. This open dialogue has supported the development of a healthy and positive attitude to sharing PGRFA, and the realization that no country has the genetic resources or the human and financial resources to have the answer to all PGRFA issues – and that this dependency exists beyond the Pacific region.

The ultimate recognition that the region fully appreciates the importance of sharing germplasm was evident from the formal placing of materials held in the CePaCT into the multilateral system of the Treaty by the Samoan Agricultural Minister on behalf of Pacific Ministers of Agriculture and Forestry at the 3rd session of the Governing Body of the Treaty in Tunis, 2009. The Minister emphasized the importance of agriculture to Pacific Islands and the need to protect biological diversity to ensure food security, especially in light of climate change and natural disasters. He said that the region's significant diversity is not enough to deal with future challenges, and welcomed access to global diversity through ratification of the Treaty.

What are the issues faced by the Pacific with regards to the implementation of the Treaty?

In September 2009, a workshop on the Treaty was held in Fiji for PAPGREN members. The aim of the workshop, funded by NZAID through PAPGREN and with supporting funding from the Treaty Secretariat under the Joint Capacity Building programme, was to identify and discuss concerns from non-contracting parties regarding accession to the Treaty, and equally concerns from contracting parties regarding implementation of Treaty obligations.

The 2009 workshop allowed for very open dialogue on the Treaty and at no point did any of the non-contracting parties express any major difficulties with the Treaty itself. Problems in ratification tended to centre on the issue of logistics, such as changes in government as previously mentioned. Human resources are also a constraint, especially in the smaller countries, where one individual has responsibility for more than one thematic area, and is often required to travel to many international meetings throughout the year. Countries do not have legal expertise in PGR policy, which leaves it to the national PGR focal point to try and explain the benefits of the Treaty to the government legal office. The smaller countries are at a disadvantage due to their human resources and the lack of understanding of legal PGR issues within the government legal office, whereas the larger countries tend to suffer from overly complex procedures. It is interesting to consider the reasons why the Cook Islands, Kiribati and Samoa were the first three countries to ratify. The Cook Islands has relatively limited PGRFA diversity, but because of its exposure to the New Zealand market had a very good understanding of the importance and usefulness of diversity. Kiribati is an atoll, also with limited diversity and much threatened by climate change. Its opportunities for market development are also poor, constrained by land and human resources. The experience of Samoa with TLB was sufficient justification for ratification. However, what all three countries had in common also was the relatively high position of the PGR focal point within the national system and stability of government.

Fiji and Palau both acceded to the Treaty in 2008, after national consultations were held in each country. This points to another factor that can delay the ratification process and that is the number of parties involved in the process, making national consultations almost essential, to ensure all stakeholders are involved in the decision-making process. A good example of just how complicated it can get in one island country can be found in the Federated States of Micronesia (FSM). For FSM to accede to the Treaty there would have to be not just approval at the national level, but also at the State level – and there are four States. This is not a reflection of lack of political will but more an indication of limited resources and a basic lack of capacity and knowledge. Despite efforts at the regional level to raise awareness as to the importance of the ITPGRFA, this information and knowledge does not reach all of those involved in the ratification process. For FSM to fully engage in the ratification process, consultations would be required within each of the four States which has obvious financial implications.

The contracting parties highlighted a number of areas which they felt were unclear and/or were impacting on implementation of Treaty obligations. The issue of having the capacity to develop genetic resources was raised several times during the course of the meeting, bearing in mind that the Treaty is set up to encourage the development of new varieties that can assist with adapting to climate change and for food security. Bearing in mind the diverse nature of the islands, there is an urgent need for capacity building in crop improvement, at the community level and national levels, preferably using participatory approaches to ensure sustainability.

All participants at the workshop – contracting and non-contracting parties – expressed a desire to have stronger representation at the Treaty meetings. With the current FAO designation, the Pacific Islands are grouped with Australia under the Southwest Pacific region banner. In the discussions, the legal expertise required to negotiate at these meetings was acknowledged and as such, Australia was best placed to represent the region. However, there are many instances where Australia and the Pacific islands would have differing opinions on a Treaty issue. Various approaches were discussed to address this concern, with the drafting of a regional paper prior to any meeting, considered the best option.

The Treaty is no exception to the many international frameworks that require implementation and as such the limited capacity in the Pacific region was discussed at length. The number of ways in which this could be supported was highlighted, for example, through education programmes, awareness raising and simply learning from other regions/countries' approaches to implementation. The workshop noted that the regional arrangement with CePaCT acting on behalf of the countries to import crop diversity from outside the region was working well and suggested that this same arrangement could be used to assist countries with implementing the Treaty. The smaller countries were very much in support of this idea. Therefore, it was recommended that SPC, in consultation with the Joint Capacity Building Programme, should draw up a proposal formalizing such a scheme for submission to the 2010 session of Heads of Agriculture and Forestry Services (HOAFS), after consultation at the technical level through PAPGREN. A draft agreement is currently being prepared to address this recommendation. Under this agreement SPC would act as an agent for the countries, both contracting and non-contracting parties.

Parties felt that generally awareness of the Treaty at the national level is low (see Annex 3 of this book for details on the main provisions of the Treaty). There is a perception in many countries that the genetic resources of a country should generate funds for the 'owner' of these resources. The need for more awareness was reinforced by some of the non-contracting parties, especially those countries where, prior to any agreement regarding genetic resources, consultation would have to occur at the provincial level. This is possibly a lesson to be learned by PAPGREN in that the momentum on awareness has not been maintained. This is a reminder of the importance of national consultations enabling a wide stakeholder audience to be reached. Case studies are a good tool for promoting any topic. The situation in Samoa raised the PGRFA diversity flag in the early 1990s, but since then there have been no similar situations strengthening that message. The SPC Genetic Resources team are implementing two studies, which aim to show how a fragile agricultural system and ecosystem can benefit from increased diversity, which has been made available through the mechanism of the Treaty. One study was initiated last year in Fiji, with three communities. Discussions with communities were recorded, and the PGRFA diversity for each of the sites, surveyed and recorded. Communities also discussed their observations regarding weather patterns. These results will form the baseline data with which to monitor both the introduction of diversity and its benefits in helping communities better manage climate change. A similar study has been initiated this year in Palau in the Kayangel Atoll.

Article 9: Farmers' Rights in the Pacific

Article 9 is also an area of interest and has been highlighted in Samoa where a participatory taro breeding programme has generated some good taro lines, over which the farmers feel they have ownership rights. This situation once again reinforces the need for awareness and that promoting and strengthening awareness has to be continuous. There is a need to demonstrate to the farmers that these taro lines have resulted from countries sharing their taro diversity and it is important to pursue that route for the sake of global food security. However, this does not negate the important role that farmers have played and continue to play in the conservation and development of plant genetic resources (see Chapter 13 for details on farmers' communities). SPC, with funding support from the Technical Centre for Agriculture and Rural Development (CTA) is conducting a study, the results of which will assist SPC to provide its member Pacific Island countries and territories with the appropriate tools to protect and promote traditional/indigenous knowledge, specifically within the context of the Treaty, at the regional and national levels. The scope of activities will include:

- a review of international, regional and/or national and/or local initiatives and best practices to comply with Article 9 of the Treaty;
- an assessment of SPC's responsibilities and opportunities for addressing the protection of traditional/indigenous knowledge in relation to the Treaty;
- consultations in three Treaty contracting parties (including Fiji) through in-country visits to explore their understanding, application of and concerns around Article 9 of the Treaty;
- consultations with key stakeholders at the regional and international level to identify areas for partnership and collaboration to advance the protection of traditional/indigenous knowledge relevant to PGRFA and enhance Farmers' Rights in accordance with Article 9 of the Treaty.

Conclusion

Promoting the ITPGRFA in the Pacific region has required discussions with all 22 Pacific Island countries and territories. The five Pacific territories are New Caledonia and French Polynesia (French) and Guam, American Samoa and the Northern Mariana Islands (US). The French territories cannot accede to the Treaty in their own right. The application of the Treaty must be extended to them by France. Since they have autonomy in national legislation, they can then decide to adopt their own legislation for implementation of the Treaty. The French Polynesia General Assembly endorsed the ratification of the Treaty by France, however, New Caledonia is still going through a consultation process. The US territories would be in the same position once the USA has ratified the Treaty.

To date, five countries have acceded to the Treaty. As previously stated this is largely due to the lack of capacity and knowledge in the region, which is exacerbated by the fragmented nature of the Pacific and the high cost of travel, making national consultations with wide participation a significant challenge. In addition, the Pacific region was not directly involved in the Treaty negotiations and only started active participation in 2006.

Despite the relatively low number of countries that have acceded to the Treaty, significant progress has been made in the last ten years in the area of PGRFA conservation and utilization with the establishment of the regional gene bank, CePaCT and the network, PAPGREN. These two developments have played a significant role in promoting and developing both the concept of sharing PGRFA and the understanding of the contribution PGRFA makes to food and nutritional security. They provide an excellent foundation on which to further accession to the Treaty. The 2009 workshop did much to highlight the issues with the non-contracting parties and the challenges facing the contracting parties. SPC with the Treaty Secretariat and Bioversity International are collaborating to ensure recommendations made at that workshop, such as the agreement that will endorse SPC's role as an agent acting for the countries in the implementation of the Treaty, are acted upon. The climate-ready collection established by CePaCT demonstrates to the countries the importance of accessing PGRFA from outside the region, with its significant number of sweet potato accessions from the International Potato Center (CIP). Activities such as these, and the ongoing case studies, will continue to reinforce the need for crop diversity.

Notes

1 The Pacific region is commonly divided into three sub-regions.
2 Australian Agency for International Development.
3 New Zealand Agency for International Development.
4 Australian Centre for International Agricultural Research.
5 Department of Agriculture, Fisheries and Forestry.
6 Commonwealth Scientific and Industrial Research Organisation.

References

Lebot, V., Prana, M. S., Kreike, N., van Hech, H., Pardales, J., Okpul, T., Gendua, T., Thongjiem, M., Hue, H., Viet, N. and Yap, T. C. (2004) 'Characterization of taro (*Colocasia esculenta* (L) Schott) genetic resources in southeast Asia and Oceania', *Genetic Resources and Crop Evolution*, vol 51, pp381–392

Mace, E. S., Mathur, P. N., Izquierdo, L., Hunter, D., Taylor, M. B., Singh, D., Delacy, I. H., Jackson, G. V. H. and Godwin, I. D. (2006) 'Rationalization of taro germplasm collections in the Pacific Island region using simple sequence repeats (SSR) markers', *Plant Genetic Resources*, vol 4, no 3, pp210–220

SPC (2003) 'Policy issues relating to plant genetic resources in the Pacific: A guide for researchers and policymakers', Secretariat of the Pacific Community, Nouméa, New Caledonia

SPC (2005) 'Report of the First Regional Conference of Ministers for Agriculture and Forestry', Ministries of Agriculture and Forestry Services, 9–10 September 2004, Suva, Fiji Islands

SPC-CRGA (2008a) 'Food security in the Pacific', Paper presented at the 38th meeting of the Committee of Representatives of Governments and Administrations, Nouméa, New Caledonia, 13–16 October 2008

SPC-LRD (2008b) 'Land Resources Division Strategic Plan 2009–2012', Secretariat of the Pacific Community, Nouméa, New Caledonia

Part II

Perspectives on the Treaty by Stakeholders in the World Food Chain

Chapter 10

International Non-governmental Organizations

The Hundred Year (or so) Seed War – Seeds, Sovereignty and Civil Society – A Historical Perspective on the Evolution of 'The Law of the Seed'

Patrick Mooney

A half-century lapsed between 1911 when Nikolai Vavilov joined the Bureau of Applied Botany in St Petersburg and when Erna Bennett and Otto Frankel convened the first international technical conference on plant genetic resources in 1961. Twenty years after that, crop genetics suddenly grew into a political intergovernmental debate during an FAO conference that, two years afterwards, created the International Undertaking and Commission on Plant Genetic Resources (IU). It took another couple of decades before the voluntary IU became a legally binding Treaty. When the Treaty's Governing Body convened in Bali to assess its progress in 2011, it had an entire century, 'a 100 Year Seed War', for review and reflection.

Most of this past century is a story of scientists and policy makers – 'courageous and farsighted leaders' like Nikolai Vavilov and his Russian colleagues, Harry and Jack Harlan, Erna Bennett, Pepe Esquinas, Melaku Worede, Fernando Gerbasi, Tewolde Berhan Gebre Egziabher, Jan Borring, and some others less courageous (some downright cowardly) best unnamed and forgotten.

The place and the perspective of civil society, in this century-long history, are less certain. I can only offer this account as a personal remembrance of the past 35 years or so full of the 'mismembering' and myopia of one witness. It is a human weakness that we tend to see ourselves always at centre stage and we forget who was standing there alongside us. My apologies for all of these weaknesses.

It is tempting to outdo Vavilov by beginning the story with a Polish-American farmer, David Lubin, and his almost single-handed construction of the International Institute for Agriculture in Rome in 1905. Angered by the grain cartels of that era, Lubin marched off his California farm back to Europe where he somehow arm-twisted the King of Italy into convening the world's first international intergovernmental agricultural meeting. Certainly, Lubin's story is as gigantic and heroic as Vavilov's, but he died in 1919 and there is no evidence in his poorly studied memoirs to indicate that he knew anything at all about plant genetic resources. David Lubin, however, knew something about Farmers' Rights; would easily understand food sovereignty; and would have no difficulty identifying the new integrated multinational cartels that dominate food and agriculture today. Throughout the decades of colourful controversy (from the Green room to the Red room to the Blue room in FAO's building A over the initial IU and later Treaty), David Lubin's legacy has been all around us and most especially in the library named after him on the ground floor of building A. If Lubin were alive today he would be a member of the '*Via Campesina*' and he would be preparing to fight for Farmers' Rights and food sovereignty in Bali.

However, in the mid-1970s, when Cary Fowler first told me about crop genetic erosion, there was no '*Via Campesina*'. When civil society's food researchers first met together in Saskatchewan's Qu'Appelle Valley, in November 1977, there were no farmers among us, and the topic of seeds seemed alien to the much greater interest in monitoring the grain trade, ocean fisheries, the expansion of the dairy industry in Asia, and the campaigns against infant formula. Only Cary and I wanted to talk about seeds. Through his research on 'Food First', Cary had figured out genetic erosion and the links to mergers between seed and pesticide companies. Following his trail, I stumbled on plant breeders' rights. Others did not seem to think it was important.

In March 1979, Erna Bennett herself came out to the Saskatchewan prairies to confront the seed trade; do battle against intellectual property over seeds; and advocate for plant genetic resources conservation. Until then, Cary and I had only talked with her on the telephone. No one who attended the packed meeting in Regina will forget her Irish eloquence.

By the summer of 1979, I had tortured a long pamphlet into a small book titled *Seeds of the Earth* but reluctantly concluded that my original target, FAO's World Conference on Agrarian Reform and Rural Development later that year would not yield a sympathetic audience and opted instead, to take the book for its unveiling to the UN Conference on Science and Technology for Development in Vienna. The book 'launch' was singularly unmemorable. I did, however, press a copy into the hands of M. S. Swaminathan who – as Independent Chair of FAO – raised the issue in his speech at FAO's conference a few months later. Knowing that Indira Gandhi would address the next FAO conference in 1981, we opted to try again pushing 'seeds' at FAO.

In the summer of 1981, Cary Fowler and I were subcontracted via Art Domick at American University (and an old admirer of Erna Bennett's) to do some work on food policies for the Mexican government. That gave me an opportunity to go to Mexico City in September and meet with government officials to propose that Mexico take up 'seeds' at the upcoming FAO conference. A former Mexican minister of agriculture, Oscar Brauer, who had moved on to FAO had already contracted me (through ICDA – the International Coalition for Development Action) to write a report on the implications – if any – of my book for FAO seed policy. Brauer's support probably helped us with the Mexican government.

History records that FAO's 1981 conference agreed to consider the formation of a body to study plant genetic resources. The contentious paragraphs were to be considered by the COAG (Committee on Agriculture) at its 1983 meeting and would then be passed on to the next FAO conference in November 1983 (see Annex 1 of this volume for the list of all Commission and Treaty meetings). However, that is getting ahead of things. My own memories of the 1981 FAO Conference are more kaleidoscopic. Cary and I had managed to convince allies at ICDA to join us in Rome for the campaign. We met outdoors in the café across from FAO on the '*Aventino*' before the opening session and prepped for the unfamiliar encounter ahead. We were being followed everywhere by a Japanese film crew and when we took our seats in the Observer section of the Blue room for Gandhi's speech, we gathered embarrassing attention. The battery of cameras, trained on the speaker's podium, was interrupted by the Japanese crew's singular focus on our little civil society group off in the corner. The Japanese film crew gave us our first global media coverage. They were unrelenting. Before Rome, they ventured to the ICDA offices overlooking Covent Garden (in the cheap days before the restoration) in London only to find the office door absent and the lone filing cabinet empty. In 1981, we were not impressive.

The champion of the 1981 conference was the Mexican delegation led by the very pleasant and charming son of Mexico's former president Luis Echeverría. However, the delegation was intellectually strengthened by Francisco Martínez Gómez (Pancho) who took on the issue as a personal 'cause célèbre'. As civil society, we intervened in the debates as best we could but spent much of the time wringing our hands and anxiously passing notes to the Mexican delegation. Much more effective, I am sure, was Pepe Esquinas who – as a member of the FAO Secretariat in the International Board for Plant Genetic Resources (IBPGR) – seemed to know everybody in Latin America and had his own clandestine avenues. I had met Pepe at FAO either earlier that year or perhaps even the year before – while being berated by Trevor Williams (then, the head of IBPGR) in his office doorway at the time; however, we had not had many opportunities to talk. Most of my links to the internal machinations of the FAO Secretariat were through Erna Bennett who was in the process of being fired. It was only after she left that we realized how strong and important Pepe Esquinas was as an ally and leader.

Also in 1981, the IBPGR hosted another International Technical Conference on Plant Genetic Resources at FAO. Cary and I were determined to attend and were made to feel distinctly unwelcome. Erna Bennett had sent me an interoffice memo from Trevor Williams to Dieter Bommer, the ADG for agriculture, warning that I was planning to come and advising that I would not be allowed to enter the building. When I entered, I was confronted by an official who told me I would not be allowed in. I showed him my copy of the memo and pointed out that the meeting was public and that I would go immediately to the media if I was kept outside. I was allowed in. On reflection, it probably would have been more fun to stay outside although we were entertained by Trevor Williams' discourse on the various venues for a world gene bank: the arid south of Argentina, the basement of FAO, or on the frozen island of Svalbard. His best shot: a cupboard on a space station.

In March 1983, ICDA scraped together enough funds for me to return to Rome to attend the Committee on Agriculture where the 1981 decision was to be debated again. COAG had set aside one or two hours for the discussion on a Thursday afternoon. Long conversations with Pepe Esquinas persuaded us that we needed to press for three things: the formation of an intergovernmental committee to take on the politics and practice of plant genetic resource conservation at FAO; the formation of a global fund to collect and conserve plant genetic diversity (we thought around $350 million); and (this was at Pepe's insistence) the construction of a World Gene Bank as a backup to other national and regional gene banks.

I was the lone NGO observer but, unbeknownst to governments, I had a secret weapon – one of the original IBM PCs. A young high school student named Beverly Cross (whose farm was near my own) painstakingly typed in the entire IBPGR germplasm databook into a spreadsheet. It was miraculous. Suddenly, we were able to identify exactly how much germplasm of which crops every country in the world had either donated or received. I was able to go to literally every delegation in Africa, Asia and Latin America and hand them a note that clearly showed how much germplasm that country had donated and how much it had received – including a list of the countries to which their germplasm had gone. Of course, the figures showed overwhelmingly that the South was a massive contributor of free germplasm and that the North was actively using the germplasm to develop new varieties protected by intellectual property. What was supposed to be a one-hour discussion on a Thursday ran through the afternoon and early evening on to all day Friday and then onward to the following Monday afternoon. Highlights: the Bolivian ambassador demanded that the UN flag be planted on every gene bank and the American delegate advised the other countries present to follow the dictum of Mark Twain ... 'if it ain't broke, don't fix it'. The North was furious. Genetic resources were a non-issue being handled perfectly adequately by existing scientific institutions. They could not understand why the South was insisting that intergovernmental control be asserted over the world gene banks. Manoeuvring in the background all the time was the Mexican delegation led by another son of another former Mexican president – José Ramon Lopez Portilo who later became the Independent Chair of the FAO Council. José Ramon was brilliantly backed by Pancho Martínez, and Pepe Esquinas was everywhere talking to everybody.

In the end, COAG produced a report that called for the creation of an intergovernmental FAO committee and Undertaking to address plant genetic resources. The report was to go to the FAO conference at the end of the year.

About the time of the COAG, I had a telephone call from Sven Hamrell, the director of the Dag Hammarskjöld Foundation in Uppsala, Sweden. Sven, who I had met once or twice since 1981, wondered if I would write an article for his journal, *Development Dialogue*, that could be published later in the year. I enthusiastically agreed knowing that the Journal was mass-distributed free to around 18,000 policy and opinion makers around the world.

Following the COAG, I devoted most of my time (leaning heavily on Cary Fowler and Hope Shand for advice and research) writing the article that evolved like the pamphlet four years earlier – into a kind of book that the Dag Hammarskjöld Foundation finally agreed to title *The Law of the Seed*.

FAO's November conference was to be the big battle. More than 20 friends from European civil society organizations agreed to join Cary and I in Rome to press for the COAG recommendation as well as for funding and for a global gene bank. I had met Henk Hobbelink earlier in the year and Henk turned into an invaluable colleague and one of the key organizers of our November campaign.

We had another secret weapon for the November meeting – Olle Nordberg, Sven Hamrell's accomplice of the Dag Hammarskjöld Foundation who flew to Rome on the opening day of the conference with boxes of *The Law of the Seed* that he managed to place directly into the box of every government delegation at FAO. Although it seemed unlikely that busy delegates would take the time to read a couple of hundred pages about the politics of genetic resources in the midst of the conference, many of them actually did. On the second day of the conference, we were invited to meet with M. S. Swaminathan who was still Chair of the FAO Council. M.S. had clearly marked out passages he wanted to discuss. Later that morning, we met Mohamed Zehni, Libya's ambassador to FAO who I think was chairing the G-77. Zehni is also a geneticist. He had a copy of the book in his hand when we had coffee and I asked him what he thought of it. He delicately offered the advice that it was 'perhaps a little rich for delegates here ...'. Nevertheless, he had read it! And so had FAO's imperious Director-General, Edouard Saouma. Later in the conference, Saouma's secretary appeared at my elbow cryptically commenting 'the director general is not unhappy with your activities'.

The events of the two-week conference are something of a blur. First we fought in the Green room, trying to enlarge the original COAG proposals and then we carried the Commission report to the Blue room where it was debated again.

I never fully understood an almost-violent encounter between José Ramon and the FAO Legal Council on the podium of the Green room, which ultimately led to the upgrading of the recommendation to create an intergovernmental committee into an intergovernmental Commission.

The plenary battle in the Blue room is probably remembered differently by different folks depending on whether you were sitting on the podium as part of the Secretariat, as was Kay Killingsworth, for example, or if you were ensconced in the NGO cheap seats on the sides (Cary, Henk and me), or if you were in the middle of the fray among the delegates like Zehni and Lopez Portilo. Pepe Esquinas – who never sat – was buttonholing delegates, writing bits of text and stalking the corridors outside – sometimes simultaneously.

My fractured memory does recall José Ramon on his feet challenging John Block, the US Secretary of Agriculture who was chairing the Conference session. Block was trying to gavel the issue away but Lopez Portilo was having none of it. The Mexicans wanted a Commission and Block wanted nothing but was prepared to go along with a lower-level committee instead. Block kept calling for a show of hands and concluding that his side had won. The Mexicans kept challenging the count. There may have been as many as six rounds of voting before Block conceded that he had lost. Before that concession, however, he actually called for a timeout, advised government delegations to consult their capitols, and darkly advised delegations to inquire into any undisclosed paragraphs of any bilateral agreements or treaties that they had signed recently. It was all a bit remarkable. At the end of the conference, Mexico had won and I flew happily to Barcelona for the annual meeting of ICDA, leaving Cary Fowler alone to track the FAO Council that immediately followed the conference and was to practically dispose of the conference decisions.

Cary called me from Rome while I was in Barcelona reporting that the fight had continued through the Council, with the US and other governments in the North trying to undermine the conference's decisions. Throughout it all, José Ramon – with a growing number of riled-up South governments – held his ground with tactical support from Cary and Pepe. Every few years since that memorable 1983 meeting, I have run into old friends who were in the room at the time. Each adds an anecdote or two and I have noticed that the anecdotes tend to become a little more dramatic and bizarre as the years go by. My own included, I am told.

Immediately following the FAO conference – and at Cary Fowler's inspiration – Hope Shand, Cary and I established the Rural Advancement Foundation International (RAFI) and formally set about working together with plant genetic resources as our one and only issue. When I left ICDA, they wisely went straight to Henk Hobbelink and asked him to take over their seeds campaign. Now, the ICDA Seeds Campaign has broadened its work and reputation enormously since then and has become Genetic Resources Action International (GRAIN) – with Henk still brilliantly at the helm. Renée Vellvé joined Henk a couple of years later and immediately became a key player in Commission negotiations.

In 1985, Cary and I were given two plane tickets to travel around the world talking to governments about the issues before the first meeting of the FAO Commission. We went first to Rome to talk with Pepe Esquinas before travelling onward through Africa and Asia.

Coming out of the 1983 conference, we had both an intergovernmental Commission and an International Undertaking. The Undertaking had some influence but no legal authority and was ambiguous in several areas including the issue of intellectual property. We knew that pressure would be on at the Commission's first meeting to accept plant breeders' rights and to insist that, what we called 'farmers' varieties', and what most scientists preferred to call 'landraces' or even 'stone-age seeds', were to be exchanged freely.

Literally en route to the first Commission meeting in Rome, we concocted the idea of Farmers' Rights which we simplistically saw as a counterweight against plant breeders' rights. We wanted to insist that farmers varieties were the product of farmer genius and should not be treated in any way as being less than varieties produced by the public or private sector. We were not quite sure what to do with the idea beyond presenting it as a threat and possibly a barrier to accepting plant breeders' rights. With Henk Hobbelink, we agreed to vet the idea at a news conference in downtown Rome early in the Commission's first meeting. We also wanted to find a way to introduce it into the intergovernmental debate. We had not had a chance to talk about the idea with the Mexican delegation or any of our other friends in other countries.

In the end, from the back of the Green room, we got the microphone and proposed Farmers' Rights as part of the IU. The lack of interest was deafening. It did not seem that anybody was going to pick up our proposal. Then, Jaap Hardon, the head of the Dutch Gene Bank and Netherlands delegate to the Commission, literally as he was preparing to leave, decided he couldn't resist and took the floor to ridicule Farmers' Rights as romantic and naive. Beside him, the Mexican delegation exploded. Suddenly Jose Ramon was on his feet staunchly defending Farmers' Rights and attacking his good neighbour, Jaap. With Mexico in full rhetorical flight, the Bolivians, Venezuelans, Cubans, Nicaraguans, Ethiopians and many others began waving their flags and championing the cause. Through a messenger, I sent Jaap a note thanking him for his intervention and I heard his hearty laughter as he raced off for his airplane. Although we have often disagreed, Jaap was then and still remains one of my heroes. For that matter, so does his hand-picked successor, Bert Visser.

The first meeting of the Commission maintained the ambiguity around intellectual property but included Farmers' Rights. We saw it more as a place marker from which we could launch other battles in the years ahead.

At most, the first four sessions of the Commission were heavily influenced by civil society. Governments were still trying to come to grips with the creature they had let others create and those of us at the back of the room still had the advantage in terms of computerized data and political strategy. In 1987, we were able to press for a Code of Conduct on Germplasm Collection and for a study of the possible impact of biotechnology on genetic resources. It seemed that whatever we suggested would be taken up and – more or less – adopted.

In 1988, the Keystone International Dialogue on Plant Genetic Resources got underway in Keystone Colorado bringing together 40 or 50 protagonists from various governments, scientific organizations, and a couple of us from civil society. Hope actually attended a preliminary discussion about the dialogue in Washington some weeks earlier but Cary and I were not invited to the first formal meeting until a week or two before it happened. It was clear that many governments in the North were not at all sure they wanted us to be there. Cary could not attend for personal reasons. Chaired by M. S. Swaminathan, the first meeting went surprisingly smoothly as we realized that none of us actually had horns or tails and we could have a decent conversation. It was even pleasant … sometimes.

In the summer of 1989, Don Duvick (the vice president for research at Pioneer Hi-Bred) and Henry Shands (of the US government's genetic resources programme) proposed a small meeting in Washington to discuss the possibility of US entry into the Intergovernmental Commission. Much to my surprise and, largely due to the Keystone dialogue process, I was invited to join along with Pepe (representing FAO), Jaap and Melaku Worede. Camila Montecinos (who now works with GRAIN in Chile) also attended at my specific request. Camila is one of the toughest people I know and I did not want to be the sole NGO at the small meeting.

Don Duvick was clearly the 'mover and shaker' with the US government and his big concern was that if the North were to join and to eventually provide funding, the South had to guarantee to make all of its germplasm available. There was no way that the South was going to – or should – agree to this. However, in the far-from-perfect IU there was the notion that public and private researchers could identify a category of germplasm that they hoped eventually to commercialize that could remain exclusive. Companies argued that they might have material in the nursery trial stage that was a generation away from being commercialized that should not be just taken by somebody else at the last minute. We argued that the same held true for the South. For example, if Ethiopia has naturally occurring caffeine-free coffee trees that it knows to be invaluable but is still some years away from entering the international coffee trade, it would be unfair to force Ethiopia to surrender such obviously invaluable material. We were all sitting out in Henry Shands' yard when we made our case. Don looked at us, and nodded. The battle was over – hardly even engaged – before it started. We typed up a half page statement and took it to the US undersecretary of Agriculture the next day for his agreement. Don did the talking and the deal was done. The USA joined the Commission. I knew it was not good to have the US 'inside' at that point in the development of the Commission but I could not see how to keep them out. If CSOs had not been there, a deal would have been reached that would have let the United States come in and would not have been in any way to the advantage of the South. As it turned out, both sides were left with 'plausible denial', for virtually any germplasm they wanted to argue was 'still under development'. I received a cheque for the reimbursement of my plane ticket to Washington and my hotel stay from the US Department of Agriculture. That will never happen again, I thought. And it has not.

The Keystone process had many important moments as we met in larger or smaller groupings from Colorado to St. Petersburg (then Leningrad) to New York, Madras, Ottawa, Rome and finally Uppsala and Oslo. The process created bonds of cooperation and, sometimes, comradeship that have held up over the years. It did not really cause people to change positions so much, but to at least be able to understand one another's positions and find common ground where common ground was occasionally useful.

Three anecdotes stand out: Melaku Worede, Jaap Hardon, Don Duvick, Henry Shands and I were all in the car somewhere in the countryside beyond St Petersburg. It was hot and we had run out of petrol and were waiting impatiently for a Russian host to solve the problem. We had been talking a lot and suddenly Don accused me of not being interested in plant genetic resources at all but just wanting to bash multinational corporations. He was angry. Jaap leaned forward from the back seat and calmly said that whatever I felt about multinational corporations, I was dedicated to diversity. Don liked the answer. We got along much better afterwards. A year or so later in Madras, Don and I had been asked to write anonymously about different approaches to funding plant genetic resources. The papers had been circulated to the group a few days before the meeting. At one point in my proposal, I had written that the seed industry's arguments, that farmers should happily give up their own varieties in return for commercial varieties, was like saying that the Greeks should give up their claim on the Elgin marbles in return for the Rolling Stones. Don announced to the room that he was the author! A few minutes later Jaap, who learned nothing from his assault on Farmers' Rights five years earlier, attacked our opposition to the word 'landrace' by insisting that no one named their cars after people either. John Peano jumped in with one word, 'Volkswagen', and I followed with 'Land Rover' and Jaap did what he does best, dissolve into laughter. At another encounter (either Madras or Oslo, I forget) Cary, Pepe and I walked away from a long drafting session where we'd left John Deusing, a lawyer with what was then Ciba-Geigy, to clean up the text for presentation to the whole group the following morning. The sun was already coming up when we realized we had left the final delicate wording to our corporate 'enemy'. We shrugged – knowing that we trusted him to complete our task fairly. The morning proved us right.

Of the 1980s, there are still tales that probably should not be told. We all felt sometimes like Jedi warriors taking on the Evil Empire – variously identified as IBPGR, Monsanto or the US government. When Erna left FAO, she shipped us boxes of papers that took weeks to cipher. Other documents were got through US Freedom of Information requests and a few more appeared mysteriously under hotel room doors, behind mirrors in FAO washrooms, pushed across a table during a furtive airport meeting, or passed openly and anonymously via smiling messengers in the Red room. Most of the best information came, however, from Hope Shand's number-crunching through germplasm collections, seed catalogues and plant patent lists. Throughout it all, Pepe Esquinas was an amazing presence – a hybrid somewhere between Don Quixote and Machiavelli (with a pinch of Rasputin), challenging and charming. I have emblazoned in my memory, Pepe, very very late at night in the semi-darkness of his office after a day of Commission drafting trying to cajole a nuance out of the Oxford dictionary that the stuffy volume just could not conjugate. Among us, Pepe Esquinas was 'Wiley Quixote'. Even at his 'wiliest', however, Pepe remained passionately loyal to the loftiest principles of the United Nations and FAO.

'Us' in the early days, was a small group. Throughout these years, 'civil society' included both Henk Hobbelink and Renée Vellvé at GRAIN, Camila Montecinos (then at CET now at GRAIN), Rene Salazar at Searice, Vandana Shiva (wherever she wanted to be at), Andrew Mushita at CTDT (Community Technology Development Trust) and Cary, Hope and I at RAFI. In addition, many friends we could call upon if things got tough. Around the time of the Leipzig Technical Conference, 'us' expanded wonderfully to include Patrick Mulvany, Liz Hoskins, Neth Dano, Edward Hammond and many many others.

At one point, visiting IBPGR as part of the Keystone process, Dick van Sloten expressed disbelief when I mentioned that I had not been to their offices since they moved to the old cheese factory a few kilometres from FAO. 'Well, not in daylight, anyway', I added. I think he took me seriously!

When the curtain came down on the Keystone Dialogue in 1991, the clearly unfinished business was intellectual-property. Jaap Hardon – a glutton for punishment – approached Henk Hobbelink and me about the formation of a second dialogue when we were all in Zimbabwe together in late 1992. A few days later, the three of us were in Nairobi at a CGIAR meeting involving Geoff Hawtin. The final shape of what became known as the 'Crucible Group' was formed in the bar late one night while Geoff and Henk danced on a tabletop secured by Jaap and me. It was Jaap's idea but I claim the name and I spent much of the next several years explaining what a crucible is. Over most of a decade, the Crucible Group produced three books but not much progress. Perhaps because we had already gone through the Keystone Dialogue, Crucible did not have the same spin-off effects.

In the almost-intuitive move from Undertaking to Treaty, the 1993–1994 CGIAR stripe reviews of genetic resources suddenly became important. I was invited to join the review and Henry Shands became its Chair. The big change was IBPGR (en route to becoming IPGRI (and, now, Bioversity International) where the palace coup had led to the selection of Geoff Hawtin as the organization's second-ever director. By any definition, Geoff was/is the CGIAR systems best advocate and smartest strategist. He was a breath of fresh air in FAO Commission meetings and became a critical ally (and, sometimes, opponent) from 1991 onward. At Geoff's quiet insistence (from the sidelines), the stripe review came up with the remarkable conclusion that the CGIAR's gene banks should be placed under the auspices of the Undertaking and that gene bank policies should be guided by its Commission. The report was presented to the CGIAR mid-term meeting in New Delhi in May 1994. I attended the New Delhi meeting as a member of the review panel. It was Ismail Serageldin's first meeting as Chair of CGIAR and, of course, as a Vice-President of the World Bank. I was furious when Henry presented our report and then stepped aside from his role as Chair to advise that maybe the CGIAR should rethink the key recommendation of surrendering policy control to FAO. I immediately wrote to Serageldin urging him to move quickly to implement the review's principal recommendation. My letter was followed by a month of silence.

Then, as I passed through RAFI's Ottawa office en route to Uppsala (for Sven Hamrell's retirement party at the Dag Hammarskjöld Foundation) and then Nairobi for an organizational meeting of the newly created Convention on Biological Diversity, Bev Cross handed me a fax from Serageldin. I read it standing in the doorway and realized that the World Bank vice-president was saying that it would be 'foolhardy' for the CGIAR to implement the stripe review's recommendation and that he wanted to talk with lawyers at the Bank about other possibilities. I faxed the letter to Henk at GRAIN and to Geoff Hawtin at IPGRI and then headed for the airport. At Sven's party in Uppsala, I met up with Carl-Gustaf

Thornström and showed him the letter. He was alarmed and asked to make a copy. On my onward flight from Stockholm via Amsterdam to Nairobi I encountered Norway's Jan Borring and several other delegates flying to the same meeting and handed out copies of the letter. Everybody was shocked. In Nairobi, Henk Hobbelink and I grabbed Geoff Hawtin and persuaded him to attend a news conference on the topic that had been hastily arranged by GRAIN. It is a testimony to Geoff Hawtin's integrity that he agreed to attend.

In the intergovernmental biodiversity convention meeting, Sweden and Malaysia joined forces to accuse the World Bank of the 'dawn raid' on the CG's gene banks and of trying to take over the banks to orchestrate access to germplasm for multinational seed companies. Geoff Tansey wrote up the story for the *Financial Times* and *New Scientist*, blasting the Bank for the attempted coup. Within two days, Geoff Hawtin read out a letter from Ismail to the Nairobi meeting announcing that he would personally sign the policy turnover to the FAO Commission on behalf of each of the 11 gene banks by the time the CGIAR held its annual meeting in Washington in October.

Did the World Bank really intend to take over the CG gene banks? The sequel to the story played out in August 1994 when Serageldin invited me to Washington for lunch to talk about our differences. In a preparatory phone call, it was clear to me that senior CGIAR staff had not bothered to actually review the fax that I had received signed by Serageldin. I was even told that the fax I had received was not the fax they had sent. When I invited them to reread the critical paragraphs, there was a pause on the line as they looked for a copy, and then the quiet comment, 'I can see how you might have formed the impression' from the deeply chastened official. I am not sure if the coup was intended. I am sure that if we had not acted quickly, the agreement between FAO and CGIAR would not have been signed. I am also sure that it would have left the door open to other forces inside the bank and out, and that it might have understood the potential value of the gene banks and sought to use them in other ways. The bottom line is that FAO's weak and voluntary Undertaking and Commission suddenly had high profile responsibility for the world's 11 most important gene banks. The logic of moving from Undertaking to Treaty was becoming more apparent.

Many of the most dramatic events had nothing to do with CSOs. Dick van Sloten's own courageous efforts to restore order at IBPGR – a palace coup in fact – remains for others to tell. Rene, Hank and I sat dumbfounded another time as the Brazilian Ambassador accused his American counterpart of 'terrorism'. She broke into tears. There are other stories, I am sure, that we, in civil society, never heard of.

If not sooner, the 1991 Commission meeting was certainly the last that was dominated by civil society. By the time governments met again in 1993, the Commission was thoroughly institutionalized and government delegations coming to Rome had marching orders from their capitals that demanded obedience. We could still cajole and tease but we could not decide.

Pepe Esquinas consulted widely over the idea of turning the IU into a legally binding treaty. He had the idea that the negotiation of the Treaty could be done in tandem with negotiations leading to a new International Technical Conference on Plant Genetic Resources including a State of the World report and rolling Plan of Action for genetic resources work. It was a bold and complex agenda. I was enthusiastic about the Plan of Action and saw the negotiation of the Treaty as a problematic but useful way to maintain a high profile political agenda for genetic resources work. Cary Fowler, my old comrade-in-arms, with his razor wit and laser focus on genetic resources, had moved to Norway to work with Noragric around the end of the Keystone Dialogue, and was the logical person to take on the Technical Conference and Plan of Action.

Cary's so-called 'technical' conference in Leipzig in 1996 brought together the largest number ever of civil society organizations – South and North – in support of plant genetic resources. By then, however, we were more cheerleaders than controllers and we accepted our more traditional role of acting as clarifier's of issues and supporters of the more progressive positions of South negotiators. Our overall influence was modest although our involvement was appreciated. For many at FAO, CGIAR and in governments, we were hard to categorize, since we had the sometimes-unnerving capacity of being 'spoilers' – able to make or break a move or idea or to turn a minor into a major issue unexpectedly. Nevertheless, as useful or concerning as this role is, we were not in the driver's seat anymore.

By the time of the 1998 Commission meeting, neither GRAIN nor RAFI were sure we should even be present. All of a sudden, however, two major developments changed at least RAFI's view. First, in March, Hope Shand discovered a joint USDA/Delta and Pine Land patent granted on what they described as a 'Technology Protection System' that rendered GM seeds sterile at harvest time. We quickly called the new technology, 'Terminator'. We wanted FAO and the Commission to condemn the technology. Second, friends inside the CGIAR told us of two plant breeders' rights claims made by Australian agencies on CGIAR gene bank accessions. The two claims presented the first clear examples of 'biopiracy' concerning gene banks. We took both issues to the Commission and eventually got strong support for both. The Australians quickly abandoned their patents and Jacque Diouf, FAO's Director-General, roundly condemned Terminator technology. With enormous help from Geoff Hawtin and Cary, CGIAR also announced that it would never touch the suicide seeds.

NGOs were invited into the closed Treaty 'contact group' negotiations because governments wanted industry present and they could not really invite industry without inviting their civil society watchdogs. Ultimately, seed companies would either be required – or 'volunteered' – to pay a proportion of their profits, royalties or revenues so OECD states demanded that they be at the table and the global South knew that no consensus was possible without industry acquiescence. We had no illusions – but participation gave us the opportunity to blow whistles and apply pressure to both our friends and foes around the table.

The need to be present was made painfully evident in one of the first meetings of the contact group. Before we – or industry – were invited, the North moved to sideline Farmers' Rights by imposing a 'chapeau paragraph' that rendered the strong affirmative language beneath almost irrelevant. We had always understood that Farmers' Rights would be sacrificed by the South as a bargaining chip but we had hoped it would be better used and carefully positioned for post-Treaty negotiations. I was sharing a hotel room with René Salazar who found himself as an NGO on the Philippine delegation. Returning to the room very late that night, René woke me up, alarmed by the last-minute manoeuvres in the contact group. Only Norway and Poland – and the Philippines – expressed concern over the sudden text changes. Very early in the morning, we both knocked on hotel room doors trying to convince our allies to return to the issue. They all claimed innocence or ignorance and they all advised us not to worry. The deal was done. René – who was powerless to stop it – was heartbroken.

Either Silvia Ribeiro (who joined us at RAFI in 1999) or I attended the contact group negotiations. They were usually the worst meetings of our lives.

Here and there, we were able to use our civil society independence to speak bluntly and give clarity to points and positions that governments dared not say publicly. This clarifying role was especially helpful during the Spoleto negotiation where the Commission's Chair, Fernando Gerbasi of Venezuela, managed a breakthrough making the final Treaty possible (see Annex 3 of this book for details on the main provisions of the Treaty).

Following the adoption of the Treaty at the FAO Conference in 2001 (see Annex 2 of this volume for the contracting parties per FAO regional groups), I was happy to accept Fernando Gerbasi's invitation to a celebration party at his home in Rome. The room was filled with old friends and old enemies but the times had changed – I felt less like Darth Vader and more like Art Deco standing in the corner.

What role did civil society really play? Henk and Hope and Camilla and Rene and Renée and Cary and I could debate this to a draw among ourselves. It is a matter of perspective. If I had been in the audience – as most governments were most of the time – I think I would have seen us on the stage all right – stage left, I hope – clowning and conspiring, sometimes loud and pontificating, sometimes in the shadows, often tangled in the curtains or messing with the lighting, and sometimes mischievously inserting text into the teleprompters of other actors.

Cary, the inspirational architect behind the now-famous Doomsday Vault, invited me to the Vault's opening in Svalbard at the end of February 2008. There, I began to feel more comfortable with the changes. It was an emotional occasion. I picked up a box of the International Center for Tropical Agriculture (CIAT) bean seeds (appropriate, given our shared legal action with CIAT defending the Mexican yellow beans – first evidence that the FAO/CGIAR agreement could have legal weight) with Clive Stannard and walked down into the vault to place them in storage. Many of us carrying the boxes had tears in our eyes. Ditdit Pelegrina (who had replaced Neth Dano who had replaced Rene Salazar as head of Searice) and I had an opportunity to speak at the seminar that preceded the formal opening of the vault and I recalled our three objectives when civil society first came to FAO pressing the seeds issue back in 1981: we wanted an intergovernmental organization to address the issues; we wanted $350 million a year for genetic resource conservation; and we wanted a World Gene Bank. With the Governing Body of the

Treaty (and the new enlarged FAO Commission on Genetic Resources for Food and Agriculture), creation of the Global Crop Diversity Trust, and the establishment of the Svalbard Vault, we had come a fair way to achieving our original goals. Not all the way – but some ways. Amid the good feelings, remains the feeling that we had not asked for enough in the first place. There are other seed wars looming – some bigger than any we have seen before.

As a civil society organization, ETC Group (we changed our name from RAFI in 2002) is back to where it was in the late 1970s/early 1980s. We are outsiders once again – with a new agenda that neither FAO nor most governments understand. Our concerns about genetic resources now cover everything from mammals to microbes and our concern about multinational corporations – the ones we've loved to call 'Gene Giants' – and their efforts to monopolize seeds has moved on to include Synthetic Biology, the effort to monopolize biomass, and our opposition to the new 'Biomassters'. We are not only concerned about fighting intellectual property monopolies but also fighting biological and technological monopolies. With climate change, the biggest battle of all is to support the efforts of peasant producers around the world to use the genetic diversity at their fingertips to respond to the changes ahead. There is lots to do. It feels like old times.

However, the real change – a century in coming – is David Lubin's legacy. He is no longer alone – a single peasant fighting the grain trade. Now there is '*Via Campesina*' – a massive farmers' movement around the world – that is clearly in the lead in civil society and among social movements and has its own plans. *Via Campesina* has moved us firmly from our narrow focus on Farmers' Rights to food sovereignty. The seed wars have new seed warriors!

Chapter 11

International Research Centres

The Consultative Group on International Agricultural Research and the International Treaty on Plant Genetic Resources for Food and Agriculture

Gerald Moore and Emile Frison

Introduction

The Consultative Group on International Agricultural Research (CGIAR) is a strategic alliance of 64 members comprising governments, international organizations and private foundations that support a common mission: to achieve sustainable food security and reduce poverty in developing countries through scientific research and research-related activities in the fields of agriculture, forestry, fisheries, policy and environment. It was set up in 1971 under the co-sponsorship of the World Bank, the Food and Agriculture Organization of the United Nations (FAO) and the United Nations Development Programme (UNDP) to mobilize science to benefit the poor.

The CGIAR supports 15 international agricultural centres (CG Centres), whose tasks are, inter alia, to conserve genetic resources for food and agriculture, to develop improved varieties and to promote the sustainable utilization of those genetic resources. The CG Centres maintain collections of plant genetic resources for food and agriculture (PGRFA) numbering over 650,000 accessions, whose importance for food and agriculture has been recognized in the International Treaty on Plant Genetic Resources for Food and Agriculture (the Treaty).[1] A major interest of the CG Centres has been to ensure that PGRFA can continue to be available for research, breeding and training for food and agriculture for the benefit of developing countries, within a stable international system that allows for the equitable sharing of benefits arising from the use of those resources.

This chapter examines the history of the involvement of the CG Centres in the conservation and sustainable utilization of PGRFA, and the role played in the negotiating of the Treaty (see Annex 1 of this volume for the list of all Commission and Treaty negotiating meetings).

The nature of the CGIAR and its centres

The first international agricultural research centres[2] were established in response to concerns that food resources in developing countries would be insufficient to meet the needs of their growing populations and the need to seek improved varieties to increase food production. A number of new centres were set up within the CG system during the 1970s and 1980s,[3] and yet other existing centres brought within the CGIAR during the 1990s,[4] thus bringing the centres to its present total of 15 (see Table 11.1 below). The need for a more focused system-wide approach within the CG system has been the driving force for reforms over the last few years. The first such reform was the establishment of the Alliance of CG Centres in 2004 as a means of providing a collective unified voice for the centres on matters requiring a common position. Under the Alliance procedures, decisions could be taken by majority vote that would bind all centres. More far reaching reforms have recently been instituted, involving the establishment of a Consortium of CG Centres, with its own legal personality, together with a Fund Council, which, it is hoped, will provide more direction and funding stability for the CG system.[5]

The work of the initial centres, and, in particular, that of plant breeders like Norman Borlaug of CIMMYT, bore spectacular results, heralding the birth of the so-called 'Green Revolution'. At the same time, the introduction of new improved varieties tended to supplant existing local varieties and threatened the very biodiversity on which the original green revolution, and future crop improvements, depended. The CG system responded to this new threat by increasing its efforts to collect and conserve endangered PGRFA.

In 1974, the CGIAR set up the International Board on Plant Genetic Resources (IBPGR) hosted by FAO with the task of coordinating an international plant genetic resources programme, and organizing collecting missions as well as building and expanding gene banks at the national, regional and international level. Over the period 1974–1980, IBPGR collected and conserved over 65,000 accessions from over 70 countries. The material collected through these missions, which were for the most part carried out jointly with institutions in the countries concerned, were deposited for conservation in some 52 gene banks. These gene banks included both national or parastatal institutions such as EMBRAPA in Brazil, the Kenyan Agricultural Research Institute (KARI), CGN in The Netherlands, VIR in Russia, and the Rural Development Administration of Korea, as well as international institutions such as the CG Centres, AVRDC and CATIE. In total 8 CG Centres[6] formed part of the nascent network.

Table 11.1 *The 15 Centres[**] of the Consultative Group on International Agricultural Research*

Abbreviation and name of CGIAR Centres	*Former name*	*Dates of founding*	*Location of headquarters*
AfricaRice – Africa Rice Centre*	WARDA –West Africa Rice Development Association	1971	Cotonou, Benin
Bioversity International*	IBPGR – International Board for Plant Genetic Resources; then IPGRI –International Plant Genetic Resources Institute merged with INIBAP	1974	Rome, Italy
CIAT – International Centre for Tropical Agriculture*		1969	Cali, Columbia
CIFOR– Centre for International Forestry Research		1993	Bogor, Indonesia
CIMMYT – International Maize and Wheat Improvement Centre*		1963	Mexico City, Mexico
CIP – International Potato Centre*		1971	Lima, Peru
ICARDA – International Centre for Agricultural Research in the Dry Areas*		1975	Aleppo, Syrian Arab Republic
ICRISAT – International Crops Research Institute for the Semi-Arid Tropics*		1972	Patancheru, Andhra Pradesh, India
IFPRI – International USA Food Policy Research Institute		1979	Washington DC,
IITA – International Institute of Tropical Agriculture*		1967	Ibadan, Nigeria
ILRI – International Livestock Research Institute*	Previously existing as two institutions: ILCA – International Livestock Centre for Africa; and ILRAD – International Laboratory for Research on Animal Diseases	1973	Nairobi, Kenya
IRRI – International Rice Research Institute*		1960	Los Baños, Philippines
IWMI – International Water Management Institute	IIMI – International Irrigation Management Institute	1984	Colombo, Sri Lanka
World Agroforestry Centre*	ICRAF – International Council for Research in Agroforestry	1978 1991	Nairobi, Kenya
WorldFish Centre	ICLARM – International Centre for Living Aquatic Resources Management	1977	Penang, Malaysia

* Article 15 of the International Treaty on Plant Genetic Resources for Food and Agriculture states that contracting parties to the Treaty call upon the International Agricultural Research Centres (IARCs) to conclude agreements with the Governing Body of the Treaty with regard to ex situ collections. The centres with an asterisk close to the name signed agreements with FAO on 16 October 2006 for the inclusion of their ex situ collections within the purview of the Treaty and to make PGRFA listed in Annex I available in accordance with the provisions set out in Part IV of the Treaty. Since 2007, those centres also distribute PGRFA other than those listed in Annex I of the Treaty and collected before its entry into force with the standard material transfer agreement (SMTA) and an interpretative footnote was endorsed by the Governing Body for the use of a unique SMTA.

** There were 16 centres until 2004. ISNAR – the International Service for National Agricultural Research – based in The Hague, The Netherlands was founded in 1980 and ceased to exist in 2004. Some competencies were transferred to the Knowledge, Capacity, and Innovation Division of IFPRI

During the period 1980–2004, the number of new accessions into the CG Centres as a whole has ranged from a high of almost 35,000 per year in 1984 to a low of 5000 in 2004 (Halewood and Sood, 2006). Today, the CG gene banks contain a total of over 650,000 accessions. While this represents only about 12 per cent of the total accessions held in ex situ collections worldwide, the CG collections are particularly valuable in light of the high proportion of landraces and wild relatives. They are also well maintained and documented (Moore and Tymowski, 2005). Improvements in the conservation and maintenance of the CG collections have recently been introduced through the Global Public Goods Project financed by the World Bank.[7] And the financial security of the collections is being secured through a series of long-term funding arrangements with the Global Crop Diversity Trust (GCDT), a new endowment fund set up to ensure the long-term conservation and availability of plant genetic resources for food and agriculture.[8]

The CG system firmly believes that the true value of plant genetic resources lies in their use. In so far as possible, both unimproved and improved PGRFA are treated as international public goods, and distributed as freely and widely as possible to breeders and farmers throughout the world. Since 1 January 2007, the CG Centres have been distributing PGRFA of crops listed in Annex I to the Treaty under the SMTA adopted by the Governing Body at its First Session in June 2006. At its second session in November 2007, the Governing Body also authorized the Centres to distribute PGRFA of non-Annex I crops collected before the entry into force of the Treaty under the same SMTA. The early experience with use of the SMTA indicated that most of the material distributed consists of improved materials.[9]

In describing the role of the CG system in general and the CG Centres in particular in the negotiation of the Treaty, it is important first to understand their legal status.

The original CGIAR system[10] was made up of the Consultative Group[11] itself (the CGIAR), an independent Science Council[12] and 15 International Agricultural Research Centres.[13] Neither the CGIAR system nor the CGIAR[14] itself had any independent legal personality of their own, either under international law or indeed under any system of national law. The International Agricultural Research Centres (IARCs) on the other hand each have their own independent legal personality. Initially there were doubts as to whether some of the CG Centres as originally established had legal personality under international law as opposed to national law. Most if not all of these doubts have been resolved through agreements concluded in the last 15 years explicitly recognizing the international legal personality of the centres concerned.[15] Today, the international legal personality of the CG Centres holding ex situ collections of PGRFA has been recognized and forms the basis of the agreements signed by the Centres with the Governing Body of the Treaty, as mandated in Article 15 of the Treaty.

The International Undertaking and the Convention on Biological Diversity (CBD)

As noted above, the CGIAR and the individual CG Centres have always been committed to ensuring the maximum availability of PGRFA, including both unimproved and improved materials, as a means of promoting agricultural research and breeding for the benefit of farmers in developing countries and elsewhere. Due to the special nature of PGRFA, and the spread of PGRFA across country and continental borders over the centuries, all countries and regions are now highly dependent on PGRFA from other countries and regions to sustain and develop their agriculture and food security (Moore and Tymowski, 2005).

This interdependence was recognized in the International Undertaking on Plant Genetic Resources (IU) adopted by the FAO Conference in 1983,[16] which was based on the 'universally accepted principle that plant genetic resources are a heritage of mankind and consequently should be made available without restriction'.[17] Its stated objective was to 'ensure that plant genetic resources of economic and/or social interest, particularly for agriculture, will be explored, preserved, evaluated and made available for plant breeding and scientific purposes'.[18]

The IU also called for the development of international arrangements then being initiated by FAO and the International Board for Plant Genetic Resources, the predecessor of IPGRI, now Bioversity International, to develop a global system for plant genetic resources, including 'an international network of base collections in gene banks, under the auspices or jurisdiction of FAO, that have assumed the responsibility to hold, for the benefit of the international community and on the principles of unrestricted exchange, base or active collections of the plant genetic resources of particular plant species'.[19]

A series of discussions were held in the FAO Commission on Plant Genetic Resources during the latter part of the 1980s on the legal arrangements that would be appropriate to establish the international network, which as noted above, covered both national and international institutions. In the end, the Commission decided to go ahead only with the establishment of agreements (the so-called In Trust agreements of 1994) with the CG Centres and gene banks holding the Coconut Genetic Resources Network (COGENT) collections. However, a number of the institutions in the original network as promoted by the IBPGR in addition to the CG Centres are being considered for financial assistance by the GCDT as part of an efficient and sustainable global system of ex situ collections.

In the international climate generated by the IU, it is not surprising that the highest rates of germplasm acquisition and distribution by the Centres as a whole were achieved in the years 1983–1985.

During the subsequent years, the concept of free availability of PGRFA started to be eroded. From the side of the plant breeding industry came the push to recognize the rights of formal breeders and researchers over the products of their breeding and research (see Chapter 12). From the side of developing countries providers of PGRFA came a countervailing movement for the recognition of the sovereign rights of countries over their natural resources, including PGRFA. The result was the adoption of a series of Agreed Interpretations of the IU recognizing on the one hand that plant breeders' rights, as provided for under the UPOV Convention, were not incompatible with the IU, as well as the rights of farmers arising out of their contribution to the conservation and development of plant genetic resources. The Agreed Interpretations, on the other hand, also recognized the sovereign rights of countries over their plant genetic resources.

The concept of sovereign rights over genetic resources and the right of national governments to determine access to those resources in accordance with their own national legislation became a cornerstone of the CBD which was opened for signature in 1992. The CBD provided for access to genetic resources to be subject to the prior informed consent of the country of origin providing the resources and to be on the basis of mutually agreed terms. While there is nothing in the CBD that requires that prior informed consent and mutually agreed terms be on a bilateral basis, this was the way in which the Convention was in practice implemented, at least until the negotiation of a set of mutually agreed terms for access to some PGRFA under the Treaty.

In this atmosphere of intense national concern over the sovereign rights of nations over their patrimony, it is hardly surprising that the rate of acquisition of new materials by the CG Centres dropped to an all time low (Halewood and Sood, 2006). New acquisitions in 1993 dropped to under 10,000, almost a quarter of the total in 1984. Although the rate rose again in 1994, this was due more to transfers between centres, or to transfers from developed country gene banks, such as the United States Department of Agriculture (USDA), rather than to new acquisitions from collecting missions in countries of origin (Halewood and Sood, 2006).

The need to find a more appropriate system of access to PGRFA, given the dependence of all countries on easy and effective access to PGRFA from other countries and regions, coupled with a lack of clarity over the legal status of the ex situ collections acquired before the entry into force of the CBD, led directly to the conclusion of the In Trust agreements between the CG Centres and FAO in 1994, and the renegotiation of the IU. Both were the subject of Resolution 3 adopted by the Diplomatic Conference that adopted the CBD in 1992. Resolution 3 called for the development of complementarity and cooperation between the CBD and the FAO Global System for the Conservation and Sustainable Use of Plant Genetic Resources for Food and Sustainable Agriculture, and recognized the need to seek solutions to outstanding matters concerning plant genetic resources with the Global System, including, in particular, access to ex situ collections not acquired in accordance with the CBD, and the question of Farmers' Rights.

In Trust agreements of 1994

A study prepared by the Legal Office of FAO in 1987[20] pointed out the uncertainty that surrounded the legal status of many of the existing ex situ collections forming part of the international network, including those of the CG Centres. The uncertainties related, in particular, to the legal status of the institutions holding those collections, ownership over the accessions in the collections, and the rights of the host governments over the collections. As noted above, these uncertainties were left outstanding by the CBD, which expressly did not cover ex situ collections of genetic resources acquired before its entry into force.

The CGIAR system responded in a number of ways to this situation. The first response was to develop the concept of the 'in trust' status of the ex situ collections held by the CG Centres. Collections held by the CG Centres were not the property of individual nations, nor were they the property of the CG Centres themselves, but were held by the centres 'in trust' for the international community. This concept was first set out in a CGIAR Policy on Plant Genetic Resources adopted in 1989.[21] As can be seen, the concept drew, to a large extent, on the notions set out in the IU. The concept is still referred to in Article 15 of the Treaty.

The second response was to clarify the international legal status of the individual centres holding ex situ collections.[22]

The third response was to clarify once and for all the status of the collections in agreements with FAO representing the international plant genetic resources community. This was achieved through the signature on 26 October 1994 of a series of agreements[23] between FAO and the 12 CG Centres then holding ex situ collections of germplasm. Under the agreements the centres formally placed their collections of designated germplasm under the auspices of FAO as part of the International Network of ex situ collections provided for under the IU. The agreements also clarified the status of the designated germplasm as being held in trust by the centres for the benefit of the international community. The centres undertook neither to claim legal ownership over the material nor to seek intellectual property rights over it or related information. They also undertook to manage the designated germplasm in accordance with internationally accepted standards and to make samples of it available to users for the purpose of scientific research, plant breeding and genetic resources conservation without restriction. The centres were to ensure that where material is transferred to the recipient and subsequent recipients, these recipients are also bound by the same conditions. Perhaps most significantly, the centres recognized the intergovernmental authority of FAO and its Commission on Plant Genetic Resources in setting policies for the International Network, and to give full consideration to any policy changes proposed by the Commission. The In Trust agreements were to remain in force for four years and be subject to automatic renewal for further periods of four years unless terminated by either party.

The In Trust agreements had a double significance for the centres. In the first place they clarified the status of the collections held by the centres. Second, by recognizing the intergovernmental policy authority of FAO and its Commission, they brought the centres on board in the process of renegotiation of the IU. This latter aspect was as important to the centres in safeguarding the future of the collections, as it was to the renegotiations in ensuring that these important collections would be brought within the purview of the new international instrument being negotiated.

The CGIAR and the Negotiation of the Treaty

The CG system was represented at all stages of the negotiations on the Treaty, including in the sessions of the contact group which spearheaded the final negotiations. It was perhaps this continuous presence, coupled with the steady stream of timely, relevant and reliable technical inputs and the political neutrality of the CG system that contributed most to its influence on those negotiations. The role of the CG system, represented primarily by IPGRI,[24] which had the mandate to represent the CG system as an observer in the negotiations, was necessarily limited, given the intergovernmental nature of the negotiations. Nevertheless, it did play a significant part in promoting the concept of a multilateral system (MLS) for PGRFA, and in providing the necessary scientific and technical information that allowed for its acceptance (see Annex 3 of this book for details on the main provisions of the Treaty). It was particularly effective in providing technical information on the current state of gene flows and the interdependence of all countries, including, in particular, developing countries, on access to plant genetic resources for their own agricultural development. In an atmosphere of technical uncertainty that characterized the early stages of the negotiations, this provision of impartial and reliable scientific information was particularly helpful in bringing about a consensus.

The role played by the CG system in the negotiation of the Treaty has been examined at some length in an article published in 2003 (Sauvé and Watts, 2003). In the article, the authors found that the CG system 'exerted influence on the issue of the multilateral system of access and benefit ... [and] the level of influence it exerted on this specific can be deemed important, but not critical'. It also exerted critical influence to ensure that access for conservation as well as utilization was included in the scope of the MLS. IPGRI (and the CG system) also had an influence on the scope of the MLS, canvassing successfully for the expansion of the MLS to cover most of the CG mandate crops, although they were unsuccessful in achieving complete coverage of those crops in Annex I to the Treaty. They also argued successfully for the coverage of in trust collections held by the CG Centres in a specific article (Article 15) of the Treaty, and for the legal mechanism finally adopted by the Treaty of bringing those collections within the purview of the Treaty by means of separate agreements between the centres and the Governing Body of the Treaty, in recognition of the international legal personality of the individual centres. Most important, however, was the general role of IPGRI and the other CG Centres 'as a leading source of scientific and technical information to delegates, through studies, seminars, formal interventions during the negotiations and personal contacts'. The study had revealed that IPGRI was seen as 'a consistent and reliable presence throughout the negotiations', [had] 'consistently promoted the concept of the Multilateral System', and 'had improved the general understanding of the issues being dealt with in the negotiations and that it shed light on the nature of the interlinkages between issues, especially between the issues of access and benefit-sharing'.

The ex situ collections and the Treaty

The Treaty contains one article dedicated to ex situ collections held by the CG Centres and other relevant international institutions.[25] In Article 15, the contracting parties recognized the importance of the collections held in trust by the CG Centres and called on the centres to sign agreements with the Governing Body placing those collections within the purview of the Treaty. As noted above, this approach was necessitated by the fact that the centres, for the most part, possess their own independent international legal personality but are not States and thus can neither be bound by the Treaty itself nor become parties to the Treaty in their own right.

Article 15 sets out the main terms and conditions that are to be contained in such agreements. Annex I PGRFA held by the centres are to be made available in accordance with the same conditions as applicable to collections held by contracting parties – that is, they are to be made available under the SMTA.

The conditions under which non-Annex I material is to be made available depend on the date when it was collected.

Material collected before the entry into force of the Treaty were to be made available in accordance with the MTA then being used by the centres under the In Trust agreements of 1994. This MTA was to be amended by the Governing Body no later than its second session to bring it into line with the relevant provisions of the Treaty, including, in particular, the provisions relating to facilitated access and benefit-sharing. In fact a decision was taken at the second session of the Governing Body that the centres should use the SMTA itself for transfers of non-Annex I material as well as for Annex I material. The Governing Body agreed to the addition of an explanatory footnote to the SMTA clarifying its application to Annex I as well as non-Annex I material (ITPGRFA, 2007). The centres are to periodically inform the Governing Body about the MTAs entered into in accordance with conditions established by the Governing Body,[26] are to make samples of PGRFA collected in in situ conditions available to the contracting party where they were collected without an MTA, and to take appropriate measures, in accordance with their capacity, to maintain effective compliance with the conditions of the MTA and promptly inform the Governing Body of cases of non-compliance.

Non-Annex I material collected after the entry into force of the Treaty, on the other hand, is to be made available for access on terms consistent with those mutually agreed between the centres receiving the material and the country of origin of those resources, or the country that acquired them in accordance with the CBD or other applicable law.

Under Article 15 the contracting parties agree to provide centres that have signed agreements with the Governing Body with facilitated access to Annex I PGRFA. They are also encouraged to provide centres with access on mutually agreed terms to non-Annex I material that is important to their programmes and activities.

Article 15 also includes general provisions drawn from the former in trust agreements, including: recognition of the authority of the Governing Body to provide policy guidance relating to the collections held by them; the collections to be administered in accordance with international accepted standards; and for technical support and assistance with the evacuation or transfer of threatened collections to the extent possible.

The agreements between the CGIAR Centres and the Governing Body of the Treaty

The agreements provided for in Article 15 of the Treaty were signed by FAO on behalf of the Governing Body and the 11 CG Centres holding ex situ collections on World Food Day (16 October) 2006. The agreements repeat almost verbatim the relevant provisions of Article 15.

At the same time, the CG Centres issued a statement regarding their interpretation of the agreements, on much the same lines as the joint statements issued at the time of the signature of the In Trust agreements with FAO in 1994. The statement clarified the centres' common understanding of certain provisions of the agreements and indicated some actions that the centres would be taking to implement them.

On the issue of availability of the germplasm held in trust by the centres, the centres clarified their understanding that while the agreements talked only in terms of making samples of PGRFA available to contracting parties, this would not prevent the centres from also making germplasm available to non-contracting parties, using the SMTA in the case of Annex I materials and the MTA (now the SMTA with footnotes) for non-Annex I material. The centres also voiced their understanding that the agreements did not preclude them from making PGRFA also available to farmers for direct cultivation, as was the case with material made available under the earlier In Trust agreements.[27]

The Statement also clarified the steps that the centres would take to promote compliance by recipients with obligations under the MTA, including requesting explanations in respect of perceived violations, informing the Governing Body and taking action with national authorities for violations involving intellectual property rights.

In much the same way as the centres had done in their earlier joint statements regarding the implementation of the In Trust agreements, the centres further indicated the way in which they would be implementing the provisions of the agreements regarding the obligation to make PGRFA available. In this respect, the centres made it clear that while they would do their best to respond to all requests as quickly as possible and free of charge, sound management practices as well as practical or even biological constraints (such as seed availability or the health status of a sample) may at times limit the ability of centres to provide PGRFA, and that centres would have to use some discretion in determining the size and number of samples to be provided at any given time to a particular recipient. In some cases, such as for woody species, multiplying and supplying accessions can involve very time-consuming and expensive procedures. In such circumstances it would be unreasonable to expect that centres could guarantee unlimited quantities or immediate availability of all germplasm. At their discretion, centres might request that users cover all or part of the costs involved in multiplication.

The CGIAR Centres' experience with the implementation of the Treaty

The agreements with the Governing Body entered into force in January 2007, and the centres chose to implement them in full as from 1 January 2007. In the first 19 months of operation (1 January 2007 to 31 July 2008) the Centres distributed approximately 550,000 samples of PGRFA under the SMTA. Of these, almost three quarters were materials that the centres had been involved in improving. The overwhelming majority of the samples were sent to developing countries (74 per cent) and countries with economies in transition (6 per cent).[28] In only three cases in the first seven months of implementation did potential recipients refuse to accept materials under the SMTA. There were no instances of refusal during the period 1 August 2007 to 1 August 2008. There were, however, a number of queries and concerns raised over the SMTA, particularly during the earlier stages of implementation, including, in particular, its length and complexity. Many of the questions raised are being responded to in a series of Frequently Asked Questions on the websites of the individual centres. Other questions of a more complex nature are being referred to an ad hoc technical advisory committee on the SMTA and the MLS set up by the Governing Body at its Third Session in 2009.

On the whole, however, the experience of the centres with the implementation of the Treaty has been positive; even more so since the decision of the Governing Body at its second session to authorize the centres to use the same SMTA for both Annex I and non-Annex I material. This simplifies considerably the task of the centres in making PGRFA available and reduces the administrative costs involved. Even more streamlined procedures for the distribution of germplasm will inevitably come about with the introduction of the computerized one-stop ordering system for the CG system.

Conclusions

The CGIAR system has always been committed to ensuring the conservation of PGRFA and the wide availability of both unimproved and improved materials, as a means of promoting agricultural research and breeding for the benefit of farmers in developing countries and elsewhere. It has also been concerned to ensure that a stable global system is in place that would allow the centres to continue to play their part in conserving and promoting the sustainable use of PGRFA as a means of achieving food security. It was with these interests in mind that the CGIAR system has played a significant role in the development of the Treaty, and is now working with contracting parties to ensure its full implementation.

Notes

1 ITPGRFA, Article 15.1.
2 The International Rice Research Institute (IRRI) was established in 1960, the Centro Internacional de Mejoramiento de Maíz y Trigo (CIMMYT) in 1964, the International Institute of Tropical Agriculture (IITA) in 1967 and the International Centre for Tropical Agriculture (CIAT) in 1969.
3 International Crops Research Institute for the Semi-Arid Tropics (ICRISAT) (1972); Centro Internacional de la Papa (CIP) (1973); International Laboratory for Research on Animal Diseases (ILRAD), now incorporated with ILCA into the International Livestock Research Institute (ILRI) (1973); International Food Policy Research Institute (IFPRI) (1979); International Livestock Centre for Africa (ILCA) now incorporated with ILRAD into ILRI (1974); West Africa Rice Development Association, now called Africa Rice Center (WARDA) (1975); International Center for Agricultural Research in the Dry Areas (ICARDA) (1975); International Service for National Agricultural Research (ISNAR) (1980).
4 International Council for Research in Agroforestry, now called World Agroforestry Centre (ICRAF) 1977 (1991); International Irrigation Management Institute (IIMI) now called International Water Management Institute (IWMI) 1984 (1991); International Centre for Living Aquatic Resources Management (ICLARM) now called WorldFish Centre 1977 (1992); International Network for the Improvement of Banana and Plantain (INIBAP) now merged with Bioversity International 1984 (1992); Center for International Forestry Research (CIFOR) 1993 (1993).
5 See www.cgiar.org/changemanagement/index.html.
6 CIAT, ILCA, ICRISAT, CIMMYT, IITA, CIP, IRRI and ICARDA.
7 See http://sgrp.cgiar.org/?q=node/583.
8 So far, long-term funding agreements have been concluded in respect of 13 collections of global significance, including collections held by seven CG Centres and two collections held by the Secretariat of the Pacific Community (SPC) .
9 Out of 542,493 samples distributed during the first 19 months of operation of the SMTA, 372,170 (over 68.6 per cent) were of improved material. See 'Experience of the Centres of the Consultative Group on International Agricultural Research (CGIAR) with the implementation of the agreements with the Governing Body, with particular reference to the Standard Material Transfer Agreement', FAO Docs. IT/GB-2/07/Inf. 11 and IT/GB-3/09/Inf. 15, reports submitted to the Second and Third Sessions of the Governing Body of the Treaty, October/November 2007, and June 2009.
10 The CGIAR system is now in the process of reform. The new system will now consist of a CGIAR fund and a consortium of CGIAR Centres now being established as a legal entity, See www.cgiar.org.
11 The Consultative Group is composed of 47 country members and 17 international or regional organizations.
12 The Science Council, which is an independent scientific body of the CG system consisting of a Chair and six members appointed by the CGIAR on the recommendation of its Executive Council.
13 WARDA; Bioversity International (formerly IPGRI); CIAT; CIFOR; CIMMYT; CIP; ICARDA; ICRISAT; IFPRI; IITA; ILRI; IRRI; IWMI; World Agroforestry Centre; WorldFish Center.
14 The CGIAR itself is described in the Charter of the CGIAR system as an informal association of public and private sector members. The CGIAR system is described as

a loosely connected network of components.

15 See, for example, the Agreement for the Recognition of the International Legal personality of the International Potato Center (CIP) of 1999; Agreement Recognizing the International Legal Personality of the International Rice Research Institute (IRRI) of 1995; Agreement on the Establishment of the International Plant Genetic Resources Institute (IPGRI) of 1991; Agreement between the Center for International Forestry Research (CIFOR) and the Government of the Republic of Indonesia regarding the Headquarters seat of the Centre, of 1993.

16 FAO Conference Resolution 8/83.

17 International Undertaking, Article 1.

18 International Undertaking, Article 1.

19 International Undertaking, Article 7.

20 Legal Status of Base and Active Collections of Plant Genetic Resources, FAO doc. CPGR/87/5.

21 The Policy stated that 'it is CGIAR policy that collections assembled as a result of international collaboration should not become the property of a single nation, but should be held in trust for the use of present and future generations of research workers in all countries throughout the world'.

22 See note 15 above.

23 For a copy of the agreement and the statement made by the CG Centres at the time of signature, see Booklet of CGIAR Centre Policy Instruments, Guidelines and Statements on Genetic Resources, Biotechnology and Intellectual Property Rights, at www.sgrp.cgiar.org/?q=publications.

24 IPGRI (the International Plant Genetic Resources Institute) was set up as an international organization in 1991 as a successor to the International Board on Plant Genetic Resources. Reflecting the fact that the mandate of the organization now covers all forms of biodiversity, it has been operating under the name of Bioversity International since 2006, although the legal name remains unchanged.

25 While most of the provisions of Article 15 apply directly to the collections held by the CG Centres, Article 15.5 also provides that the Governing Body will seek to establish agreements with other relevant international institutions. So far such agreements have been concluded in respect of the COGENT coconut collections, the CATIE Collection, the FAO/IAEA mutant germplasm collection, the cacao network collections held by the University of the West Indies and the ex situ collections held by the Centre for Pacific Crops and Trees of the Secretariat of the Pacific Community.

26 At its 3rd session in 2009, the Governing Body decided that reports should be submitted on a biennial basis. See Resolution 5/2009, para 15.

27 For a copy of the Statement see www.sgrp.cgiar.org/?q=publications.

28 See 'Experience of the Centres of the Consultative Group on International Agricultural Research (CGIAR) with the implementation of the agreements with the Governing Body, with particular reference to the Standard Material Transfer Agreement', FAO Docs. IT/GB-2/07/Inf. 11, and IT/GB-3/09/Inf.15, reports submitted to the Second and Third Sessions of the Governing Body in 2007 and 2009.

References

ITPGRFA (2007) Report of the Second Session of the Governing Body, para 68

Halewood, M. and Sood, R. (2006) 'Genebanks and public goods: Political and legal challenges', Paper prepared for the 19th session of the Genetic Resources Policy Committee, 22–24 February 2006.

Moore, G. and Tymowski, W. (2005) 'Explanatory guide to the International Treaty on Plant Genetic Resources for Food and Agriculture, Environmental Law and Policy Paper No. 57, International Union for Conservation of Nature, Gland, Switzerland and Cambridge, UK

Sauvé, R. and Watts, J. (2003) 'An analysis of IPGRI's influence on the International Treaty on Plant Genetic Resources for Food and Agriculture', *Agricultural Systems*, vol 78 (2003), pp307–327

Chapter 12

The Seed Industry

Plant Breeding and the International Treaty on Plant Genetic Resources for Food and Agriculture

Anke van den Hurk[1]

Introduction

Plant breeding started about 9000 to 11,000 years ago when man started with the domestication of wild plants. Farmers and growers tried to improve their crops with desired traits through trial and error. The evolutionary theories of Darwin and the genetic experiments of Mendel that were developed at the end of the 19th century gave a further impulse to plant breeding and made it more efficient. During the 20th century breeding science was further improved through knowledge of genetics, plant pathology and entomology (Bruins, 2009).

The development of hybrids (starting around 1920) was the first technology in plant breeding to offer better plant varieties to growers and farmers. The new varieties were not only uniform but also often performed better than their parents due to the heterosis effect of hybrid vigour. The increasing use of seed treatment from the 1960s onward further improved yields, as the use of the plant protection products was more precise and therefore more effective. The latest step of innovations to further widen the opportunities plant breeding offers is the use of biotechnology. On the one hand biotechnology is used to better understand genetics and enables quicker interference in the breeding process with tools like markers. On the other hand, the precise introduction of genes through genetic modification, in particular for the major crops, has been a major breakthrough for plant breeding. Genetic modification led to an increase in yield, a reduction of the use of insecticides and an increase of income for farmers (Bruins, 2009).

Professional breeders

Commercial seed industry started around the 1740s with the earliest known seed company, Vilmorin in France. This company was quickly followed by more companies in France, The Netherlands, the United Kingdom and Japan. As indicated above, the plant breeding science became more and more sophisticated and an increasing number of specialized breeding companies were established. It needs to be noted, however, that in the last decennia, consolidation of seed companies took place starting in field crops, now being followed by vegetables and flowers. The global seed market increased from US$12 billion in 1975, to around US$20 billion in 1985 and was estimated at US$36.5 billion in 2007 (Bruins, 2009).

All in all plant breeding has become a highly developed science of how to combine desired traits of plants in one variety. Yield has increased, resistances to biotic stress and tolerance to abiotic stress have been incorporated and various qualitative characteristics like taste, earliness, size, nutritional value and so on were improved (Bruins, 2009).

Relationship between genetic resources and plant breeding

Plant breeding would not be possible if biodiversity did not exist. The recombination of required traits in a plant variety is the essence of plant breeding, whatever plant breeding methodology is used. Hence, for the recombination of the required traits, genetic variation is required. For the development of modern varieties plant breeders mainly make use of existing modern varieties that consist of sets of genes that are desirable for agriculture. Through recombination it is hoped to create even better varieties. When specific traits cannot be found in related varieties, other genetic resources like landraces, wild relatives and/or related species may be used. The latter happens at a level of 5–10 per cent at most.

Plant breeding is done by many players, by farmers, small- and medium-sized companies and multinationals. It also takes place in different regions of the world, even though the method of breeding and capabilities may differ. These activities do not take place independently, as plants from different users and different regions are constantly intermingled. This means that genetic resources are continuously moved around the world. This flow is essential for the future of plant breeding as it assures breeders that they can utilize the desired sets of genes.

Plant breeding is a continuous process of improvement, in which genetic resources are both an input and an output. The genetic resources that have been developed will be input for new breeding processes. Through plant breeding new variation, new diversity may be created (Van den Hurk, 2009). Lang and Bedo (2004) showed a great increase in genetic diversity of the Hungarian wheat varieties registered over the last 50 years. This is the result of breeders using a wide range of genetic resources to come to new wheat varieties. Moreover, farmers use a wider choice of varieties at present than in the past.

Van de Wouw et al (2010) demonstrated in two studies that reduction of biodiversity through the modernization of agriculture could be observed in the 1960s when diversity in the crops researched was low. However, diversity was rising again from then on until the end of the century. These trends over the last decades demonstrate that plant breeding has a positive influence on the biodiversity at the genetic level.

Recombination and use of genetic resources are not limited to one plant breeder and one region. Plant breeders made, make and will make use of genetic resources from each other, from different countries and backgrounds. Plant species have, for example, moved around the world and may have grown into important species in other parts of the world. It is believed, for instance, that Papua New Guinea and the surrounding region is the centre of origin for sugar cane. From there it moved to northern India, where a secondary centre of origin developed. Then it moved further around the world. Currently Brazil is the top producer (Willy Degreef, personal communication).

It is not only that plant species move around the world, but also that those species may be used for different objectives and therefore gain importance. Sugar cane, for example, is not only used as a sweetener, but has also become important for ethanol production. Furthermore, crops may adapt to different climatological conditions and move to new regions. Maize growing, for example, has shifted to northern Europe, while sugar beet has been adapted for tropical circumstances (Van den Hurk, 2009).

From the above it can be concluded that no plant breeder, no nation is completely independent in terms of genetic resources. Both developed and developing countries have come to rely on non-indigenous crops for their food, feed and fibre supplies. A study assessing the degree of a country's dependence on non-indigenous crops (measured in terms of calorific contribution to nutrition contributed by crops whose centre of diversity is outside the country in question) has shown that all countries grow or import crops that come from distant lands (Palacios, 1998).

The International Treaty on Plant Genetic Resources for Food and Agriculture

From the above it can be concluded that genetic resources and plant breeding are closely intertwined. As the flow of genetic resources was at stake, it was important for the plant breeding sector to actively participate in the negotiations of the International Treaty on Plant Genetic Resources for Food and Agriculture (ITPGRFA) (see Annex 1 of this volume for the list of all Commission and Treaty negotiating meetings). The focus of input has been on the multilateral system, access, benefit sharing and Farmers' Rights (see Annex 3 of this book for details on the main provisions of the Treaty).

Plant breeding sector and the negotiations of the ITPGRFA

With the entry into force of the Convention on Biological Diversity (CBD) at the end of 1993, genetic resources were no longer freely available due to States' sovereign rights. At the time, the International Seed Federation (ISF) was of the opinion that the restrictions probably would have the most effect on public research, small breeding companies and developing countries poor in genetic and financial resources. Large companies and the developed world would be less affected as they have already collected materials from all over. Therefore, ISF supported the development of a multilateral system as proposed in the Global Plan of Action on Plant Genetic Resources for Food and Agriculture. This would leave as much freedom to operate as possible for the breeders (Coupe and Lewins, 2007).

The multilateral system should include all genetic resources of importance to present and future food security, and/or agriculture in general, at the level of genera and species: food crops, including vegetables and fruits, forage crops and mixed industrial/food crops. For each genus and species, the genetic resources should comprise wild relatives, landraces, obsolete varieties, and commercial varieties that are in the public domain (ASSINSEL, 1999) (for the list of genera and species, see CGRFA/IUND/4, Rev.1, pp40–43). Unfortunately, the final negotiated list of the multilateral system of the ITPGRFA was limited due to political reasons. Some main food crops like soy bean are missing. Furthermore, most important vegetable species are also missing from the list. Smaller crops like asparagus and strawberry are on the list while important vegetables like tomato, pepper, sweet pepper and onion are not on the list. This means that in the daily practice of plant breeders it is unclear how to deal with access to genetic resources that are not part of the list – leading to limited or no access at present.

In the negotiations of the ITPGRFA it was important to explain the interdependence of this treaty and the Union for the Protection of New Varieties of Plants (UPOV) Convention and the plant breeders' rights defined in it, which is so important for innovation and further improvement of varieties in the breeding industry (ASSINSEL, 1999).

Breeders' rights provide protection to a genome of the species, but on a specific individual plant variety in the development of which the breeder has invested. It is only limited in time. Moreover, thanks to breeders' exemption, the variety to which the title has been granted is freely available for further breeding and the result of such further breeding is freely marketable, as long as the newly developed variety is distinct, uniform and stable and not a simple copy of the initial variety. In fact, the obligation to avoid plagiarism favours biodiversity (ASSINSEL, 1999). Hence the breeders' exemption was a benefit on its own and ISF is positive that this was recognized in the final text of the Treaty.

The breeders' exemption is not applicable in patents, which means that new improved patented material is not immediately available for further breeding. For this reason, ISF members indicated to be ready to study the possibility of balancing the resulting lack of immediate availability. When the results of a breeding/research programme which includes genetic resources provided by in situ or ex situ gene banks, are patented, they agreed to participate in a fund to be established by governments, as decided in FAO resolution 3/91, and implicitly acknowledged in the Global Plan of Action. A material transfer agreement (MTA) that is linked to the multilateral system would be necessary to legalize access and benefit sharing. This ISF position, stated at the 5th extraordinary session of the Commission on Genetic Resources in June 1998, was necessary to finally get to a breakthrough in the negotiations of the benefit-sharing arrangements of the ITPGRFA (Cooper, 2002).

Whilst preferring a broad multilateral agreement, ISF acknowledged the need to keep open the possibility of bilateral agreements in exceptional cases (ASSINSEL, 1999). This could be, for instance, appropriate when a small number of countries have, or need, access to genetic diversity of a particular species or group of species, and/or when highly expensive and specialized research gives a strong competitive advantage to a single or limited number of institutions. Such conditions could prevail in the case of some industrial crops as, for example, rubber. In addition, bilateral agreements could be tailored to the needs of the parties; they could be created for specific purposes and then dissolved without the need of heavy structures; they could offer greater confidentiality (ASSINSEL, 1999).

Plant breeding sector and the negotiations of the standard material transfer agreement

To implement the multilateral system and make it effective, a standard material transfer agreement (SMTA) had to be developed. ISF supported the fast development of such an agreement and offered its experience with preparing contracts. From the start in 1998 of the negotiations on the agreement that later became the SMTA, ISF already defined several issues which should be considered in the agreement (ASSINSEL, 1999).

Breeders' exemption includes benefit sharing

First of all, the material supplied should be available without any restrictions for the recipient for breeding and research purposes. According to ISF the recipient should neither claim legal ownership nor apply for intellectual property protection over the germplasm received, per se. However, it should be possible to protect plant varieties developed from the material, if the criteria of protectability are met, by plant breeders' rights, or any other sui generis system consistent with the UPOV Convention, or by patent, according to national law. This also meant that cells, organelles, genes or molecular constructs isolated from the material may be protected by the recipient through patents, if the criteria for patentability are met (ASSINSEL, 1999). This approach of ISF was generally accepted and is incorporated in the SMTA (see Articles 6.2 and 6.10).

Another important criterion for ISF, which existed as early as 1999 in its official position and occurs for protection under UPOV-like systems, considers that the free access to the new varieties for further research should be recognized as a contribution to benefit-sharing (ASSINSEL, 1999). As stated before, ISF could see that when the results of the research are patented, the recipient should pay to the multilateral agreement fund (the Resolution 3/91 or 3/91-like fund) a certain amount of royalties, to be accepted on a contractual basis (ASSINSEL, 1999).

In the negotiations on the SMTA the part dealing with benefit sharing has taken a fair amount of time and the seed sector has been active to participate and provide relevant information. The benefits of the breeders' exemption have been recognized in the final text of the SMTA. According to Article 6.7 of the SMTA there is no obligation for any further benefit sharing; however, voluntary contributions are welcomed according to Article 6.8.

Benefit-sharing requirements

To agree to the benefit-sharing requirements, be they obligatory or voluntary, several bottlenecks needed to be dealt with. The first point to discuss is the contribution of the germplasm to the final product. This may be divided into two parts, the amount of work and research that needs to be done to get to a variety, and the contribution of the germplasm to the final product.

In the negotiations it became clear that a good balance between the work of the plant breeder and the contribution of the germplasm should be sought. A breeding process takes at least ten years and often longer, in particular, when wild relatives or landraces are used. Moreover, plant breeding companies spend 10–15 per cent of their turnover on research and development.

In the discussion on the contribution of a genetic resource the plant breeding sector suggested the following elements to be considered for benefit sharing. First of all, it would look to the amount of DNA that was incorporated in the final product. The plant breeding sector was of the opinion that benefit sharing should only take place when a great part of the genetic resources could be found back in the final product; a minimum of 25 per cent should be incorporated. In addition, benefit sharing should be able to be triggered when an identifiable trait of value or essential characteristic of the genetic resource was incorporated. Secondly, the amount of available knowledge on the genetic resource could be considered relevant. The more you know on the genetic resource, the less risk you need to take to work with the material.

The amount of DNA incorporated has been discussed in a great detail. It was not possible to come to an agreement though on what part of the DNA of the genetic resource should be incorporated. The main concerns were the traceability and control of such a system. Therefore, it had been agreed that any incorporation of a genetic resource should trigger benefit sharing be it voluntary or obligatory.

The amount of knowledge known beforehand has not been debated any further in the context of triggering benefit sharing. In fact, it was decided that information that was not confidential that became available from research of the germplasm should be shared. Knowledge sharing is considered an important form of benefit sharing.

Once it was decided what would be the trigger point for benefit sharing, the benefit sharing itself should be discussed. The seed sector pointed out that the percentage of the profit to be agreed upon cannot be very high, as any incorporation is a trigger point. Furthermore, they found it important that the percentage would be taken of the net sales, meaning that (i) discounts, customary in trade, (ii) amounts repaid or credited by reason of rejection or returns, (iii) any freight or other transportation costs, insurance, duties, tariffs and sales and excise taxes based directly on sales or turnover or delivery of products and (iv) any licence fees, should be subtracted from the gross income. Especially the developing world was concerned about the net sales; they expressed their concern on transparency on what would be subtracted from the gross income and what would not be. On the other hand it was understood that payment should not be settled on income that the breeder did not receive. To overcome the problem, the plant breeding sector estimated that the income losses of the above mentioned points are around 30 per cent (Le Buanec and Noome, personal communication). In conclusion it had been decided that the obligatory benefit sharing would be 1.1 per cent of the gross income minus 30 per cent.

A third issue in the negotiation process that took time and thought was the determination of the exact moment when benefit sharing should take place. In this discussion ISF indicated that double payment should be avoided. Furthermore, ISF was of the opinion that a reasonable point in the development chain should be found for the benefit-sharing moment (Le Buanec and Noome, personal communication). In other words, it would be important that the user of genetic resources is not forced to follow his product to the final consumer. The point of commercialization of a product was finally defined as the moment that a recipient and/or its licensee sell a product on the open market.

A fourth issue to be dealt with was the multiple uses of genetic resources in breeding programmes. To simplify the benefit-sharing system and make it work it was agreed that only one payment should be made, even if more genetic resources under more SMTAs were involved.

The fact that benefits are only created after a long time was another potential problem. This would mean that benefits would be shared only in a later stage. To circumvent this situation another option for benefit sharing had been designed. Recipients of genetic resources could opt for a lower percentage of the sales, but then on all the sales of a certain product whether germplasm was incorporated or not and/or whether the product would be available for research and breeding or not. The seed sector was involved in the discussions of this option, and saw opportunities in this approach. However, the percentage 0.5 per cent that was finally agreed upon is considered too high.

Other contractual issues

A dispute settlement was agreed upon. For the seed sector it was important that a dispute could only be initiated by either the provider or the recipient. Later, it became also relevant that the third party beneficiary would also be able to initiate this. The seed sector could support this (Le Buanec, personal communication; ESA, 2005).

The seed sector was of the opinion that terms on duration and a termination clause should be included as in any contract (Le Buanec, personal communication; ESA, 2005). During the negotiations no agreement on those items was possible and a duration and termination clause was left out.

Implementation of the ITPGRFA and the SMTA

Once the SMTA was agreed upon, it was important to implement the multilateral system of the ITPGRFA. In some countries the implementation was taken care of immediately, while in most countries it seems more difficult to implement the multilateral system and the SMTA. The exchange of germplasm continues with the countries and regional and international institutes that have implemented the SMTA. In countries where the SMTA is not implemented, no or perhaps limited exchange is taking place. For vegetables no examples of bilateral agreements are known, except if they are based on the SMTA.

If no agreements are made, benefits are not shared. Therefore, the plant breeding sector stresses the need for an effective implementation of the ITPGRFA, putting the genetic resources into the multilateral system and making them available under the SMTA. Only then, materials may be used in a sustainable manner and can be conserved.

The conditions of the SMTA are sometimes also used for non-Annex I crops. This is strongly supported by the seed sector as this creates a level playing field between Annex I and non-Annex I crops. Moreover this may assist in the support for extension of the list of Annex I. That the conditions of the SMTA are useful as a benefit-sharing tool is demonstrated through some collection missions that have been carried out recently by the Dutch gene bank CGN, financially supported by the Dutch breeding companies. In negotiations with Uzbekistan and Tajikistan it was agreed that collection missions on wild spinach could be carried out under the following conditions: the mission would be paid for by The Netherlands and the materials would be shared between the countries and CGN. Moreover, the collected materials can be given out by CGN under the conditions of the SMTA. A similar mission for wild *Allium* species has been agreed upon with Greece.

With regard to the use of the SMTA some bottlenecks need to be further discussed among parties and with stakeholders. The number of SMTAs that are being signed is increasing and the administrative burden may cause unnecessary inconveniences for both user and provider.

Another issue for consideration is the passing on of SMTAs to future users and the information that needs to be provided to the third party beneficiary. As the extensive administration may be cumbersome this may limit further distribution of genetic resources.

Putting genetic resources into the multilateral system needs to be stimulated and facilitated. Issues of concern for the seed sector doing so are several; the burden of administration is one. Secondly, it is not clear if it is the accession that becomes part of the system or the genetic constitution/information. This may be relevant as seed lots may be split in several parts: one for the multilateral system, one for own use and another for further distribution under own conditions or so. Lastly, it may be important to make arrangements so that providers of genetic resources to the multilateral system that for any reason lost their own part of the accession can obtain a copy of their accession though the multilateral system without signing an SMTA; they in the end have brought it in. Finally, the seed sector recognizes that using the SMTA to access genetic resources of Annex I that are maintained in situ is also important and should require more attention.

Link to the access and benefit-sharing system of the CBD

Currently an international regime on access and benefit sharing is being negotiated under the CBD. This regime is dealing with all genetic resources for all uses. The seed sector is of the opinion that the ITPGRFA should be recognized in those negotiations and should be excluded from this general regime. Moreover, it would be good to obtain recognition for the system as a workable system for access and benefit sharing that, in particular, suits industries, like the seed sector, that deals with a continuous flow of genetic resources. In other words it may be useful to extend the rules of the ITPGRFA to the whole breeding sector and possibly other sectors that deal with a continuous flow of genetic resources.

Farmers' Rights

The negotiation of the ITPGRFA was not only focused on a multilateral system, but also on Farmers' Rights. This was felt necessary to recognize the contribution of farmers to the conservation of genetic resources. ISF could support recognition for the farmers; and finds it also important to recognize the contribution of the plant breeding sector.

During the negotiations, ISF explained that plant breeders' rights do not have any negative impact on the activities and work of farmers and, in particular, subsistence farmers. Furthermore, ISF stressed that UPOV and, in particular, the section on farm-saved seeds would not be undermined.

As far as farm-saved seeds are concerned, Article 15 of the UPOV Convention clearly states (i) that the breeders' rights shall not extend to acts done privately and for non-commercial purposes and (ii) that each contracting party [to UPOV] may, within reasonable limits and subject to the safeguarding of the legitimate interests of the breeder, restrict the breeders' rights in relation to any variety in order to permit farmers to use for propagating purposes, on their own holdings, the product of the harvest they have obtained by planting, on their own holdings [a] protected variety.

The plant breeding sector was and is of the opinion that any rules on Farmers' Rights should be implemented at a national level as all jurisdictions have different systems to involve stakeholders including farmers in policy development, and benefit-sharing arrangements. ISF, therefore, could support the text of Article 9 of the Treaty, as long as all elements are recognized and respected.

Concern about the implementation of Farmers' Rights has been expressed by several parties. ISF supports the fact that Article 9 of the Treaty should be implemented and used to call upon the parties to assume their national responsibility.

In their latest position paper on Farmers' Rights adopted in 2009, ISF provided information on the importance of plant breeders' rights for both farmers and plant breeders and also explained the coherence between plant breeders' rights and Farmers' Rights (ISF, 2009).

To encourage the continuous and substantial investments required to support breeding and the large-scale characterization and conservation of germplasm undertaken by the commercial sector, ISF is of the opinion that breeders – whether companies or individuals – must have the opportunity to protect their new varieties through intellectual property rights in order to obtain fair remuneration. Therefore, ISF strongly supports plant breeders' rights based on the UPOV 1991 Convention as it provides an adequate protection of plant varieties against inappropriate exploitation by others.

In relation to Farmers' Rights, it is important to note that this protection is combined with free access and use for further breeding purposes (breeders' exemption) and also the compulsory exception of acts done privately for non-commercial purposes allowing subsistence farmers in developing countries to save and use seed from their own harvests (ISF, 2009).

Most national laws recognize and protect intellectual property. They allow protection of new plant varieties created by breeders through years of breeding effort and significant economic investment to the exploration, characterization and development of germplasm as intellectual property. The Treaty does so too. Even as Article 9 calls for Farmers' Rights it does not exclude the intellectual property of commercial plant breeders. Article 9.3 expressly acknowledges that implementation of a system that allows farmers to 'save, use, exchange and sell farm-saved seed' rests with national governments 'subject to national law and as appropriate'. The Treaty recognizes that each contracting party has its own domestic needs and priorities, and recognizes that a contracting party may also have obligations under other international agreements and conventions it adheres to.

Farmers are the primary market for new varieties developed and protected by commercial plant breeders. Free and unlimited use of farm-saved seed that is harvested from protected varieties developed by plant breeders destroys the economic incentive for those breeders to continue to conserve, characterize and develop the available genetic resources in important food and feed crops. If farm-saved seed of protected varieties is permitted and used, breeders should receive fair remuneration for that use. Failure to respect and protect the property newly created by breeders will eventually restrict the release of genetically diverse and improved varieties to the detriment of farmers and to society as a whole. However, farmers still have the opportunity to freely use seeds of landraces and seeds of varieties that are not or no longer protected, independently of the consent of the breeder (ISF, 2009).

Conclusion

For the plant breeding sector it is important to have sufficient freedom to operate to carry out their breeding activities and have the necessary access to plant genetic resources. This means that a flow of genetic resources should continue to take place. It is important to realize that access is required both in developed and developing countries. In the latter it may become even more important as the plant breeding sector is expanding. Moreover, genetic resources should be available for all type of users, be they small, medium-sized or large enterprises.

The value of the breeders' exemption should be continuously recognized. This guarantees a continuous flow of genetic resources of which plant breeders and thus farmers and economies will benefit.

As Annex I is only limited, it is important to consider the extension of Annex I. Many important food crops are still not on the list, while they are important for feeding the world and providing the necessary variation in diet. The seed sector would like to invite the member countries to implement the multilateral system, and make access and benefit sharing possible. As long as Annex I is not extended the seed sector urges parties to still make use of the conditions of the SMTA for the exchange of plant genetic resources that are used in plant breeding. This has proven the most successful in recent years.

So, it is important not only to implement the ITPGRFA as broadly as possible, but also to take care that the system as such is recognized and respected by the negotiators of the access and benefit sharing regime of the CBD.

With regard to Farmers' Rights the plant breeding sector can support Article 9 of the Treaty as long as it is implemented nationally. The Treaty text states the same and therefore should not be changed and/or interpreted differently. It should be realized that the needs of stakeholders and socio-economic context differ from region to region and from country to country.

The fact that Farmers' Rights and plant breeders' rights can coexist needs to be recognized and respected in the implementation of Article 9. Contrary to what is argued by some sections of society, Article 9.3 does not provide any legitimacy to save, use and sell farm-saved seed. The ITPGRFA remains consistent in recognizing existing obligations arising out of national legislation on farm-saved seed. Therefore, the special dispositions authorizing the use of farm-saved seed that States have implemented – as part of their national legislations on plant breeder's rights – can remain unchanged.

Notes

1 I thank Ms Monique Krinkels for her contribution in the writing and finalization of this chapter.

References

ASSINSEL (1999) 'A multilateral agreement for plant genetic resources for food and agriculture: An ASSINSEL viewpoint', Background documentation provided by the International Association of Plant Breeders for the protection of plant varieties, ftp://ftp.fao.org/ag/cgrfa/cgrfa8/8-i9-e.pdf

Bruins, M. (2009) 'The evolution and contribution of plant breeding and related technologies in the future', *Responding to the Challenges of a Changing World: The Role of New Plant Varieties and High Quality Seed in Agriculture*, Proceedings of the Second World Seed Conference, pp18–31

Cooper, D. (2002) 'The International Treaty on Plant Genetic Resources for Food and Agriculture', *Reciel*, vol 11, no 1, ISSN 0962 8797

Coupe, S. and Lewins, R. (2007) *Negotiating the Seed Treaty*, Practical Action Publishing, Rugby

ESA (2005) 'ESA position on the Standard Material Transfer Agreement (SMTA) within the FAO International Treaty on Plant Genetic Resources for Food and Agriculture', available at www.euroseeds.org/position-papers/PP2005/ESA_05.0427.4.pdf

ISF (2009) 'Farmers' rights', available at www.worldseed.org/cms/medias/file/PositionPapers/OnSustainableAgriculture/Farmers_Rights_2009.pdf

Láng, L., and Bedo, Z. (2004) 'Changes in genetic diversity of the Hungarian wheat varieties registered over the last fifty years', *Genetic Variation for Plant Breeding*, Proceedings of the 17th EUCARPIA general congress, Tulln, Austria, 8–11 September 2004

Palacios, X. F. (1998) *Contribution to the Estimation of Countries' Interdependence in the Area of Plant Genetic Resources*', Food and Agriculture Organization Commission on Genetic Resources for Food and Agriculture

Van den Hurk, A. (2009) 'The importance of plant genetic resources for plant breeding: Access and benefit sharing', *Responding to the Challenges of a Changing World: The Role of New Plant Varieties and High Quality Seed in Agriculture*, Proceedings of the Second World Seed Conference, pp62–70

Van de Wouw, M., van Hintum, T., Kik, C., van Treuren, R. and Visser, B. (2010) 'Genetic diversity trends in twentieth century crop cultivars: A meta analysis', *Theoretical and Applied Genetics*, vol 120, pp1241–1252

Van de Wouw, M., Kik, C., van Hintum, T., van Treuren, R. and Visser, B. (2010) 'Genetic erosion in crops: Concept, research results and challenges', *Plant Genetic Resources: Characterization and Utilization*, vol 8, pp1–15

Chapter 13

Farmers' Communities

A reflection on the Treaty from Small Farmers' Perspectives

Wilhelmina R. Pelegrina and Renato Salazar[1]

Farmers and the International Seed Treaty

The provisions of the Treaty and its implications for smallholder farmers are yet to be substantially 'processed' by farmers and their communities. Civil society organizations (CSOs) with knowledge about the Treaty can only directly reach a very limited number of farmers. Although limited in number, these farmers have substantial understanding of the implications of the Treaty. However, while these farmers are informed and are often involved in the discussions of the Treaty, the issues covered by the Treaty have to compete with other more pressing issues like agrarian reform, access to markets, seed regulations, irrigation concerns and human rights violations.

Laws and regulations developed and implemented by national governments have far greater impact on small farmers than international treaties and conventions. The nation states that negotiated, signed and adopted the Treaty (see Annex 2 of this volume for the list of contracting parties per FAO regional groups), and not the Treaty itself, are the ones that can really affect farmers by enacting and implementing laws on seeds, plant varieties, access and benefit sharing, intellectual property rights, commercial regulations development of research and extension programmes. Farmers do not feel part of international policy processes and agreements, but feel very close to national policies and laws that can affect them.

It is not surprising therefore that despite attempts to invite farmers and farmer organizations for the negotiations, only a few actively participated during the first Governing Body meeting of 2006 in Madrid. This partly improved in

Rome in 2007 and in Tunisia in 2009 with the presence of farmers from *La Via Campesina* and internationally unaffiliated farmers groups and communities from Asia, Africa and Latin America attempting to make their voices heard. Farmers, from local and national organizations in developing countries, who participated in the Governing Body meetings often wonder whether their presence mattered, as they cannot follow the discussions nor find that their interventions were heard. While there is an openness and good will from the Governing Body and the Treaty Secretariat to allow farmers to attend the negotiations, supportive mechanisms and processes, as well as financing, have yet to be set up to allow for a vibrant and constant engagement with farmers and their organizations. Primarily there is a need to support farmers to 'process' the content of the Treaty, its mechanisms and its implications for their lives. Farmers' participation to the Governing Body is an essential element to ensure that their perspectives and positions on the finer points of the Treaty are heard and deliberated as part of a healthy democratic process, advancement in the global discourse and as an embodiment of one of the Treaty's core components – the recognition of Farmers' Rights.

Farmers and Farmers' Rights

For small farmers, the most important provision is, therefore, the article on Farmers' Rights. Farmers' Rights, as a phrase, is immediately loaded with all the possible 'rights' that farmers are supposed to enjoy. However, the current and most prominent international deliberation and use of Farmers' Rights is limited to issues related to plant genetic resources (PGR) for food and agriculture.

In 2003, farmers and farmer groups in the Philippines defined Farmers' Rights to comprise 38 elements covering socio-political, economic and cultural rights (CBDC Network, 2009). In 2007, the Community Biodiversity Development and Conservation Network[2] facilitated a discussion among farmers and farmer groups in Asia (Lao PDR and Philippines), Africa (Malawi and Zimbabwe) and Latin America (Brazil, Chile, Cuba and Venezuela) on their views about Farmers' Rights. Farmers from these countries likewise defined Farmers' Rights comprehensively to include access and rights to land, agricultural resources (water, information, other inputs), appropriate technology, market, the right to organize and participate in policy decision-making processes. For farmers, Farmers' Rights is a bundle of rights. Although most farmers acknowledge the focus of the Treaty on PGR, for them, Farmers' Rights as stipulated in Article 9 of the Treaty cannot be meaningfully realized unless other entitlements are guaranteed. Farmers point out the interrelationship between seeds and land, water, energy, culture, social fabric, household and individual well-being. Farmers and farmer groups are aware that all of these rights (forming the bundle of rights) have their own arenas and institutional locations where these are deliberated and where specific 'battles' are fought, but this does not stop them from looking at the potential of Farmers' Rights, as stipulated in Article 9, to uphold their collective rights. This creates an impression that farmers and farmer groups are merely being rhetorical, contributing to further misunderstanding with negotiators, academic institutions and even CSOs who, in turn, are trying to concretize Farmers' Rights to be limited to seeds.

Farmer groups and some CSOs argue that while the Treaty recognized the rights of farmers to save, use, exchange and sell farm-saved seeds in its Article 9, the Treaty did not limit Farmers' Rights to this set of rights. The responsibility to recognize what constitutes Farmers' Rights is subject to decisions of national governments. Farmers and farmer groups recognize the challenges for a legal recognition of their identified entitlements as such articulation may be viewed as a direct challenge to the status quo, rather than necessary measures to ensure national and global food security. At the national level, there are class struggles and structural problems, which will colour the interpretation of Farmers' Rights. This is where the international and global community can play a role, by working to put forth the necessity of recognizing Farmers' Rights as a cornerstone of the country's food security and the security of the global food system. In a way, there are farmers and farmer organizations that see the utility of the Treaty and the spaces provided in the ongoing negotiations to assist them in ensuring legal entitlements to their collective rights.

Article 9 of the Treaty allows for a 'human rights based approach'. Farmers' Rights to plant genetics resources is a right that small farmers are to enjoy. However, in the real world and especially in less developed countries, rights are not handed down on a silver platter but are fought for and won. The moral and ethical high ground that underpins the rights of farmers to PGR is meaningless unless the structures and institutions that are responsible for providing this right are confronted. Small farmers who are usually among the poorer and weaker sectors of a country are keenly aware of this reality. For instance, farmers and farmer groups will continue to exercise their customary practices with or without legal recognition. For some, farming and seed saving have become an everyday form of resistance; for others it is simply their way of life; for most, it is the most practical way to survive and produce food for the family and for the community. Article 9 assists small farmers using this approach. Thus, the realization of Farmers' Rights, with the Treaty as a guide, needs to be a result of the assertion of small farmers to enjoy this right. This right should not be a gift patronizingly given to farmers by those who are rich and powerful. Gifts, even good ones, strengthen dependency and weaken the poor. Farmers' assertion of their rights will build confidence and critical learning. This will help address the 'behavioural' poverty of the poor that includes dependency and the lack of understanding of the structures that make them poor.

While there are farmers and farmer groups who have started expounding on Farmers' Rights, a large number of farmers and their organizations have yet to identify themselves with this 'social construct' (Kneen, 2009). No one can teach farmers about Farmers' Rights because it is imbedded in them and it is the role of governments and other stakeholders to ensure that farmers can continue with what they have been doing or strengthen their knowledge and skills for global public good. How these different views will play out at national and international negotiations remains to be seen, as the full potential of the Treaty as an instrument of benefit to farmers has yet to be felt and assessed by farmers themselves.

Recognizing the dynamic farmers' seed systems

Farmers exchange seeds, as genetic material, freely. Traditional agriculture depended on the constant exchange and movement of PGR to manage different biotic and abiotic stresses and to provide for the different needs of farming communities. These natural and farmers' selection pressures developed the plant genetic diversity that the world inherited today. This system of management of PGR becomes even more important as climate change is making the weather, pest and disease resurgence become unpredictable. Diverse, free and democratic management of PGR will allow greater options for climate adaptation. The farmers' system of PGR management will play an important role as they are at the frontline of changing rain pattern and stresses. The right of farmers to save, use, exchange and sell seeds is one of the most basic foundations of the farmers' system of PGR management. This is how PGR diversity is maintained and created.

It is clear that traditional agriculture has been altered or modernized as farmers react to market opportunities and as they changed from extensive to intensive agriculture. Consequently, farmers' varieties that fitted the traditional system of production were replaced by new cultivars bred for systems that are more intensive.

However, while more modern cultivars are often used, this did not stop the farmers' system from creating diversity, this time also using introduced cultivars as raw materials for their selection. Thus, new types of varieties or populations emerged, selected from modern cultivars, landraces and local varieties. For example, farmers in North Cotabato, the Philippines, developed 120 farmer rice varieties in 6 years, in contrast to the national release of only 55 inbred lines in 10 years from public research institutions. In the Mekong Delta of Vietnam, there are more than 100 farmer varieties covering more than 100,000 hectares of rice area. In the North and Central parts of Vietnam, farmers have developed more than 150 new farmer varieties. Due to traits that fit the market and intensive systems that most farmers now practice, their new rice varieties are also non-photosensitive, of short to medium duration, and are no longer tall. Furthermore, these new varieties carry adapted traits that fit the farming conditions of different macro and micro ecosystems. Saving, using, exchanging and selling seeds among themselves helped create these new cultivars. All traditional or introduced varieties constitute raw materials to be developed and adapted. If the rice varieties were protected with intellectual property rights that discouraged farmers exchanging and selling among themselves, these varieties would not have emerged. This evidence is the moral reason why farmers should be allowed to save, use, exchange and sell seeds among themselves. They already provided all their PGR to the world for free. The materials they continue to create are also free.

When the negotiations of the Treaty started (see Annex 1 of this volume for the list of all Commission and Treaty negotiating meetings), there were few evidence-based studies about the ability of farmers to develop new farmer varieties. In prime irrigated areas, modern varieties have replaced landraces, but not the ability of farmers to undertake crop breeding (Salazar et al, 2006). When a germplasm gets into farmers' hands, it enters into an endless process of experimentation – from adaptation to local conditions and farm practices, seed production and distribution (including marketing trials). Farmers are primarily concerned about their livelihood, the return on their inputs and hard labour in the form of sufficient (preferably with surplus) food supply and income. It is natural for farmers to test and innovate as part of risk management measures to ensure their livelihoods. Although there is a mention in Article 9 of the Treaty about the materials that farmers will continue to make, the focus remains on the PGR and not on this dynamic technology development process by farmers. Thus, it appears that the Treaty was negotiated based on the idea that farmers have active roles in conserving local and traditional landraces and varieties, but are not actually innovating. What is being emphasized is farmers' traditional knowledge over traditional resources. This disparity between farmers' practices and realities with that of the prevailing interpretations of the Treaty by negotiators, their advisers and even civil society groups and farmer groups may be an impediment to putting into operation the core components of the Treaty on conservation and sustainable use, Farmers' Rights and the multilateral system of access and benefit-sharing (see Annex 3 of this book for details on the main provisions of the Treaty). At the national level, this disparity in interpretation is apparent in existing seed policies. Those who were involved in drafting the existing seed laws in Bhutan, Laos and Vietnam, for example, admitted that, when they drafted and passed the laws, they were not aware of these realities. In the end, it is not just about protecting the germplasm materials, but the farmers' dynamic and collective system of technology development and diffusion through every season of research, experimentation, knowledge and skill sharing with other farmers and even with public and private entities.

Challenges ahead

The Treaty is not perfect even though it was negotiated out of the collective goodwill of the global community. It is this imperfection of the Treaty that allows for flexibility in the interpretation and negotiation at different levels but which constitute a constant source of frustration for farmers and their organizations. Most do not want to engage in the Treaty discussions and instead use their (time and human) resources towards more direct work in their fields or in their organizations. Farmers have yet to see concrete results out of the Treaty, as translated into national policies (e.g. seed rules) and programmes on PGR. While there are efforts by CSOs to facilitate opportunities for farmers to participate in the deliberations of the Treaty, it may not be enough to ensure farmers' participation and for them to articulate on their own, their views about the elements of the Treaty and their implications. There may need to be institutionalized mechanisms and processes to encourage farmers to participate. In addition, some concrete gesture has to be made. If there is a Global Crop Diversity Trust Fund, why not a Global Fund for Farmers and farmer groups to support their work on on-farm conservation and crop development? While the benefit-sharing fund is a step in this direction, it is too early to tell whether it is sufficient to support what farmers envision. The Governing Body can call on its members to develop mechanisms or compel seed banks to link with farmers, whereby any ex situ conservation efforts should be linked to on-farm conservation work. At the national level, we need to continue to provide space for farmers' discussions and deliberations, so that they can come out with their own take as to what the Treaty and its components mean to them. This can be supplemented with an information campaign to enrich the area of understanding among different stakeholders. Finally, the Governing Body can ensure farmers' right to participate in decision making related to plant genetic resources for food and agriculture, by supporting farmers and farmer communities to engage in the Governing Body process through their own platforms of consultations and processes as they review matters related to the Treaty and beyond. The support and recognition provided by the Convention on Biological Diversity (CBD) for CSOs and indigenous peoples can be a template, which the Governing Body can adopt to support farmers and their organizations. The CBD supports parallel processes and capacity building processes of CSOs and indigenous peoples, even developing platforms for joint publication and regular communication. The Governing Body and the Secretariat can start with a global farmers' conference on the Treaty, particularly on Farmers' Rights to pave the way for smallholder farmers to 'process' the Treaty within their own context.

Notes

1 The authors are not representative of farmer groups and lay no claim to speak on behalf of farmers. This article is borne out of the authors' insights from years of working directly with farmers and farmer groups in their different communities through community-based initiatives and policy advocacy work on conservation and the sustainable use of plant genetic resources for food and agriculture. What is written, are personal reflections on the issue.

2 The Community Biodiversity Development and Conservation (CBDC) Network was one of the first global networks to have put forth farmers as central actors in the conservation and sustainable use of plant genetic resources for food and agriculture in international discourses, by concretely showing the contribution of farmers to on-farm conservation and crop improvement (Southeast Asia), local and national seed systems (Africa) and farmers' role in ecosystem conservation and restoration, tapping on indigenous cosmology (Latin America). The network is composed of organizations in 21 countries.

References

CBDC Network (2009) *Farmers' Rights:Vision and Realization. Report of Farmers Consultation Process in Africa,Asia and Latin America*, Community Biodiversity Development and Conservation Network, Manila, Philippines
Kneen, B. (2009) *The Tyranny of Rights*, The Ram's Horn, Ottawa, Canada
Salazar, R., Louwaars, N. P. and Visser, B (2006) 'On protecting new varieties: New approaches to rights on collective innovations in plant genetic resources', Capri Working Paper Number 5, available at www.capri.cgiar.org/pdf/capriwp45.pdf

Chapter 14

Gene Bank Curators

Towards Implementation of the International Treaty on Plant Genetic Resources for Food and Agriculture by the Indian National Gene Bank

Shyam Kumar Sharma and Pratibha Brahmi

Introduction: PGRFA diversity in India

The Indian subcontinent is very rich in biological diversity, harbouring around 49,000 species of plants, including about 17,500 species of higher plants. The Indian gene centre holds a prominent position among the 12 mega-gene centres of the world. It is also one of the Vavilovian centres of origin and diversity of crop plants. Two out of the 25 global hotspots of biodiversity, namely the Indo-Burma and Western Ghats are located here. India possesses about 12 per cent of world flora with 5725 endemic species of higher plants belonging to about 141 endemic genera and over 47 families. About 166 species of crops including 25 major and minor crops have originated and/or developed diversity in this part of the world. Further, 320 species of wild relatives of crop plants are also known to occur here.

Presently, the Indian diversity is composed of rich genetic wealth of native as well as introduced types. India is a primary as well as a secondary centre of diversity for several crops, and also has rich regional diversity for several South/Southeast Asian crops such as rice, black gram, moth bean, pigeon pea, cucurbits (like smooth gourd, ridged gourd and pointed gourd), tree cotton, capsularis jute, jackfruit, banana, mango, *Syzygium cumini*/jamun, large cardamom, black pepper and several minor millets and medicinal plants like *Rauvolfia serpentina* and *Saussurea costus*. It is also a secondary centre of diversity for African crops like finger millet, pearl millet, sorghum, cowpea, cluster bean (transdomesticate), okra, sesame, niger and safflower; tropical American types such as maize, tomato, muskmelon/*Cucumis* species, pumpkin/*Cucurbita* species, chayote/chou-chou, chillies and *Amaranthus*; and it is a regional (Asiatic) diversity centre for crops like maize, barley, amaranth, buckwheat, proso millet, foxtail millet, mung bean/green gram, chickpea, cucumber, bitter gourd, bottle gourd, snake gourd and some members of the tribe *Brassicae.*

The major share of food comes from cultivated species such as rice, wheat, maize, sorghum, barley, sugarcane, sugar beet, potato, sweet potato, cassava, beans, groundnut, coconut and banana. Crops like chickpea, pigeon pea, pearl millet and other minor millets, cotton, sunflower, soybean, sugarcane, rapeseed-mustard, vegetable and horticultural crops have their regional importance (from the social and economic security view point) for the farming community. Besides, spices, condiments and beverages are obtained from cultivated and wild plant resources.

Major crops

Crop diversity is well represented as developed cultivars, landraces or as folk varieties in different phytogeographical regions of India among diverse crop(s)/crop-group(s). The western Himalayan region (including cold arid tracts) comprising Kinnaur, Lahul and Spiti and Pangi valleys, Ladakh and adjoining areas of Jammu and Kashmir and Uttarakhand hold rich diversity in wheat, maize, barley (hull-less types), proso millet, buckwheat, amaranth, chenopods, field peas, lentil, rice, French bean, *Cicer*, leafy Brassicae, pome, stone and nut fruits, medicago/clover, medicinal and aromatic plants. The most extensively cultivated grains in the country are rice, wheat and maize. In rice, both annual and perennial types occur particularly in the eastern and the central peninsular region including north-eastern plains. *Oryza nivara*, *O. perennis*, *O. officinalis*, *O. granulata*, *Porteresia coaractata* species and wild forms of *O. sativa* are fairly evenly distributed. Diversity in scented, deep water, cold and salt tolerant paddy types occur in various parts of the country. Considerable polymorphism is still found to exist in crops like wheat (*Triticum aestivum*, *T. dicoccum* and *T. durum*) and barley (*Hordeum vulgare*) in northern states in the Himalayan region. Maize has rich diversity in the peninsular tract, western Himalayas and north-eastern states. Fifteen distinct races and three sub-races of maize were recognized in India. *Chionachne*, *Polytoca*, *Trilobachne* and *Teosinte* also occur in this region. Millet crops have been dominant components of rain-fed agriculture on a regional basis in India. Millets are small grained, annual, warm-weather cereals of the grass family that includes 8000 species within 600 genera, of which 35 species comprising 20 genera have been domesticated. Millet used to be cultivated in an area of 35–37m/ha in India, reduced to 20–22m/ha during the past decade. The word millet was used to connote the following eight crops: great millet (*Sorghum bicolor*), pearl millet (*Pennisetum typhoides*), finger millet (*Eleusine coracana*), foxtail millet (*Setaria italica*), proso millet (*Panicum miliaceum*), little millet (*Panicum miliare*), barnyard millet (*Echinochloa colona*) and kodo millet (*Paspalum scrobiculatum*). Their adaptation to harsher environments and diverse cultural and agro-climatic situations is well known. The International

Crop Research Institute for Semi-Arid Tropics (ICRISAT) located in India, maintains 44,822 accessions of sorghum, 21,191 accessions of pearl millet and 3460 accessions of small millets.

The tribal-dominated areas of North-eastern region and the Eastern Himalaya, such as Mizoram, Meghalaya, Tripura, Manipur, Arunachal Pradesh, parts of Nagaland, north Bengal and Sikkim, are extremely rich in variability in rice, maize (including primitive popcorn), barley, wheat, buckwheat, *Chenopodium*, amaranth, soft shelled form of *Coix*, foxtail millet, finger millet, rice bean, winged bean, adzuki bean, sem, black gram, sword bean, soybean, peas, vegetables (cucurbits like *Cucurbita*, *Cucumis*, *Momordica*, *Cyclanthera*, *Luffa*, *Lagenaria*, *Benincasa*), fruits (*Citrus*, *Musa*, pineapple), oilseeds (*Brassica* spp., *Perilla*, niger, sesame), fibre crops (tree cotton, jute, mesta and kenaf), tuberous/rhizomatous types as taro/yam, and bamboos.

The eastern peninsular region, particularly the tribal belt of Orissa and Chhotanagpur plateau, holds rich crop diversity in rice, sorghum, finger millet, foxtail millet and proso millet, *Dolichos bean*, rice bean, chickpea, pigeon pea, horse gram/kulthi, brinjal, chillies, cucurbitaceous crops, mango, niger, sesame, linseed, Brassicae and castor. These areas hold tremendous variability in rice. Western arid/semi-arid region, including Rajasthan, Gujarat as well as Saurashtra region, possesses rich diversity like pearl millet, sorghum, wheat (drought and salinity tolerant types), guar, moth bean, cowpea, black gram, mung bean, Brassicae, sesame, chilli, cucurbitaceous vegetables, minor vegetables and fruits (*Capparis aphyla*, *C. deciduas* ber), *Citrus*, forage grasses/legumes and spice crops (coriander, fenugreek, ajwain, garlic).

The central tribal region covering Madhya Pradesh and adjoining tract of Maharashtra are rich in diversity of crops like wheat, rice, sorghum, small millet, grain legumes (chickpea, pigeon pea, black gram, green gram, cowpea), oilseeds (niger and sesame, Brassicae), chilli and cucurbitaceous vegetables. The western peninsular region including the Western Ghats has enormous diversity in tuber crops like *Dioscorea*, *Colocasia*, okra, eggplant, chilli and cucurbits, banana and rhizomatous types like *Curcuma*, ginger, spice crops (black pepper, cardamom, nutmeg), forage legumes and grasses, and areca nut.

Through the introduction of high-yielding varieties in major crops (rice, wheat, maize etc.), local landraces of many coarse grain cereals (particularly minor millets), are under cultivation only on a limited scale or have disappeared from their native habitats. Although rice diversity, at a local level, appears to have sustained owing to food preferences and social security of the farmers growing rice, diversity in major cereals/millet crops like wheat, pearl millet, sorghum is decreasing at the local level.

Conservation programme in India

The ex situ conservation approach requires systematic long-term conservation of viable propagules of collections outside the natural habitat of species. Realizing the

Table 14.1 *Details of various ex situ conservation sites for PGRFA in India*

Type of conservation	*Nodal Ministry/ Number of Department facilities*
Seed gene bank (long-term collections, −18°C)	Ministry of Agriculture, ICAR 1
Seed gene bank (medium-term collections, 4°C)	Ministry of Agriculture, ICAR 28
Seed gene bank (short-term collections at around 10°C)	Ministry of Agriculture, ICAR 13
Botanical gardens	Ministry of Environment, BSI 150*
In vitro conservation (4°C to 25°C)	Ministry of Agriculture, ICAR 5
Field gene bank	Ministry of Agriculture, ICAR, SAU 25
Cryopreservation [using liquid nitrogen in vapour phase (−170°C) or liquid phase (−196 °C)]	Ministry of Agriculture, ICAR 2

* National Report of MoEF for CBD (2005)

importance of collecting and conserving PGRFA, India has taken strategic steps for their ex situ conservation using appropriate approaches, especially in the last three decades. A majority of this work is carried out under the Indian Council of Agricultural Research (ICAR) by the National Bureau of Plant Genetic Resources (NBPGR), New Delhi, which is the nodal organization for ex situ management of PGRFA. Additionally, several economically important plant species are also conserved in botanic gardens of various plant science based institutes, most of which come under the jurisdiction of the Botanical Survey of India (BSI), Ministry of Environment and Forests (MoEF). The various types of components that constitute the ex situ conservation of PGRFA in India are listed in Table 14.1.

The NBPGR has been entrusted with the responsibility to plan, conduct, promote, coordinate and take the lead in activities concerning the collection, characterization, evaluation, conservation, exchange, documentation and sustainable management of diverse germplasm of crop plants and their wild relatives with a view to ensuring their availability for use over time to breeders and other researchers. The NBPGR, with its ten regional stations/base centres/quarantine centres over different phytogeographic zones of the country (Figure 14.1) has an active collaboration and linkages with over 57 National Active Germplasm Sites (NAGS), situated at different crop-based ICAR institutions and state agricultural universities (SAU) and various other crop improvement programmes. Through this network, NBPGR has been spearheading the national activities on PGRFA management. The base collection of germplasm is kept in long-term storage by NBPGR in its National Gene Bank, which is linked to numerous crop-specific active collections that are maintained at appropriate locations. The National Gene Bank of NBPGR has three types of storage facilities – seed gene bank, cryogene bank and in vitro gene bank. The seed gene bank was first established in 1986 and expanded in 1996 and presently has 12 long-term storage modules that are kept at −18°C. There are also six medium term modules maintained at 4–10°C. In addition to seed conservation, other ex situ conservation methods, such as in

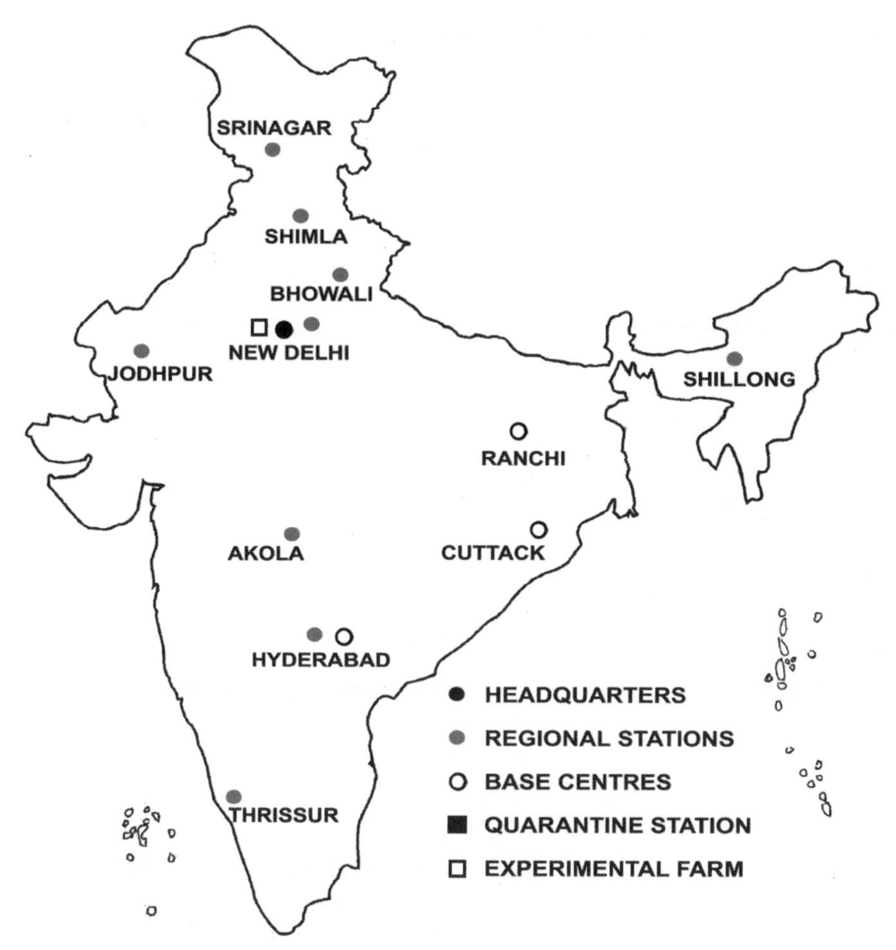

Figure 14.1 *Location of NBPGR headquarters and its ten regional stations in India*

vitro conservation and cryopreservation, have been employed to conserve species, predominantly having non-orthodox seeds (seeds which lose their viability when dried below critical moisture content and are sensitive to low temperature storage) and vegetative propagated species. The cryobank comprises six extra-large capacity (180 litre) cryotanks that store samples in the vapour phase of liquid nitrogen (from −160 to −180°C), and three smaller cryotanks (30–60 litre) where samples are held in the liquid phase (−196°C). The in vitro gene bank has four culture rooms at 25°C for maintenance of slow-growing cultures. The National Gene Bank of NBPGR has currently over 381,032 accessions of germplasm belonging to nearly 1969 species (Table 14.2).

Active germplasm collections are maintained at NBPGR regional stations and the NAGS situated at different crop-based ICAR institutions and SAU, which are held in modules maintained at 4–10°C. Eighteen medium-term storage modules (7 at NBPGR centres and 11 at NAGS) are used for the storage of active

Table 14.2 *Status of base collections in National Genebank of India (30 May 2010)*

Crop group	*Present status*	*Crop group*	*Present status*
Cereals		**Vegetables**	
Paddy	91199	Brinjal	4010
Wheat	39051	Chilli	2011
Maize	7656	Others	18141
Others	11236		
Millets and forages			**Fruits**
Sorghum	19912	Custard apple	59
Pearl millet	8137	Papaya	23
Minor millet	21252	Others	448
Others	5162		
Pseudo cereals		**Medicinal and aromatic plants**	
Amaranth	5450	Opium poppy	350
Buckwheat	858	Ocimum	402
Others	348	Tobacco	1467
		Others	4171
Grain legumes		**Spices and condiments**	
Chickpea	16867	Coriander	590
Pigeon pea	11148	Sowa	91
Mung bean	3672	Others	2150
Others	25414		
Oilseeds		**Agro-forestry**	
Groundnut	14346	Pongam oil tree	395
Brassica	10153	Others	2038
Safflower	7605		
Others	22890		
Fibre crops		**Duplicate safety samples of CG centres**	
Cotton	6423	Lentil	7712
Jute	2909	Pigeon pea	2523
Others	2173		
		Total	**381,032***

*The figure includes 3666 released varieties and 1869 genetic stocks

Note: Number of crop species conserved – 1580

collection of seed propagated crops. These centres also manage the field gene banks of clonally propagated crops. The directory of various NAGS, together with the germplasm accessions maintained, is presented in Table 14.3. In addition, there are ten more medium-term storage facilities maintained by other institutions belonging to different public and private organizations.

Table 14.3 *National Active Germplasm Sites for active collections of PGRFA in India (as in 2006)*

S. No.	Crop(s)	Institute/ AICRP*/ NRC	Seed bank	Field bank	In vitro/ Cryo
Agricultural crops					
1.	Amaranth, buckwheat, guava, mango, moth bean, rice bean	G.B. Pant University of Agriculture and Technology, Pantnagar, Centre for Plant Genetic Resources			
2.	Cotton	Central Institute for Cotton Research (CICR)	8879	0	0
3.	Cotton*	AICRP on Cotton, Coimbtore	–	–	–
4.	Crops of north-east region	ICAR, Research Complex, NEH Region, Shillong	867		
5.	Chickpea*	AICRP on Chickpea, IIPR, Kanpur	–		
6.	Fodder crops	Indian Grassland & Fodder Research Institute (IGFRI), Jhansi	6267		
7.	Forage crops*	AICRP on Forage Crops, IGFRI, Jhansi			
8.	Field crops of hills	Vivekanand Parvatiya Krishi Anusandhan Shala	–		
9.	Groundnut	NRC on Groundnut, Junagadh	8963	84	–
10.	Jute and allied fibres	Central Research Institute for Jute and Allied Fibres (CRIJ&AF), Barrackpore	3226	1427	
11.	Maize*	Directorate of Maize Research (DMR), New Delhi	2500		
12.	MULLaRP*	AICRP on MULLaRP, IIPR, Kanpur	–		
13.	Oilseeds (sunflower, safflower, castor)	Directorate of Oil Seeds Research (DOR), Hyderabad	10550	1329	–
14.	Pearl millet*	AICRP on Pearl Millet	3100		
15.	Pigeon pea*	AICRP on Pigeonpea, IIPR, Kanpur	–		
16.	Pulses	Indian Institute of Pulses Research (IIPR), Kanpur	6395		
17.	Rapeseed and mustard	NRC on Rape Seed and Mustard, Bharatpur	8082		
18.	Rice	Central Rice Research Institute (CRRI), Hyderabad	24000		
19.	Rice and Lathyrus	Indira Gandhi Krishi Vishwa vidhyalaya (IGKVV), Raipur	15000		
20.	Sesame and niger*	AICRP on Sesame and Niger, Jabalpur	–		
21.	Small millets	AIC Small Millets Improvement Project, Bangloru	13290		
22.	Sorghum	NRC on Sorghum, Hyderabad	7366		
23.	Soybean	NRC on Soybean, Indor	2500		

Table 14.3 *continued*

S. No.	*Crop(s)*	*Institute/ AICRP*/ NRC*	*Seed bank*	*Field bank*	*In vitro/ Cryo*
24.	Sugarcane	Sugarcane Breeding Institute (SBI), Coimbtore	5861		
25.	Sugarcane*	AICRP on Sugarcane, Lucknow	–		
26.	Underutilized crops	NBPGR Headquarters	199		
27.	Wheat and barley	Directorate of Wheat Research (DWR), Karnal	7000		
Horticultural Crops					
28.	Agroforestry cpp	NRC on Agroforestry	40		
29.	Arid fruits	Central Institute on Arid Horticulture, Bikaner	1319	1229	
30.	Banana, plantain	NRC on Banana, Tiruchirapalli		907	
31.	Cashew	NRC for Cashew, Puttur		519	
32.	Citrus species	NRC on Citrus, Nagpur		150	
33.	Floriculture*	AICRP on Floriculture, IARI, New Delhi			
34.	Grapes	NRC for Grapes, Pune		600	
35.	Leechi, bael, aonla and jackfruit	NRC on Leechi, Muzaffarpur		2426	
36.	Medicinal and aromatic plants	NRC on M & AP, Anand		190	
37.	Mango, Guvava, lichi	Central Institute for Subtropical Horticulture (CISH), Lucknow		848	
38.	Subtropical fruits*	AICRP on Subtropical Fruits, CISTH, Lucknow	–		
39.	Mulberry	Silkworm and Mulberry, Hosur		806	
40.	Oil palm	NRC on Oil Palm, Pedavegi, A.P.		103	
41.	Onion and garlic	NRC for Onion and Garlic, Pune		1066	
42.	Orchids	NRC for Orchids, Pakyang, Sikkim		1500	
43.	Ornamentals and non-traditional crops	National Botanical Research Institute, (NBRI), Lucknow			
44.	Plantation crops	Central Plantation Crops Research Institute(CPCRI), Kasargod		522	
45.	Potato	Central Potato Research Institute (CPRI), Shimla	457	22342	1471
46.	Spices	Indian Institute of Spices Research (IISR), Kozhikode		5695	
47.	Spices*	AICRP on Spice, Calicut		6055	
48.	Tea	Upasi Tea Research, Foundation (TRF), Vellaparai,		400	
49.	Temperate horticulture crops	Central Institute of Temperate Horticulture (CITH), Srinagar		780	

Table 14.3 *continued*

S. No.	*Crop(s)*	*Institute/ AICRP*/ NRC*	*Seed bank*	*Field bank*	*In vitro/ Cryo*
50.	Temperate horticulture crops	CITH, NBPGR RS, Shimla		908	
51.	Tobacco	Central Tobacco Research Institute (CTRI), Rajamundry		2359	
52.	Tropical fruits	Indian Institute of Horticulture Research (IIHR), Bangalore	1983	1754	
53.	Tropical fruits*	AICRP on Tropical Fruits, Bangalore	–		
54.	Tuber crops	Central Tuber Crops Research Institute (CTCRI), Thiruvanathapuram		3871	
55.	Tuber crops*	AICRP on Tuber Crops, CTCRI, Tiruvanathapuram	–	5432	
56.	Vegetables	Indian Institute of Vegetable Research (IIVR) Varanasi	16139		
57.	Vegetables	IARI, Regional Station, Katrain			

*AICRP: All India Coordinated Research Project to assist in evaluation of germplasm; NRC: National Research Centre

Numerous botanic gardens managed by the BSI and several other organizations help in ex situ conservation of economically important as well as endangered, threatened and rare plant species. The tradition of setting up botanic gardens in India dates back over 200 years when large spaces within major cities in India were set aside for the purpose. The Indian Botanic Garden at Calcutta was established in 1787. It now spreads over an area of 110 hectares and has around 15,000 plants belonging to 2500 species. Presently there are 150 organized botanic gardens or large parks in India, of which 33 gardens are managed by the government, 40 by universities and the rest are managed by state departments or civil society organizations (CSOs). The Government of India has also recently initiated establishment of a National Botanical Garden in NOIDA in Uttar Pradesh. In all, about 150,000 live plants belonging to nearly 4000 species (including 250 endemic species), are conserved in these botanic gardens.

For germplasm registration, there is a system operating at the NBPGR, New Delhi. This system is completely different from the registration of plant varieties of the Protection of Plant Varieties and Farmers' Rights Act (PPVFRA). This registration of germplasm is not a system of protection per se but a safeguard of material, developed by a breeder through publication and documentation in the public domain. This germplasm registration can be used as evidence in documentary or other forms to create and establish 'Prior Art'. Germplasm which can be registered at NBPGR could be any good performance material for specific and/ multiple traits (may not be yield superior), mutants or with a different plod level than the normal, with academic/scientific importance, parental lines of inbreds, promising experimental material or landraces and traditional varieties. The procedure and forms are available at NBPGR website (www.nbpgr.ernet.in). Efforts are being made to get more and more.

Information management

For information management, database development and its maintenance, upkeeping of the Local Area Network (LAN), computer hardware and software, statistical analysis of PGR experimental data and guidance to the researchers for the experimental designs, the Agricultural Research Information System (ARIS) cell was established at NBPGR in 1997. This cell also takes consultancy related to database management in the plant genetic resources. In addition, this cell also imparts computer training in relation to database management of genetic resources.

Many on-line databases related to plant genetic resources, and plant varieties have been developed and are in use by researchers in India. Two databases namely IINDUS (Indian Information System as per the DUS Guidelines) and NORV (Notified and Released Varieties of India) are in use at the Protection of Plant Varieties and Farmers' Rights Authority for the purpose of registration of extant and new varieties.

The website of NBPGR is hosted on its own web server at www.nbpgr.ernet.in and is updated/maintained by ARIS cell regularly. This website has information related to all the important activities of the NBPGR, on-line application for 'permit to import seed/planting material/Transgenics/GMOs (for research purpose)', 'Material Transfer Agreement (MTA)', 'Guidelines for Registration of Plant Germplasm', 'Guidelines for Documentation and Conservation of Folk Varieties', 'Guidelines for Submission of Seeds/Propagules with National Genebank', 'Approved Fee Structure for Import of Germplasm Material', 'Guidelines for Filing Application of Plant Varieties for Registration under PPVFRA, 2001' and 'Format of Passport Data Sheet for Allotment of IC No'. In addition, all the announcements for the conferences, training programmes and meetings are regularly updated in the website.

Regulatory and protection mechanisms for access to plant genetic resources

Many international developments during the last two decades have directly or indirectly affected the genetic resource management programmes (see Annex 1 of this volume for the list of all Commission and Treaty negotiating meetings). Plant breeders have traditionally relied on open and free access to PGR for developing new, high-yielding crop varieties. With the adoption of the Convention on Biological Diversity (CBD), which advocates national sovereignty over the biological resources, the authority for access to genetic resources rests with the national governments and this access is subject to prior informed consent of the providing country on mutually agreed terms. This led to enactment of the Biological Diversity Act (BDA) for India which governs access to all genetic resources of India and encompasses provisions for equitable benefit sharing. The International Treaty on Plant Genetic Resources for Food and Agricultural (ITPGRFA) is another legally binding treaty which has provisions for facilitated access to 64 crops and forage species, under a multilateral system of access and benefit-sharing (MLS) (see Annex 3 of this book for details on the main provisions of the Treaty). This exchange is under the conditions of a standard material transfer agreement (SMTA). NBPGR, being a single-window system for the exchange of small samples of plant germplasm meant for research, has developed a suitable MTA for providing access to PGR both within and outside the country. After operationalization of the Treaty in India and harmonization of the provision of the BDA with the obligation of the Treaty, the exchange of the PGRFA would be operated through NBPGR. There are some issues as listed further in the chapter, which need to be looked into before such an arrangement is expected to be put in practice.

In addition, under the GATT/ WTO/ TRIPs regimes, restrictions have been imposed on free trade in commodities, including the agricultural products. Countries are required to adopt patenting or enact effective *sui generis* system or a combination of both, for the protection of plant breeders' rights. As a national obligation for the TRIPS Agreement of WTO, the new legislations namely, the Protection of Plant Varieties and Farmers Right Act (PPVFRA) 2001, and Geographical Indications of Goods (Registration and Protection) Act 1999, were enacted and suitable amendments made in other existing intellectual property rights (IPR) legislations, which have a bearing on the product, processes and technologies developed. The Indian plant variety protection is unique in providing equal rights to the farmers as breeder and conserver of genetic resources of local importance. To facilitate this activity with identification of distinctiveness of newly developed varieties, ICAR has provided the requisite support to PPVFR by developing guidelines for distinctiveness, uniformity and stability (DUS). To date, DUS guidelines for 35 crops have been developed and notified. Plant variety protection, under the PPVFRA, currently covers 17 crops; others are to be notified soon. NBPGR facilitates the submission of applications for plant variety registration under the PPVFRA. Over 700 applications of ICAR/SAUs have been submitted through NBPGR.

Implementation of the Treaty in India

During the long negotiation phase of the Treaty (see Annex 1 of this volume for the list of all Commission and Treaty negotiating meetings) and later as a contracting party to the Governing Body (see Annex 2 of this volume for the list of contracting parties per FAO regional groups), representatives of the Indian Ministry of Agriculture, mainly Joint Secretary (Seeds), Department of Agriculture and Cooperation (DAC) and India Council of Agriculture Research (ICAR) actively participated in the meetings. Their interaction with the Middle East and South Asian countries as part of the group and at individual level with the GRULAC (Latin American) group helped India to understand and consolidate their views and stands taken during these negotiations. This interaction was focused specially on the list of crops to be included in Annex I, the conditions of the SMTA, the role of FAO as the third party beneficiary and on the funding strategy of the Treaty. India has now designated Joint Secretary (Seeds) DAC, Ministry of Agriculture, as the nodal point for implementation of the Treaty. Regarding the obligations of the Treaty, there are various issues which still need to be worked out before effective implementation of the Treaty in the country. Some of these are discussed here.

National implementation of the Treaty

In India, access to genetic resources to outsiders is governed by the provisions of the BDA 2002, which was the outcome of the implantation of CBD in the country. It is the umbrella legislation to govern access to India's genetic resources including the PGRFA. The Treaty also in its Article 12.3(a) provides for access to PGRFA, subject to national laws. Therefore, there is a need to harmonize the provision of the BDA 2002 with the provision of the access to PGRFA under the multilateral system of the Treaty recognizing the legally binding nature of the Treaty. A notification to this effect for exchange of PGRFA covered under Annex I of the Treaty and use of SMTA for such exchange needs to be brought out by the National Biodiversity Authority which is the apex body in India for implementation of BDA 2002.

Implementation authority

DAC is the focal point in India; the actual custodian of genetic resources for food and agriculture is the ICAR, working under the Department of Agriculture Research and Education (DARE). For effective implementation of the national obligation of the Treaty, greater intervention of ICAR/DARE with DAC is expected, since the PGRFA for exchange needs to be routed through NBPGR for two reasons. First, all germplasm access is being conserved and regenerated through NBPGR. Second, NBPGR is envisaged as the single window for export/import of PGRFA in the country for research purposes. It also has the authority delegated through DAC for quarantine certification of material under exchange.

Access to PGRFA

It is not clear, in India, (i) whether the material has to be accessed through a single window system or (ii) whether it has to be accessed directly from the concerned party. Similarly, when material is being accessed by individuals from IARCs, which is a part of the MLS of the Treaty, the information is not collated at one place in the country. The mechanism is not helping the focal points to have record of the material coming into the country from the MLS of the Treaty.

Designation of PGRFA for the Treaty

As enumerated above, NBPGR is mandated to conserve PGR and manage PGRFA. Most requests for PGRFA from within India and from abroad are sent to the NBPGR which (i) caters to such request by arranging the material either from its stations or from NAGS, (ii) clears the material from the quarantine angle and (iii) dispatches the material to requesters. All exchanges are recorded and always under an MTA. There are, though, difficulties experienced at each step of such supply:

- lack of enough multiplied seed for each and every accession requested for;
- passport data of all the material in the gene bank is not available;
- recorded ownership of the material especially where no passport information is available.

All these problems have delayed the discussion on the identification of the material covered by Annex I of the Treaty. Secondly, the gene bank also holds material received from other countries being a nodal institute mandated to exchange plant genetic resources for research purpose in India. Another apprehension before designation is material of Indian origin available in the IARC. The SINGER data base shows that about 10 per cent of all material available at IARCs is of Indian origin. Such material has already become a part of the MLS through the agreements between the IARCs and the Governing Body of the Treaty. These materials are being supplied by IARCs on a regular basis through SMTA. This should be recognized by the Treaty as a meaningful and substantial contribution of countries like India towards designation of their material.

Conclusion

The exchange of PGRFA is crucial for crop improvement programmes and ultimately the food and nutritional security of the world. India has contributed its share of genetic resources to the world through various national and international exchange programmes and would continue in the future also under the new ITPGRFA regime. NBPGR, being the nodal organization in India entrusted with exchange of PGRFA, would be working towards the operationalization of the Treaty under the national exchange guidelines. However, the procedure is taking its own course. A regulation for implementation of the Treaty at the national level is under way and soon the Treaty will be fully operational in India.

Bibliography

Ambasta, S. P., Ramachandran, K., Kashyapa, K. and Ramesh, C. (eds) (1986) *The Useful Plants of India*, Publications and Information Directorate, Council of Scientific and Industrial Research, New Delhi, India

Arora, R. K. (1991) 'Plant diversity in Indian gene centre', in R. S. Paroda and R. K. Arora (eds) *Plant Genetic Resources Conservation and Management Concept and Approaches*, IBPGR Regional Office for South and South East Asia, New Delhi pp25–54

Arora, R. K. and Nayar, E.R. (1984) 'Wild relatives of crop plants in India', *Sci. Monograph No. 8*, National Bureau of Plant Genetic Resources, New Delhi, India, p90

Arora, R. K. and Pandey, A. (1996) *Wild Edible Plants of India: Diversity, Conservation and Use*, National Bureau of Plant Genetic Resources, New Delhi, India

Bisht, I. S., Rao, K. S., Bhandari, D. C, Nautiyal, S. and Maikhuri, R. K. (2006) 'A suitable site for in situ (onfarm) management of plant diversity in traditional agro-ecosystems of western Himalaya in Uttaranchal state: A case study', *Genetic Resources and Plant Evolution*, vol 53, pp1333–1356

Das, T. and Das, A. K. (2005) 'Inventorying plant biodiversity in home gardens: A case study in Barak Valley, Assam, North East India', *Current Science*, vol 89, no 1, pp155–163

Dhar, U., Manjkhola, S., Joshi, M., Bhatt, A., Bisht, A. K. and Meena, J. (2002) 'Current status and future strategies for development of medicinal plant sector in Uttaranchal, India', *Current Science*, vol 83, no 8, pp956–964

Gadgill, M., Singh, S. N., Nagendra, H. and Subhash Chandran, M. D. (1996) *In situ Conservation of Wild Relatives of Cultivated Plants: Guiding Principles and a Case Study*, Food and Agricultural Organization of the United Nations and Indian Institute of Science, Bangalore, India

Ganga Prasad Rao, N. (2006) 'The rise and decline of millets in Indian agriculture and an outlook on future research and development', in *Strategies for Millets Development and Utilisation*, Society for Millets Research, NRC for Sorghum, Hyderabad pp11–31

NBPGR (2007) 'State of plant genetic resources for food and agriculture (1996–2006)', a Country Report, National Bureau of Plant Genetic Resources, New Delhi-110012, India Protection of Plant Varieties and Farmers' Rights Act (2001) available at www.plantauthority.gov.in

Chapter 15

Plant Breeders

The Point of View of a Plant Breeder on the International Treaty on Plant Genetic Resources for Food and Agriculture

José I. Cubero

Introduction to plant breeders

Plant genetic resources (PGR) are the most important tool for plant breeders. Access to these resources was free since the beginning of agriculture. Only in the last 50 years, has the value of the genes, hence of the living organism carrying them, increased in astronomic proportions. This value is translated not only in their monetary price but more importantly also in their strategic and political value as they are the only way to reach food security in the future. Having always been important, food security has reached an even higher level of significance as food insecurity has acquired the unfortunate character of *endemic* at a global scale (Sasson, 2009).

Plant breeding is as old as agriculture itself. In fact, the first farmers also were the first breeders: they sowed what they spared for sowing the previous year – that is, what they selected. The only conscious method of crop improvement was what is nowadays called 'bulk selection', consisting in choosing the seeds of the best individuals, or even the best seeds in the whole harvest, and mixing them to form the sowing bulk for the next season; obviously, there were spontaneous crosses among plants of different plots, but these crosses were done by Mother Nature, not by a careful planning by the breeder-farmer. Hand-made crosses with the purpose of increasing the variation found in the varieties used by farmers was not possible until the sex in plants was scientifically demonstrated at the end of the 17th century by Camerarius (*De sexu plantarum epistola*, 1694). Other methods were added in the 20th century such as polyploidy, artificial mutation and, recently, genetic engineering (Cubero, 2003). The common practice by plant breeders consisted in applying the chosen method to a suitable variety obtained either by him or by any other breeder. It was not a written statement but a universal practice taken for granted as it can be seen in the first textbook on plant breeding (Bailey, 1895) and all the classical ones in the 20th century (for example, Davenport, 1907; Poehlman, 1959; Sánchez-Monge, 1955, 1974; Allard, 1960, 1999; Simmonds, 1979; Jensen, 1988; Hayward et al, 1993). Recent textbooks on plant breeding as those of Cubero (2003) and Acquaah (2009), already include the subject as an important topic for plant breeders.

Germplasm collections were freely exchanged and national organizations were happy to provide subsets of their collections under request. Only when the problem emerged of applying intellectual property rights (IPR) to the work performed by plant breeders, was the question of the indiscriminate use of varieties such as a source of genes for own work put on the table and more concise terms were sought to define the practice. Thus, the traditional practice followed by plant breeders had to be modified to accommodate the IPRs to their productions; the concept of *breeder's exemption* or *scientific option* was probably coined during the first meetings held on that and related topics (see below) under the Food and Agricultural Organization's (FAO) umbrella in the 1960s. This broad concept was later on defined in Article 15iii of the 1991 UPOV Convention[1] and incorporated in national laws; for example, Spanish Law 3/2000 governing the Protection of Plant Varieties, states in its Article 15 that the varieties protected in Spain may be used as initial source of variation to breed new varieties without requiring the breeder's authorization or generating rights for the owners of the protected varieties used.

Since the beginning of agriculture, farmers were accustomed to reserving a portion of the harvest as seed for the next season. Now, when varieties were produced by professional plant breeders (roughly speaking since the end of the 18th century (Cubero, 2003)) and registered, the varieties obtained reached the farmers through seed companies or official agencies, but farmers usually continued with their old practice. At the same time as it was necessary to refine the concept of *breeder's exemption* (see above), the ancestral practice of farmers had to be discussed as it confronted the implementation of breeders' property rights. The very important concept of *farmer's privilege* had precedents in consuetudinary practices in some countries as, for example, the *farm saved seed* in the US, the *semence de ferme* in France or the *landwirte vorbehalt* in Germany (the latter translates exactly as 'farmer's privilege'); it was probably introduced in the meetings in the 1960s as the counterpart of the *breeder's exemption*; it was finally accepted in the legal texts following the 1991 UPOV Act (Article 15.2) (Sánchez, 2009). The *farmer's privilege* meant that farmers can be exempt from paying royalties due to the producer of the variety, provided the farmer kept the seeds for his own use, never to be multiplied and sold in the market. The *farmer's privilege* has produced a considerable amount of literature in many fields, agronomical as well as juridical (Elena, 2007; López de Haro, 2007; Mateos, 2009).

It is not advisable to use the farmer's privilege for a long time because of the *varietal degeneration*; the best practice is always to resort (if not annually, at least periodically) to *reliable* private or official seed producers, even in the case of the most favourable materials for the farmer (self-pollinating or vegetative reproduced varieties), but in spite of that technical difficulty, farmers still save seeds for their own use.

The conflict between the traditional farmer's practice and the plant breeders' rights as a *sui generis* system of IPRs was evident and has become even more critical in recent times: valuable cultivars possessing important characteristics were released under contract and royalties were demanded by private seed companies to developing countries which revolted as many of these genes were identified in landraces or wild forms found in their territories. In many cases, developing countries prohibited germplasm recollections without special permission and under agreement of sharing the material collected. In some cases, the germplasm collections were placed under the authority of the defence ministries.

Besides, many abuses were committed under the *breeder's exemption*: to transfer a character by backcrossing is usually easy, especially between modern cultivars; the use of wild or primitive forms is much more complicated because the useful gene is generally linked to undesirable ones and to 'clean' the former requires, in the best cases, many years of painful backcrosses and selection. The temptation to transfer a useless but easily identifiable gene to an outstanding cultivar was very high. Only after a few backcrosses, a variety possessing the whole valuable genotype plus an insignificant new gene would be able to be registered in commercial lists as being *distinctive, uniform and stable* (DUS in the breeders jargon); but the true value was that of the original genotype. As a solution to these abuses, the concept of *essentially derived varieties*, to separate what was an important breeding contribution from unimportant derivatives, was launched in the UPOV Act of 1991. The main idea is to preserve the breeder's exemption but maintaining the rights of the first breeder, whose permission would be necessary to market the derived variety (CIPR, 2002). This concept is easier to understand than to put in practice, because in the 1991 Act this concept is not very well defined (for example, would a transgenic variety with a simple but very valuable transferred gene be an essentially derived variety?). This lack of precision and the fact that many countries have not yet signed the 1991 Act are causing many legal difficulties for its application.

Other serious concerns arose from IPRs concerning vegetal materials (CIPR, 2002). The Agreement of Trade-Related aspects of Intellectual Property Rights (TRIPS) established that a *sui generis* system could be applied as property rights for plant varieties, but left somewhat undefined the *sui generis* concept and, in fact, there are many possibilities to apply it. Europe favours *protection* as defined by the UPOV Convention revised in 1991, and several other countries led by the USA mainly use the *patent* system. The USA has indeed a long tradition concerning vegetative reproduced varieties (the seminal Plant Patent Act dates from 1930, later amended and modified; the USA Supreme Court has also decided on plant variety rights in favour of patents in recent times). In fact, the American concept of *patent* applied to vegetal products is not much different from the European *protection* concept for the same purpose.

Any system designed to protect breeders' IPRs over their varieties could be used provided they have the same legal enforcement. As it has just been mentioned, *sui generis* systems can be devised for that purpose (UPOV *protection* is, in fact, one of them) and there are many possibilities between the *protection* as defined by UPOV and *patent*. The important objective is to acknowledge the work of the breeders in producing new plant varieties and make available to them the same IPRs due to other innovators. Worth mentioning, *patents* are much broader than the UPOV *protection* and much better known for historical reasons by lawyers and judges, a fact that runs in their favour for the future. Besides, a fact adding difficulties to the problem is that several biotechnological innovations are being considered under the umbrella of industrial patent, such as, for example, genes modified in the laboratory by genetic engineering techniques fall under the strict concept of Industrial Property patent (CIPR, 2002).

The problem is more complex because although *cultivars* cannot be patented at present in the EU, genes or genetic constructions artificially produced or modified in the laboratory (*transgenes*) can be. The consequence is that a transgenic cultivar, that is, a cultivar whose genotype has been modified by genetic engineering, enjoys a peculiar situation: it is *protected* but the transgene it contains is *patented* in the current legal use of the concept, the result being a *hidden* or *virtual* patent running against many countries and therefore against most plant breeders. In fact, the UPOV Act of 1991 does not exclude the dual possibility of protection and patent for vegetal materials, at the same time allowing for restrictions to the traditional practice of both the *breeder's rights* and the *farmer's privilege*. The door is opened for patenting plant varieties as the limits between traditional and modern breeding techniques are more dubious every day. The *terra nullius* in this field, as in any other, is clearly to the advantage of the people first occupying it, as many court cases demonstrate in recent times.

Positions regarding the Treaty's negotiation and implementation

Genetic resources that were of free use some 50 years ago were fully controlled by the end of the last century. Restrictions were imposed by countries, private as well as public organizations, and by breeders themselves. Traditional rights, like the farmer´s privilege and the breeder's rights were or are in the way of being suppressed. It is a revolution in classical agricultural practices, a revolution concerning genetic resources and, especially, their control. Some international action seemed necessary as there were conflicts at all levels: political, geographical, economic and scientific.

In a certain sense, genetic resources go beyond strict plant breeding projects. They can be used, for example, in the recovering of degraded areas, but even in this case a breeding effort can obtain better results by improving adequate materials. The same can be said of industrial applications: chemical compounds can be obtained directly from wild plants, but the best results are always obtained through domesticated forms of any organisms: genetic resources collections are still a must. Within the industrial uses should be included the new uses of old crops for agro-fuels that is affecting the food security itself (Sasson, 2008). Old crops can be re-domesticated (for example, a forage crop as a seed crop) but there is still an unexplored wealth of genetic resources, both in collections and in the wild, not conflicting with food production: many non-food plants could be used to find new sources of agro-fuels without using staple crops such as wheat and maize for that purpose. No doubt, biotechnology can play a relevant role in this matter (Ruane and Sonnino, 2006), and not only in developed countries but in developing ones (Sasson, 1993, 1998, 2000).

Germplasm collections have acquired a high economic value, obviously related to their strategic importance – a fact bothering the breeders' work, constrained to use what is commercially available to them. Germplasm collections have to be used if they are to be conserved; they cannot be 'stamp collections'. As in many other cases, what is not used easily disappears. International restrictions on the use of germplasm collections, both in situ and ex situ, can lead to their erosion or even their loss in a short period, as their conservation in good shape is expensive and politicians are reluctant to spend even a small budget on something that will not allow them to show up every day in the news.

Free movement of germplasm would have undesirable consequences such as the introduction of pests in new environments and the erosion of local landraces and wild forms not only because of the spreading of modern cultivars but also as a result of careless collectors. It is also worth mentioning the inadequate facilities for germplasm conservation in countries that face great difficulties in sharing the collections because of the fear of someone getting valuable genes without any benefit sharing. The experience accumulated on other related challenging threats like pest introduction, land races erosion and biopiracy is also very wide. Nowadays these threats are still recurrent in spite of all the scientific and historical knowledge accumulated on the various topics and in many cases we have not been able to prevent them or minimize their consequences. This is due in part to a lack of social knowledge of the problem and also to a lack of interest in stimulating the social awareness of it.

Norms for germplasm collectors were set up to stress the need for them to respect the environment and the local traditions (for some incredible examples of collectors' misconduct see Fisher, 1989), never eroding the local plant populations, emphasizing the right of the prospected countries over their genetic resources and leaving a duplicate of the collected material in the host country. It was an ethical and not compulsory code, and in recent times bilateral agreements between developed and developing countries facilitated the collecting tasks. But it was felt by the international scientific community that this was not enough. Most breeders did accept these rules as they were always respectful of others rights. The problem was the greed of a few and the fear of many to suffer the consequences of the behaviour of the former.

Previous steps to solve the problems

The problems just outlined above concern a wide range of matters, not only those relevant to plant breeders. But all of them affect the production, release and spread of new varieties and the improvement of farming around the world.

FAO accepted the quixotic task of trying to solve these problems. Several international meetings and conferences have been held since 1965. It is impossible in the space of the present chapter to give an account of the many difficulties in order to reach an agreement valuable for all the interested parties. The matters under discussion went from idealism to pragmatism, and in spite of the great achievements, especially those established in the International Treaty on Plant Genetic Resources for Food and Agriculture (hereafter, the Treaty; see Annex 3 of this book for details on the main provisions of the Treaty), there is still a rough way ahead. Problems concerning plant breeders' versus farmers' rights have produced numerous papers, books and meetings (a recent one covering both matters in spite of its title can be seen in Anonymous, 2007) and a clear and complete review of the history leading to the Treaty is given by Esquinas-Alcázar (2005) (see also the introduction to this book and Annex 1 of this volume for the list of all Commission and Treaty negotiating meetings). Fortunately, from the 1950s until now, the positions seem to have moved from pure idealism (for example, 'natural resources are common heritage of mankind') to real pragmatism ('natural resources belong to the country where they are found'). The loss of ethical value is compensated by the necessity of agreements among countries, private and public agencies and, generally speaking, among all stakeholders. The balance has to be positive.

One of the problems in the numerous conferences held in the last 50 years is the fact that genetic resources for food and agriculture were frequently covered (many times even hidden) by the more general concept of 'natural resources' or, even better, 'biodiversity'. One of the great achievements was to separate both concepts, but this did not happen until rather recently.

The main dates, at least concerning plant breeding, can be outlined as follows:

In 1965 the FAO started the technical work on Plant Genetic Resources for Food and Agriculture (PGRFA) collection and conservation, and triggered a series of international technical conferences on the topic. Although 'for Food and Agriculture' was always present in the meetings, 'Plant Genetic Resources', as mentioned before, were frequently included in conferences on more general topics on environment, as, for example, the United Nations Conference on the Human Environment held in Stockholm in 1972.

For plant breeders, an important step was the creation in 1974 of the International Board for Plant Genetic Resources (IBPGR, later renamed the International Plant Genetic Resources Institute – IPGRI – and today Bioversity International), belonging now to the Consultative Group of International Agricultural Research (CGIAR) with the mandate of coordinating collection and conservation efforts. Very important from a legal point of view was the establishment of the International Union for the Protection of New Varieties of Plants (UPOV, see above) to defend breeders' rights; the last revision, as already mentioned, was that of 1991 although it has not been signed by several countries. By then, many countries had already restricted the access to their own genetic resources, and wide discussions between developed and developing countries were on the table. Developed countries favoured IPRs while developing ones tried to focus the discussions on the recognition of Farmers' Rights. The Commission on Genetic Resources for Food and Agriculture (CGRFA; now it includes all components of biodiversity for food and agriculture, including farm animals, forestry and fisheries) was set up in 1983 within FAO and the International Undertaking (IU) on Plant Genetic Resources was adopted, although it was non-binding (see Annex 1 of this book).

In fact, the first binding agreement on biological diversity (in general) was adopted at the Rio Conference in 1992 and is known as the Convention on Biological Diversity (CBD); agricultural biodiversity was only related to a set of subjects discussed in the Convention, but under the scope of FAO and its offshoot, the Commission on Genetic Resources for Food and Agriculture. A revision process of the IU led to the adoption of a new binding international instrument in 2001: the International Treaty for Plant Genetic Resources for Food and Agriculture (ITPGRFA). The IU is still applied in those countries that have not signed the Treaty yet (for a history of the revision of the IU, see the Introduction of this volume and Chapter 10).

Although its objective was the conservation and sustainable use of biological diversity and the equitable sharing of benefits arising out of their use (Article 1), thus not specifically referred to food and agriculture, the importance of the CBD on the topic was clear. The statements declaring states' sovereign rights over their own biological resources and those on the responsibility of humankind over the biological diversity are since then well established principles. The adequate transfer of technology was also firmly established and 'biotechnology' was defined in Article 2 as 'any technological application that uses biological systems, living organisms, or derivatives thereof, to make or modify products or processes for specific uses'. A financial mechanism (Article 21) was created by developed countries to support developing ones, but rather on philanthropic terms as it was not compulsory.

Meanwhile, between 1993 and 1996 the CGRFA developed the Leipzig Global Plan of Action on plant genetic resources and the first report on the state of the world's PGRFA. The FAO conference at Leipzig recognized the role of farmers since the very old times, hence including the indigenous and local communities, *as well as that of plant breeders*. Equally important was the Global Plan of Action for the conservation and sustainable use of genetic resources for food and agriculture adopted in 1996 in Leipzig; all countries are interdependent concerning these resources. The Leipzig conference also established important actions for breeders such as in situ and ex situ (i.e., the germplasm collections) conservation and the importance of the recovery of infra-utilized species.

Unfortunately, all these advances, as well as the financial mechanism established for genetic resources for food and agriculture were accepted concepts without any mechanism to implement them. Financial procedures were not established and, more important perhaps for plant breeders, there was a very light treatment of biotechnology at a moment when (the Leipzig conference was held in 1996) biotechnological achievements via genetic engineering were being introduced in the market; the first transgenic cultivars were in the farm and some medicines, like the 'transgenic' insulin, were already in the pharmacies. For plant breeders, statements on the essential importance of genetic resources as a base for reaching food security sounded logical. They were surprised that an International Conference to establish that obvious principle was necessary. To promote a just and equitable distribution of benefits was out of their scope and possibilities.

The International Treaty on Plant Genetic Resources for Food and Agriculture: Challenges ahead

After many years of wide conferences and consultancies at all levels, the Treaty was approved by the international community in 2001 (in force since June 2004). It was an International Treaty, hence a legal compulsory instrument in order to ensure conservation and sustainable use of genetic resources for food and agriculture, as well as the equitable sharing of benefits for all signatory countries. An essential difference with all the previous agreements was the multilateral way of access to and benefit-sharing arising out of the plant genetic resources and the establishment of a financial mechanism and a governing body to support the implementation of the agreement. Worth to be repeated, it was compulsory for all the signatories.

The Treaty has some important points from the plant breeders' point of view:

1. A genetic resource for food and agriculture is considered 'any genetic material of plant origin of actual or potential value for food and agriculture', 'genetic material' being 'any material of plant origin ... containing functional units of hereditary'. The definitions are very wide in scope as they include all the plant kingdom. This is a scientifically sound interpretation as, following the success of genetically modified organisms in agriculture, the fourth genetic pool under the Harlan and de Wet system (Harlan, 1992) contains all the living beings.
2. But recipients of genetic resources will not claim for any intellectual property right limiting the access of these genetic resources *or their genetic parts or components* in the form received by the multilateral system (Article 12.3.d). This statement is confusing and, indeed, provoked a lot of discussion and contradictory explanations between developed and developing countries.
3. The Treaty includes a list of plants included in the multilateral system that is far from complete and acceptable by breeders. Some important crops are lacking and the list, as a whole, seems to be set up more politically than technically or scientifically.

Point 2 is especially important for plant breeders. The different points of view expressed between developed and developing countries at the moment of the approval of the Treaty did not leave great room for hope. Breeders from developed countries (especially those working in private companies) consider the material received from developing countries as not included in the Treaty if it has been modified especially by biotechnological methods. The donors of landraces and wild forms, generally persons or institutions working in developing countries, think differently: they argue that they are the real owners of these valuable genotypes, the operated transformation, even by biotechnological means, being, according to them, of minor importance; thus, they defend that the plant materials they send to developed countries have to be included in the Treaty even if they are later on modified by genetic engineering. A lack of agreement in this sense would likely affect plant breeding at a global scale.

Point 3 is also very important. The feeling that it was a political issue is not helping in international relations among plant breeders. It seems as if participants in the Treaty negotiations were more concerned about restricting access to their own genetic resources than in granting access to the global gene pool that so far had been the main factor of agricultural development. This point is very important as many non-classical uses (bio-alcohol and biodiesel among many other industrial applications) will require the study and use of all kind of plant resources. Of course, the Treaty did not close all possible negotiations in 2001; the list of plants contained in Annex I can be modified in the future.

Plant breeders experience many constraints in their daily work: asking for permits to import and export his/her own productions and/or national germplasm that could benefit other colleagues in different regions of the world. This is not the traditional behaviour of plant breeders. In recent times, I was able to observe the exchange between breeders belonging to two countries at war at that very moment. It would be paradoxical for the same breeders not to be able to exchange the same materials in peacetime.

Interactions among plant breeding agencies

Farming and plant breeding came about by the same human act of sowing some wild seeds during a certain period of time to solve a problem of food scarcity. The first farmers also were the first breeders: they sowed the seeds that they spare (i.e., select) from the previous year's harvest, and repeated the practice over several years; we now call the method *automatic selection* to differentiate it from the *intuitive* (but conscious) one performed by already authentic farmers much later. But in operating in this way, they selected only a subset of the genes present in the previous generation.

Hence, the domesticated form (the cultigen) only had a minimum set of the whole amount of genes present in the wild stock. But we do not know about the nature and possibilities of those genes that were not chosen or that later on were discarded when farmers were conscious about the possibilities of their crops. Breeders need that material for their own work. The Treaty, in this sense, while offering many more possibilities than any other agreement on natural resources made up to now, is setting some limits to the free accession to these sources of genes.

The work accomplished by plant breeders in the last two centuries has produced varieties of high value, varieties characterized by genotypes not present in Nature. This work has to be recognized, but it would not be a wise practice to hinder its use by other professionals. The advance reached so far would stop. This limitation in the exchange of plant materials can bring negative consequences, especially since genetic engineering is producing new forms by integrating alien genes in plant genotypes. These new breeding forms increase the pool of genetic resources and adding new administrative and cumbersome tasks for their transfer can constitute an additional barrier for further developments.

Conclusion

For millennia, farmers were also plant breeders as they selected their seeds for the next sowing season themselves. Generally speaking, the creation of the first seed producer companies (Vilmorin being the very first one early in the 18th century) separated both professions in different persons: the farmer and the plant breeder at the service of his employer. In the countries that adopted both the industrial and the agricultural revolutions – that is, the 'developed-countries-to-be' – marketing strategies were used to sell their seeds to the farmers. Good farmers perceived the advantages of purchasing seeds of good quality for their fields, and little by little, the traditional landraces disappeared in these countries.

But the rest of the world still maintained their traditional farming practices, 'farmer' and 'plant breeder' still coexisted in one person. Interchanges were generally performed on a local basis. The amount of genetic variability in crops was still huge.

At the turn of the 20th century, the genetic erosion in crops in developed countries was manifest. Breeders such as Henry Harlan perceived the problem and started collecting barley landraces around the world (he described his voyages in *One Man's Life with Barley*, Exposition Press, New York, 1957); Nikolai Vavilov started in Russia collecting a multitude of landraces of almost one thousand crops, a work that he widened to explore most of the countries where trips could be done in those times.

But still the worst was to come. After the Second World War, travelling was easier, routes were safer, marketing techniques were much more elaborate and varieties of developed countries such as the maize hybrid cultivars were almost perfect. Powerful seed companies easily spread these varieties out through the world. The genetic erosion reached most corners in the globe. Many landraces and wild forms persisted in developing countries because of economic or trade difficulties, but in developed countries the genetic homogeneity of main crops was already a very serious problem by in the second half of the 20th century (National Academy of Sciences, 1972).

By that time, some developing countries had perceived that many modern cultivars obtained in the developed world but marketed also in their farming areas were carrying important genes transferred from their own landraces that had being collected in many cases without explicit permission from national authorities. The request by the developed agencies of royalties for using these cultivars sent a fire over all developing countries. They claimed property rights over those genes and, in general, over the vegetal materials taken, with or without permission, in their lands, and accused developed countries of malpractices ranging from abuse to biopiracy. They opposed *farmers' privilege* to *plant breeders' intellectual property rights*. One further complication was the introduction of molecular techniques in plant breeding as they rendered the use of a whole plant not necessary by using only a tiny portion of it in order to extract its DNA.

The FAO, through its Commission on Genetic Resources for Food and Agriculture, very patiently tried to aggregate both sides. It took a long time before concepts were defined and agreements started to be settled. The painful path to a solution has been described in this and other chapters in this book. The International Treaty on Plant Genetic Resources for Food and Agriculture was finally signed in Rome in 2001. It is a binding tool sharing benefits among those who are able to offer valuable plant materials and those possessing the techniques to modify them, thus increasing their biological value. But the signature has not overcome a wide reticence originated in past behaviours. Claims that the vaults of both private companies and public institutions of developed countries are full of plundered plant genotypes are still alive. Mistrusts among the signatories have not been thrown away. Besides, although the steps already achieved were unimaginable some 20 years ago, there is not yet a common reading by developed and developing countries of at least one crucial article (namely 12.3.d) of the Treaty concerning the modification, especially by biotechnological means, of the material received. It is probable that this difficulty will decrease in importance once developing countries have access to these techniques, as it is in fact the case for Brazil, India, China and several other countries.

From the point of view of plant breeders, the already mentioned difference in interpreting some specific (but important) aspects of the Treaty is an added difficulty in their work because they are interested not only in wild forms and in the old landraces produced through the millennia by local communities around the world. They are also interested in the new plant material obtained by applying all kinds of technologies, including 'biotechnology'. If additional progress is required to increase food production in the future, then facilitated access to genetic resources will always be a must.

Note

1 UPOV is The International Union for the Protection of New Varieties of Plants (the acronym follows the French wording of the name), an intergovernmental organization with headquarters in Geneva (Switzerland). It was established by the International Convention for the Protection of New Varieties of Plants 1961; the last revision is of 1991.

References

Acquaah, G. (2009) *Principles of Genetics and Breeding*, 2nd edn, Blackwell, Oxford
Allard, R.W. (1960 and 1999) *Principles of Plant Breeding*, John Wiley & Sons, Inc. New York
Anonymous (2007) Seminar on enforcements of plant variety rights, Madrid, 22–23 February, 2007, Community Plant Variety Office, Angers, France
Bailey, L. H. (1895) *Plant-breeding*, MacMillan and Co, New York
CIPR (Commission of Intellectual Property Rights) (2002) *Integrating Intellectual Property Rights and Development Policy*, CIPR, London
Cubero, J. I. (2003) *Introducción a la Mejora Genética Vegetal*, 2nd edn (1st edn, 1999), Ediciones Mundi-Prensa, Madrid
Davenport, E. (1907) *Principles of Breeding*, Gion and Co, Boston, MA
Elena, J. M. (2007) 'Community rules on farm saved seeds', Seminar on Enforcement of Plant Variety Rights in the European Community, 22–23 February 2007, Session II, Madrid, Spain
Esquinas-Alcázar, J. (2005) 'Protecting crop genetic diversity for food security: Political, ethical and technical challenges', *Nature Reviews-Genetics*, vol 6, pp946–953
Fisher, J. (1989) *The Origins of Garden Plants* (revised edition), Constable and Co, Ltd., London
Harlan, J. R. (1992) *Crops and Man*, 2nd edn, American Society of Agronomy, Madison, WI
Hayward, M. D., Bosemart, N. O., Romagosa, I. (eds) (1993) *Plant Breeding: Principles and Prospects*, Chapman & Hall, London
Jensen, N. F. (1988) *Plant Breeding Methodology*, John Wiley & Sons, New York
López de Haro y Wood, R. (2007) 'Problems of enforcement of plant variety rights'. Seminar on Enforcement of Plant Variety Rights in the European Community, 22–23 February 2007, Session I, Madrid, Spain
Mateos, C. (2009) 'El privilegio del agricultor y del obtentor', in Anonimous, *Jornadas sobre la protección de las obtenciones vegetales*, pp103–114, Consejo Superior Agrario, Ministerio de Medio Ambiente y Medio Rural y Marino, Madrid, Spain
NAS (1972) *Genetic Vulnerability of Major Crops*, National Academy of Sciences, Washington DC
Poehlman, J. M. (1959) *Breeding Field Crops*, Holt, Rinehart and Winston, New York
Ruane, J. and Sonnino, A. (2006) *The Role of Biotechnology in Exploring and Protecting Agricultural Genetic Resources*, FAO, Rome, Italy
Sánchez-Gil, O. (2009) 'El privilegio del agricultor y del obtentor', in Anonymous, *Jornadas sobre la protección de las obtenciones vegetales*, pp95–101, Consejo Superior Agrario, Ministerio de Medio Ambiente y Medio Rural y Marino, Madrid, Spain
Sánchez-Monge y Parellada, E. (1955) *Fitogenética*, Salvat Editores, Barcelona, Spain
Sánchez-Monge y Parellada, E. (1974) *Fitogenética (Mejora de Plantas)*, INIA, Madrid, Spain
Sasson, A. (1993, 1998, 2000) *Biotechnologies in Developing Countries: Present and Future*, UNESCO publishing, Paris, France
Sasson, A. (2008) *Bioenergy and Agrofuels: Relevance beyond Polemics*, Hassan II Academy of Sciences and Technology, Rabta, Morocco
Sasson, A. (2009) *The Global Food Crisis: Causes, Prospects, Solutions*, Hassan II Academy of Sciences and Technology, Rabat, Morocco
Simmonds, N.W. (1979) *Principles of Crop Improvement*, Longman, New York

Chapter 16

The Global Crop Diversity Trust

An Essential Element of the Treaty's Funding Strategy

Geoffrey Hawtin[1] *and Cary Fowler*[2]

Introduction

This chapter focuses on the International Treaty on Plant Genetic Resources for Food and Agriculture (the Treaty) and its potential impact on the ex situ conservation of plant genetic resources for food and agriculture (PGRFA), reflecting the mandate and focus of the Global Crop Diversity Trust. Other important areas covered by the Treaty (e.g. in situ conservation, sustainable use or Farmers' Rights) are covered extensively elsewhere and are not considered here.

Starting with a look at why ex situ conservation is important and the links between ex situ conservation and crop improvement, the chapter goes on to explore briefly the need for facilitated access as promoted by the Treaty. It then considers the status of ex situ conservation and why the Global Crop Diversity Trust and the Svalbard Global Seed Vault are needed, showing how the Treaty has, among other things, paved the way for both of these important and related institutional developments. Finally, the chapter looks at the relationship between the Treaty and the Trust, and ways in which the Trust is supporting the implementation of the Treaty as an essential element of its funding strategy.

Why ex situ conservation?

The demands placed on agriculture will continue to increase in the future as the human population expands towards nine billion, as climates change, as new pests and diseases are encountered and as human needs and expectations evolve.

Meeting these demands will only be possible if we continue to have access to the genetic diversity contained within crop varieties and their wild relatives. This genetic diversity underpins today's agriculture and provides the raw material that enables farmers and professional plant breeders to develop the new crop varieties needed for agriculture to adapt and adjust to changing circumstances. There is a growing consensus among agriculturalists that the development of new varieties will be critical for successful adaptation to climate change and hence ensuring food security in the future.

Conserving genetic diversity ex situ is vital if plant breeders are to have ready access to the traits and genes they need to do their work. It would be impossibly complicated and expensive if new materials had to be freshly collected from the wild or from farmers' fields, often in far away countries, every time a plant breeder needed new genetic diversity.

While many individual breeders maintain their own collections of the germplasm they are likely to need in the short term, there are clearly considerable efficiencies to be gained through the collective effort underway around the world, mostly supported by governments, to maintain more comprehensive collections for use over the longer term, in more centralized gene banks operating at the national, regional or international level. The value of maintaining collections in such gene banks is considerable; for example:

- Having invested in collecting plant material from the wild or from farmers' fields – an expensive exercise – the cost of maintaining it in a gene bank is often small by comparison.
- Samples are available from gene banks throughout the year, unlike plants growing in the wild or on farmers' fields that can generally only be collected in certain periods of the year such as at harvest time.
- Gene banks are generally able to supply adequate quantities of good quality seed for research and breeding purposes. It is often difficult to collect adequate numbers of seeds of good quality from plants growing in the wild.
- Gene banks are generally able to supply seed samples that are free from pests and diseases; it is much harder to guarantee the health of seed collected in the wild without going through expensive indexing and cleaning processes.
- Collections maintained in well-run gene banks have minimal genetic drift and remain stable over time, unlike varieties maintained by farmers or populations maintained under in situ conditions. This facilitates research and the generation of reliable information about samples, which, in turn, encourages their use in breeding programmes.
- Gene banks offer a 'one-stop' shop for acquisition. Breeders are able to access a large range of diversity, often from many different countries, with a single request.
- Well run gene banks have the facilities, administrative systems and experience not only to maintain samples but also to distribute them nationally and internationally.

- Ideally, ex situ collections have reliable and readily available accession-level passport, characterization and evaluation data, and, increasingly, data at the molecular level. Such data are critical to the ability of users to make informed choices about which materials to request.
- Over time, collections become ever more valuable as the data on the accessions in them become more comprehensive. Useful comparative data can be built up and made available for sets of accessions grown across multiple environments.
- Ex situ collections provide a 'safety net' – a last resort – that enables locally adapted varieties and/or unique traits to be reintroduced back into farming systems after they have been lost due to natural or human-induced disasters, changing production systems, or as a result of their replacement by new varieties.

Facilitated access

Historically there were few barriers to prevent plant breeders from acquiring the genetic diversity they needed for their breeding work. However, over time, and particularly in the 1980s and 1990s, the expanding use of intellectual property protection measures to protect crop varieties, especially though the increased use of patents, resulted in countervailing measures being taken by some countries to restrict the free availability of the raw materials of plant breeding – the varieties and landraces developed by farmers. Accusations of 'biopiracy' were rife.

In parallel with this, the increasingly influential environmental movement took action, resulting in the Convention on Biological Diversity (CBD), to counter the threats to the existence of biodiversity and the unequal ability of developing and developed countries to exploit it. Other measures were taken by individual countries or groups of countries and the overall net effect was that the 'rules of the game' became increasingly unclear (Louwaars, 2007) and it became ever more difficult for countries to collect or obtain genetic resources for plant breeding from abroad and even, in some cases, from within the country itself. There is also evidence of a slowdown in flows of materials from gene banks in the 1990s and 2000s (Visser et al, 2000) although this does not appear to have been the case with the distribution of germplasm from the International Agricultural Research Centres (IARCs) of the Consultative Group on International Agricultural Research (CGIAR) (FAO, 2009). Recognition of this situation and concerns about future access to genetic diversity for crop improvement were key motivations for many countries to become involved in the negotiation of the Treaty.

While the Treaty has laid the ground rules for accessing PGRFA and sharing the benefits resulting from its use (see Annex 3 of this book for details on the main provisions of the Treaty), there are still a number of issues to be ironed out and the Treaty's impact on promoting increased flows of genetic materials is still uncertain (Byerlee and Dubin, 2010). If germplasm flows are to be further facilitated, it is important that the Treaty build on its positive start and continue to develop ever more effective mechanisms for facilitating access to PGRFA and promoting its use. This might be achieved, for example, through expanding the list of crops in Annex I; ensuring efficient and rational conservation systems are in place and that accurate information on the conserved resources is readily available; appropriate technology is transferred; effective institutions and regulations are in place at the national level; and that there are adequate and effective national and international funding mechanisms.

In spite of the need for further development, the Treaty has had the effect of taking some of the political heat out of the debate and as described below, this has paved the way for the creation of new institutions and funding mechanisms aimed at providing greater security and promoting increased use of PGRFA.

Access to genetic resources is likely to become ever more important in the future as zones of crop adaptation shift and new crops and varieties are needed to combat evolving pest and disease spectra, different temperature and rainfall regimes and other predicted impacts of climate change (Lobell et al, 2009).

Status of ex situ conservation

As pointed out above, the diversity contained within collections of PGRFA is critical for underpinning crop genetic improvement. However, many collections are in very poor shape and in urgent need of attention.

According to the draft Second Report on the State of the World's Plant Genetic Resources (FAO, 2009) there are currently more than 1750 gene banks worldwide, of which about 130 hold more than 10,000 accessions each. They are located on all continents, but there are relatively few in Africa compared to the rest of the world. While it is estimated that about 7.4 million accessions are maintained globally, it is probable that at most only between 25 and 30 per cent of these (or 1.9–2.2 million accessions) are distinct, with the remainder being duplicates held either in the same or a different gene bank. Clearly there is a need for greater rationalization within and among collections.

While the majority of collections are maintained nationally, international collections are critically important for their size and coverage, the availability of information on them and the ease of obtaining samples. Eleven of the CGIAR Centres manage germplasm collections on behalf of the world community and of these, the collections maintained by CIMMYT, ICARDA, ICRISAT and IRRI, each comprises more than 100,000 accessions. Collectively, the centres maintain a total of about 685,000 accessions of 3145 species of 508 different genera. National gene banks housing more than 100,000 samples include those of Brazil, Canada, China, Germany, Japan, India, Russia, South Korea and the USA.

In spite of the large number of gene banks and collections around the world, many of them, especially in developing countries, are unable to guarantee the safety of the material they house and valuable collections are in jeopardy because their storage conditions and management are suboptimal. As pointed out in the draft Second Report on the State of the World's Plant Genetic Resources (FAO, 2009), much remains to be done.

The report states, for example:

- While many countries recognize the importance of collecting, conserving, regenerating characterizing, documenting and distributing plant genetic resources, they do not have adequate human capacity, funds or facilities to carry out the necessary work to the required standards.
- Greater efforts are needed to build a truly rational global system of ex situ collections. This requires, in particular, strengthened regional and international trust and cooperation.
- While there are still high levels of duplication globally for a number of crops, especially major crops, much of this is unintended and many crops and important collections remain inadequately safety duplicated. The situation is most serious for vegetatively propagated species and species with recalcitrant seeds.
- In spite of significant advances in the regeneration of collections, many countries still lack the resources needed to maintain adequate levels of viability.
- For several major crops, such as wheat and rice, a large part of the genetic diversity is now represented in collections. However, for many other crops, especially many neglected and underused species and crop wild relatives, comprehensive collections still do not exist and considerable gaps remain to be filled.
- To better serve the management of collections and encourage an increased use of the germplasm, documentation, characterization and evaluation all need to be strengthened and harmonized and the data need to be made more accessible. Greater standardization of data and information management systems is needed.
- In situ and ex situ conservation strategies need to be better linked to ensure that a maximum amount of genetic diversity is conserved in the most appropriate way, and that biological and cultural information is not lost inadvertently.
- Greater efforts are needed to promote the use of the genetic resources maintained in collections. Stronger links are needed between the managers of collections and those whose primary interest lies in using the resources, especially for plant breeding.

A study was published by Imperial College Wye comparing data from 99 countries collected by FAO in 2000 to similar data from 151 governments collected in 1996. It found that in 66 per cent of countries the number of accessions held in collections had increased over this period, however, in 60 per cent of the countries gene bank budgets had remained static or had been reduced. More than half of developing countries and 27 per cent of developed countries reported an increase in the number of accessions in urgent need of regeneration.

The report concluded that:

> *... it is time to think about how to mobilize global resources to meet a global challenge. New and imaginative means of support must be found. Until now, gene bank funding has largely been dependent on annual disbursements from national budgets, which can vary from year to year. However the need to keep crop diversity collections safe exists in perpetuity. To let it lapse even one year may mean the sacrifice of irreplaceable crop genetic resources. Therefore, funding must be stable and forever.*
>
> *To garner these resources, the world community must look beyond the annual budgets of individual countries or donor organizations. Resources can be pooled into one global fund – an endowment for the future of agricultural diversity and a foundation for food security.*
>
> *A substantial endowment would match the perpetual need for crop diversity conservation with a perpetual source of support for the world's national and international plant genetic resources collections. It could support the maintenance needs of the world's most critical collections and help to build the capacity of under-funded collections. An endowment could help realize the ideals of the International Treaty on Plant Genetic Resources by taking as its starting point conservation of the 35 priority food crops and 80 forages listed under the Treaty. Over time, it could grow in size and scope to encompass additional gene bank collections and crops.* (Imperial College Wye, 2002)

Two recent institutional developments

As pointed out above, the Treaty has enabled a number of key institutional innovations to take place that were not possible earlier (see Annex 1 of this volume for explanations on all Commission and Treaty negotiating meetings). Two very significant developments have been the creation of the Global Crop Diversity Trust and the Svalbard Global Seed Vault. With respect to these two institutions, the draft Second Report on the State of the World's Plant Genetic Resources (FAO, 2009) states:

- The Global Crop Diversity Trust, founded in 2004, represents a major step forward in underpinning the world's ability to secure PGRFA in the long-term; and
- With the establishment of the highly innovative Svalbard Global Seed Vault, a last resort safety back-up repository is now freely available to the world community for the long-term storage of duplicate seed samples.

These two institutions are described further below.

The Global Crop Diversity Trust

The idea of establishing an endowment fund to support the ex situ conservation of PGRFA has been around for many years and, as the Imperial College study pointed out, was urgently needed. However, prior to the entry into force of the Treaty, many potential donors had expressed strong concerns that if they provided funds there might be no reciprocal access rights granted to the material conserved, or if funds were provided conditionally on the material being made available, then they feared being publicly accused of biological imperialism or the like. These fears were significantly reduced once it was apparent that the Treaty would become a reality following its approval at the 31st session of FAO Conference in November 2001 (see Annex 2 of this volume for a list of its contracting parties per FAO regional groups). Although it was not until 2004 that the Treaty actually came into force, nevertheless from 2001 it became possible to begin planning the establishment of an endowment fund to support ex situ conservation.

An extensive series of consultations with all major stakeholder groups took place between 2001 and 2003, spearheaded by Bioversity International (then the International Plant Genetic Resources Institute, IPGRI) acting on behalf of the CGIAR and FAO, culminating in the drawing up in early 2004 of a Constitution[3] and Establishment Agreement[4] for a new international funding mechanism: the Global Crop Diversity Trust. The Trust was formally established in October 2004 as an independent organization under international law, this status being conferred on it through the signing of an Establishment Agreement by seven states from five of the regions referred to in the basic texts of FAO.

The objective of the Global Crop Diversity Trust, as contained in its constitution, is to ensure the long-term conservation and availability of PGRFA, with a view to achieving global food security and sustainable agriculture. More specifically the Trust aims:

- To safeguard collections of unique and valuable plant genetic resources for food and agriculture held ex situ, with priority being given to those that are plant genetic resources included in Annex I to the Treaty or referred to in Article 15.1(b) of the Treaty;
- To promote an efficient goal-oriented, economically efficient and sustainable global system of ex situ conservation in accordance with the Treaty and the Global Plan of Action for the Conservation and Sustainable Utilization of Plant Genetic Resources for Food and Agriculture;
- To promote the regeneration, characterization, documentation and evaluation of PGRFA and the exchange of related information;
- To promote the availability of PGRFA;
- To promote national and regional capacity building.

Specific activities of the Trust, as listed in its constitution, include:

- Establishing an endowment fund to provide grants to support the maintenance of eligible collections of PGRFA that meet agreed standards of management and availability of the genetic resources, related information, knowledge and technologies, and to cover operating expenses and other expenses incidental thereto;

- Receiving funds other than funds intended for the endowment fund, to provide grants to support the holders of potentially eligible collections in upgrading their collections so that they can meet agreed standards of management in order to become eligible for maintenance grants.

In order to be able to effectively target its limited resources to supporting collections of highest priority, the Trust has sponsored the development of a set of international collaborative conservation strategies. The process of developing the strategies has brought together gene bank managers, researchers and other experts on plant genetic resources from developing and developed countries. Although commissioned by the Trust, the strategies have been developed independently by the different communities involved, and will evolve as the situation of collections around the world changes.

The strategies aim to identify:

- The collaborative arrangements necessary for efficient and effective conservation;
- The collections that are of highest priority for support by the Trust and other donors and the appropriate roles for such collections with a global system;
- Major needs in collecting, storage and maintenance, distribution and research;
- Appropriate roles for other stakeholders in the conservation, regeneration, documentation and distribution of crop diversity.

Two complementary and mutually reinforcing approaches have been taken to developing these strategies: (a) on a regional basis and (b) on a crop basis. Collectively they respond to calls from the Global Plan of Action, to '*develop an efficient goal-oriented, economically efficient and sustainable system of ex situ conservation*' (FAO, 1996) and likewise the requirement under Article 5.1.(e) of the Treaty that contracting parties 'promote the development of an efficient and sustainable system of ex situ conservation, giving due attention to the need for adequate documentation, characterization, regeneration and evaluation, and promote the development and transfer of appropriate technologies for this purpose with a view to improving the sustainable use of plant genetic resources for food and agriculture'.

By the end of 2009, conservation strategies had been developed and published for more than 30 regions and crops,[5] a major undertaking that has recently been reviewed for eight themes: regeneration, crop wild relatives, collecting, crop descriptors, information systems, user priorities, new technologies and research, and challenges to building a strategy for rational conservation (Khoury et al, 2010).

Early in its existence the Trust developed an important strategy document entitled *The Role of the Global Crop Diversity Trust in Helping Ensure the Long Term Conservation and Availability of PGRFA* which was endorsed by its Executive Board.[6] This document outlined the basic assumptions and principles that underpinned the Trust's conception of how a rational global system might be constructed. It contained an important 'decision tree' that made explicit the basis upon which the Trust would determine funding priorities.

The Trust, in accordance with its constitution, consulted with its Donors' Council and the Governing Body of the Treaty in the development of a formal fund disbursement strategy. This strategy was based on the earlier paper on *The Role of the Global Crop Diversity Trust*.[7] The fund disbursement strategy was endorsed by the Donors' Council and the Governing Body, and then adopted by the Executive Board. The strategy, while directly related to the Trust, also provides a clear and rather specific description of a rational, effective, efficient and sustainable global system, noteworthy in part due to its endorsement by the Governing Body of the International Treaty.[8]

Based largely on this constellation of strategies and formal policies, as of March 2010 the Trust has provided long-term maintenance grants to collections of aroids, banana, barley, bean, cassava, fava bean, forages, grass pea, pearl millet, rice, sorghum, wheat and yam. In addition, and in partnership with a large number of other institutions, the Trust has funded numerous projects around the world that have contributed to, inter alia:

- The regeneration of collections of priority accessions of more than 20 crops in over 50 developing countries;
- The development of a global PGRFA information system;
- The development of a freely available, multilingual gene bank data management system;
- Crypopreserving part of the world's largest banana collection and the development of cryopreservation protocols for other vegetatively propagated crops;
- Upgrading gene bank facilities, especially in southern Africa;
- Rescuing material in the Philippines National Plant Genetic Resources Laboratory following the devastation caused by Typhoon Xangsane in 2006;
- The establishment of the Svalbard Global Seed Vault (see below).

As of March 2010, the Global Crop Diversity Trust has received total pledges of support amounting to almost US$170 million and of this more than US$136 million has already been received.

Svalbard Global Seed Vault

The second new institution to be considered here is the Svalbard Global Seed Vault, a facility that aims to provide an insurance against both incremental and catastrophic loss of crop diversity held in traditional gene banks around the world. However, unlike many traditional gene banks, the Vault does not house any unique, original material, but aims to serve as a fail-safe back-up facility; a safety net for the world's germplasm collections. The ultimate goal of the Vault is to safeguard a duplicate set of as much of the world's unique crop genetic material as possible.

The idea of creating an international back-up seed storage facility has also been around for many years. In the early 1980s the Nordic Genetic Resource Centre (then the Nordic Gene Bank) identified Svalbard as a suitable location for storing seeds in the permafrost and in 1983 began to safety-duplicate its accessions there in a coal mine near Longyearbyen.

In 1989, following discussions with the Government of Norway, FAO and the International Board for Plant Genetic Resources (now Bioversity International) undertook a survey of Svalbard to identify a suitable site for an international seed storage facility. Norway offered to cover the costs of the actual construction of the facility, whilst FAO and IBPGR agreed to take care of the administrative and operating costs through the creation of a fund based on capital from external donors. In the event, however, concerns by potential seed depositors and funders over the question of access to, and ownership of any materials stored in the facility, as well as questions about the quality of storage conditions on offer (ambient conditions of about −3.5°C) and the reluctance of the international community to fund the facility, led to the idea being shelved. With the clarity and increased trust among parties that resulted from the entry into force of the Treaty, it once more became possible to consider the development of an international seed back-up facility in the permafrost. Thus, in 2004, the Norwegian Ministry of Foreign Affairs and the Ministry of Agriculture and Food reopened the subject. A group of experts was appointed to carry out a preliminary study, which strongly recommended the establishment of a storage facility on Svalbard. They recommended that storage be offered, at no cost, to all interested gene banks worldwide for them to store a duplicate set of their collections. They further recommended that the facility be located in its own dedicated facility (not in the mine) and that storage conditions meet international standards for long-term conservation.

In November 2004, the FAO Commission on Genetic Resources welcomed the proposal, and plans for the facility, named the Svalbard Global Seed Vault, were then drawn up. Construction began early in 2007 and the Vault opened in February 2008. The cost of the construction, some US$9 million, was funded entirely by the Government of Norway, which owns the facility (but not the seed stored within it) and is responsible for maintaining and administering it. Under the terms of a tripartite agreement between the Norwegian Government, the Global Crop Diversity Trust and the Nordic Genetic Resource Center (NordGen), responsibility for managing the Vault lies with NordGen, overseen by an International Advisory Council. The Global Crop Diversity Trust covers the primary ongoing operational costs of running the Vault. The Vault comprises three chambers set back more than 125 metres into the mountainside, each having the capacity to store 1.5 million seed samples. While the chambers are artificially cooled to −18°C, a large measure of security against a prolonged loss of cooling is provided by the fact that they are set deep within the permafrost at a temperature of minus 3–4°C. All the material in the Vault is maintained under 'black box' conditions; that is, with ownership and access rights to the material remaining with the depositor. This means that seed packages and boxes sent for storage cannot be opened or sent to anyone except the original depositor and that the responsibility for testing material and for any subsequent regeneration remains with the depositor.

As of March 2010, the Svalbard Global Seed Vault housed some 522,000 seed samples, deposited by 28 institutions in 24 countries. The seed samples themselves were initially sourced by these institutions (a number of them international institutions) from virtually every country in the world. Information on the material deposited can be found in a database on the Vault's website, maintained by NordGen[9] and further information on the Vault can also be found on the website of the Norwegian Ministry of Agriculture and Food[10] and the Global Crop Diversity Trust.[11]

The Treaty and the Global Crop Diversity Trust

In drawing up the Constitution of the Global Crop Diversity Trust[12] a very close relationship was foreseen between the Trust and the Treaty, which at that time had yet to enter into force (see Annex 3 of this book for details on the main provisions of the Treaty). Article 7 of the Constitution is solely concerned with this relationship and states:

1 The Executive Board shall, as soon as practicable after the entry into force of the International Treaty, enter into an agreement with the Governing Body of the International Treaty, defining the relationship of the Trust with the International Treaty.
2 The relationship agreement shall include the following:
   - recognition of the Trust as an essential element of the Funding Strategy of the International Treaty;
   - the authority of the Governing Body of the International Treaty to provide overall policy guidance to the Trust on all matters within the purview of the International Treaty;
   - reporting obligations of the Trust to the Governing Body of the International Treaty;
   - recognition that the Trust will be free to take its own executive decisions on disbursement of funds, within the general framework of the overall policy guidance of the Governing Body of the International Treaty.

Following discussions between the Trust and the Governing Body, an agreement was signed in June 2006 defining the relationship between the two parties and recognizing the Trust as an essential element of the funding strategy of the Treaty.[13]

The Governing Body is responsible for providing overall policy guidance to the Trust and, in addition to the elements of the relationship outlined above, the constitution calls for the Governing Body to appoint 4 of the 11 (or up to 13) members of the Executive Board of the Trust. At least two of these appointees must come from developing countries. The Executive Board is also obliged to consult with the Governing Body before adopting either the Trust's fund disbursement strategy or the principles upon which it will decide on the eligibility of collections, projects and activities for funding.

The Trust, as an essential element of the funding strategy of the Treaty, contributes in multiple ways to the achievement of the Treaty's objectives. In particular, it is assisting contracting parties to fulfil their obligations set out in Article 5: '*Conservation, Exploration, Collection, Characterization, Evaluation and Documentation of Plant Genetic Resources for Food and Agriculture*'.

As the Treaty further develops other aspects of its funding strategy, and in particular the benefit-sharing fund, it is anticipated that there will be many opportunities in the future for the Trust to partner with such bodies and thereby contribute further to the achievement of the overall objectives of the Treaty.

Conclusions

While a number of areas covered by the Treaty are still being discussed and developed, its coming into force in 2004 did much to bring clarity to the issues of access to PGRFA and sharing the benefits resulting from its use. This, and the consequent building of trust among the parties, has helped pave the way for some very significant institutional developments that were not possible prior to the existence of the Treaty. The creation of two such institutions is described in this chapter: the Global Crop Diversity Trust and the Svalbard Global Seed Vault.

Neither of these initiatives would have been possible without the Treaty and together they aim to make a substantial contribution to the achievement of one of the key objectives of the Treaty, namely the ex situ conservation of PGRFA and promoting their sustainable use.

With landraces and farmers' varieties continuing to be lost from farmers' fields, and crop wild relatives increasingly coming under the threat of extinction as a result of changing climates and land use patterns, it is more important than ever that existing crop genetic diversity be adequately and safely conserved. The Treaty and consequent establishment of the Global Crop Diversity Trust and Svalbard Global Seed Vault provide increased confidence that the genetic resources needed to tailor our crops to meet future challenges will continue to be available for a long time to come.

Notes

1. Senior Advisor, Global Crop Diversity Trust, Manor Farm House, 17 Front Street, Portesham, Dorset, DT3 4ET, UK.
2. Executive Director, Global Crop Diversity Trust, FAO, Viale delle Terme di Caracalla, 00153 Rome, Italy.
3. Constitution of the Global Crop Diversity Trust, 2004. www.croptrust.org/main/governance.php?itemid=5.
4. Agreement for the Establishment of the Global Crop Diversity Trust, 2004 www.croptrust.org/main/governance.php?itemid=5.
5. www.croptrust.org/main/identifyingneed.php?itemid=514.
6. www.croptrust.org/documents/web/RoleofTrustSept08.pdf.

7 ibid.
8 www.croptrust.org/documents/WebPDF/GCDT%20Fund%20Disbursement%20Strategy%20FINAL.pdf.
9 nordgen.org/sgsv.
10 www.regjeringen.no/en/dep/lmd/campain/svalbard-global-seed-vault.html.
11 www.croptrust.org.
12 Constitution of the Global Crop Diversity Trust, 2004. loc cit.
13 www.croptrust.org/main/governance.php?itemid=6.

References

Byerlee, D. and Dubin, H. J. (2010) 'Crop improvement in the CGIAR as a global success story of open access and international collaboration', *International Journal of the Commons*, vol 4, no 1, February 2010, pp452–480

FAO (2009) *Draft State of the World's Plant Genetic Resources for Food and Agriculture*, FAO, Rome, Italy

Imperial College Wye (2002) 'Crop diversity at risk: The case for sustaining crop collections', Imperial College of Science, Technology and Medicine, London

FAO (1996) *Global Plan of Action for the Conservation and Sustainable Utilization of Plant Genetic Resources for Food and Agriculture*, Paragraph 79, FAO, Rome, Italy

Khoury, C., Laliberte, B. and Guarino, L. (2010) 'Trends in *ex situ* conservation of plant genetic resources: A review of global crop and regional conservation strategies', *Genetic Resources and Crop Evolution*, vol 57, no 4, pp625–639

Lobell, D., Naylor, R., Falcon, W., Burke, M. et al. (2009) 'Climate extremes and crop adaptation', Summary statement from a meeting at the Program on Food Security and Environment, 16–18 June, Stanford, CA available at http://foodsecurity.stanford.edu/publications/climate_extremes_and_crop_adaptation/

Louwaars, N. P. (2007) 'Seeds of confusion: The impact of policies on seed systems', Wageningen University and Research Centre, Wageningen, The Netherlands

Visser, B., Eaton, D., Louwaars, N. and Engels, J. (2000) 'Transactions costs of germplasm exchange under bilateral agreements', Global Forum for Agricultural Research, GFAR-2000, Dresden, Germany, 21–23 May 2000. GFAR and International Plant Genetics Research Institute, Rome, Italy

Chapter 17

Consumers

Biodiversity Is a Common Good

Cinzia Scaffidi

Introduction

The International Treaty on Plant Genetic Resources for Food and Agriculture (ITPGRFA or the Treaty) stands as a tool of *governance* of plant resources that is, 'the genetic material of plant origin with effective or potential value for food and agriculture' designed to respond at a global level to the objectives of economic solidarity and environmental sustainability.

At first glance it could seem to be a matter between governments and farmers: in fact the Treaty, after stating in the Preamble that the contracting parties are convinced of the special nature of plant genetic resources, goes on to recognize that these resources are 'the raw material indispensable for crop genetic improvement, whether by means of farmers' selection, classical plant breeding or modern biotechnologies', affirming that 'the past, present and future contributions of farmers in all regions of the world, particularly those in centres of origin and diversity, in conserving, improving and making available these resources, is the basis of Farmers' Rights' (see Annex 3 of this book for details on the main provisions of the Treaty).

In recognizing the enormous contribution that local and indigenous communities and farmers have made, and continue to make, to the conservation, development and sustainable use of plant genetic resources, it is agreed to *realize Farmers' Rights*. In the Preamble to the Treaty this is emphasized by affirming that the rights to:

> *save, use, exchange and sell farm-saved seed and other propagating material, and to participate in decision-making regarding, and in the fair and equitable sharing of the benefits arising from, the use of plant genetic resources for food and agriculture, are fundamental to the realization of Farmers' Rights, as well as the promotion of Farmers' Rights at national and international levels.*

Community is one of the key issues. Collective interest in boosting biodiversity and introducing quality in agricultural systems switches the direction of development. Until now advances have been made by researchers using sophisticated improvement techniques or genetic manipulation. Now it is possible to support the free circulation and exchange of seeds by reducing the transaction costs rather than by offering the incentive of exclusive exploitation.

Sensitivity towards issues such as the importance, function and protection of agro-biodiversity has grown in the world in general. More specifically, all the work that went into drawing up and ratifying the Treaty has also exerted an educational and cultural impact on contemporary societies.

What became clearer and clearer is that a network is needed: governments, farmers and consumers need to take action in a consistent way to protect biodiversity. In particular, governments need to promote information and public participation. Consumers should be well informed about issues related to biodiversity and genetic resources. This would do much to support the rebuilding of a food culture which, in many wealthy countries has been eroded.

Rightly, the Treaty has focused on agriculture until now. However, it is time to involve the consumers in the defence of biodiversity. Farmers are, depending on the countries, from 60 per cent to 2 per cent of the population. Consumers constitute 100 per cent and can make the difference. This leads us to think about the importance of educational initiatives and activities aimed at consumers. This can be considered as a way to implement and truly apply the core of the Treaty. It is clear that protecting biodiversity is not possible in the absence of an educated public.

Which agriculture, which consumer?

Reductionism has influenced 'modern' agriculture and given it a highly industrial profile. The farmer involved in this kind of agriculture behaves in a way very similar to that of a worker in a factory. Moreover, this kind of agriculture tends to adopt a 'singular' approach:

- It normally involves not the whole family but the single farmer, and when more than one individual of the same family is involved, he tends to work separately on specific tasks.
- It usually encompasses only one gender, the male, women being marginalized by this model of development.

- Aiming at *the market* (not *markets*), it specializes in products that have to be as uniform and homogeneous and numerous as possible. This is why it handles few products obtained by seeds improved by commercial procedure: combinations of pure parental lines that yield uniform plants and fruit – that is, hybrids.
- It is specialized, meaning that it tends to privilege one activity or a few unconnected activities.

Industrial agriculture is a linear system: it does not reuse the output of production. In this way it leaves a heavier footprint on resources. It also overlooks many other possible products, either because it has no interest in them, or because, very often 'precisely on account of the production method' some outputs cannot be exploited. One example is manure which, in industrialized livestock farms, cannot in certain cases be used as a fertilizer insofar as it is too contaminated with antibiotics.

The only objective of this kind of agriculture is the market, which measures its success on the distance between the place of production and the place of sale, believing that 'the further the better'.

Last but not least, this kind of agriculture aims at only one kind of consumer: A consumer whose awareness is low and who takes into consideration only a few factors, the most important of them being value for money. This kind of agriculture also counts on the 'laziness', and lack of information of this kind of consumer, who doesn't want to be involved in collective decisions and follows only some criteria, forgetting the others, exactly as reductionist thought does. In conclusion, this very rigid agriculture applies the same few rules to very diverse situations and thinks it can resolve its own lack of adaptability through external inputs.

A specular analysis can be made of traditional farming, which may be described as an integrated system because it tends to reuse outputs and by-products for other production phases or to launch new products. In this way its footprint on planetary resources is lighter and thus lowers production and environmental costs:

- It involves both genders and more than one generation, since it is supported by the knowledge and skills of men and women, old and young, without stopping the flow of information that allows people to grow up respecting nature and feeling part of it.
- Insofar as the main aim of traditional farming is to harvest to feed the family, it is devoted to the cultivation of more than one species, each of which is cultivated in more than one variety; each variety is produced from traditionally improved seeds and in any given population, shows a high level of variability. All this leads to even higher adaptability to climatic conditions. Whether the season is damp or dry, whether a new or an old parasite appears, there will always be a part of the crop that won't be affected by the problem.
- It is not specialized. A traditional farm performs many activities, the most important being growing and breeding. At the same time, processing and selling are present too, along with several 'non-target' activities, such as education, landscape conservation, biodiversity protection and so on.

- It has multiple tasks: not only to reach the market, but, first and foremost, to feed the family and its animals, to keep the soil fertile, to create a pleasant environment that can attract visitors.
- It has many ways of reaching markets, almost always the closest to the production site. This can be through a farm shop, through local markets, through consumer purchasing groups or associations. It is important to consider 'short-distance' selling because this allows the farm to maintain links with the local culture, while the consumers can judge the quality of a product properly and diminish costs considerably for both sides of the equation. This contrasts with the aforementioned focus of industrial agriculture on 'the' market rather than on markets.
- It caters for several kinds of consumers: elderly people, young people looking for reliable information about food, environmentalists who want to be consistent in their behaviour, gastronomes aware that quality starts from production method. All of these people have one point in common: their appreciation for the food they buy, the importance they attach to a food's identity, their consideration of food as a means of expression, a language. They are prepared to pay a fair price because they know that an excessive low price entails many risks. And they know that the end price of a food product is the result, but also the core of a complex system in which you cannot isolate only one factor.
- It is an integrated system because it tends to reuse outputs and by-products for other production phases or to launch new products. In this way its footprint on planetary resources is lighter and thus lowers production and environmental costs.

Here we are actually referring to *man agricultures*, in constant evolution on account of the adjustments and integrations they receive from other cultures, industrial culture not excluded. Small-scale traditional or subsistence agriculture knows how to make the best use of all knowledge, refusing to apply the same model to every situation.

Traditional and industrial agriculture: Trade models, social networks and product variety

Further consideration should be given to the trade models of these two different production systems. As we have said, industrial agriculture markets its products through modern food distribution systems, namely large-scale retail. Supermarkets nowadays form part of the huge shopping malls that colonize the suburbs of towns and cities, contributing to the soil sealing process, attracting an uninterrupted flow of consumers thanks to round-the-clock opening hours. In these huge retail spaces they need to present many different food options, which is why they encompass a vast geographical territory. Local origin and seasonality are not considered in this kind of retail: on the contrary the possibility of finding whatever food in whatever season is advertised as an added value.

It is not easy to have full awareness of the changes caused by this process, at least as far as our nutrition is concerned. Every day we buy fruit, vegetables, meat and milk without any idea of their distance from their place of origin or about the incidence of transportation, methods of cultivation and the quality of the organization of the labour that has produced them. This kind of market has grown into a technical-economical space, where the need for free circulation of goods has cancelled the productive vocation of single areas, leading us to ignore objective differences in terms of quality and cultural identity.

An important role is played by the advertising system only at the end of the process. In fact, the uniformity and anonymity of industrial food would end up being totally unappealing. In order to reconnect the consumer with those forms of reassurance that the industry cannot offer (transparency of production methods, origin of ingredients, naturalness, history and so on), the trade communication system 'dresses' the product with a style that it cannot have per se.

What then is the trade model of traditional agriculture? The nearby market, which doesn't necessarily mean the *short supply chain*? That is an ambiguous idea, since it only takes a part of the problem into consideration. It focuses on the *number* of transactions that take place between production and final purchase, and on the *quantity* of time that passes from production to sale, thus ultimately adopting a reductionist approach. You can have a mozzarella from Naples in a NYC restaurant in 24 hours with a direct contact between buyer and seller, but the farmers' market concept is much more than this.

As with the supermarkets, it may be useful to consider ideas of time and space. The farmers' market *invests* in time instead of trying to save it; it *takes care* of space instead of trying to have huger and huger amounts of it. Time is invested in social relationships, in information, in education: the possibility of a direct contact between the producer and the consumer (who is active and curious, not lazy and indifferent like his supermarket cousin!) gives both new opportunities for learning how to play their respective roles better. From what the consumer asks, the producer learns how to best satisfy him or her; from what the producer answers, the consumer learns about nature, about the labour that goes into food and also 'how to evaluate that food' and 'what a fair price for it should be'.

As for space, farmers' markets do not need a lot of it. They are at the service of the surrounding area: the urban centre that receives economic and other benefits from their presence and rural districts that likewise receive attention and consideration, as well as economic benefits. Space should be seen not as surface area, but as the place in which many different kinds of exchange go on, revitalizing channels of social, economic, cultural and natural life that would otherwise risk being totally forgotten. It is in the matter of exchanges and relationships that one of the most important differences between the two production systems becomes evident. Because sustainable and ecological agriculture has one more function: the permanent and mutual educational process that involves farmers, consumers, cooks, school, institutions and research (Petrini, 2009).

Scarcity and abundance

As Anderson (2009) writes, the most common definitions of economy share the same annoying element: a privileged attention to scarcity, especially to the allocation of limited resources. It is difficult to overcome the importance the concept 'that you cannot have everything for free' has in the economy. The whole discipline is focused on the study of the exchanges and of the conditions in which they happen.

Thanks to new technologies, markets have multiplied, as have the potential buyers of each producer and the potential suppliers of each consumer. The number of actors potentially involved in any kind of relationship, dialogue, bargain and creation has grown.

Somehow, thanks to the development of innovations, a new era of abundance has opened, but it is unable to fit into the normative patterns set by an economy born and developed in a context of scarcity.

But abundance is older than technologies. In fact, not only the world that new technologies have allowed, but also the one based on natural laws, fall under the realm of abundance. This is exactly why the legal solution, which plays a prominent role apropos the use of genetic resources, cannot regulate the realm of life or creativity. In a situation of scarcity, the use of genetic resources would bring about the extinction thereof, which is why it has to be regulated. Yet genetic resources belong to nature, where scarcity is not considered.

Farmers learn from nature the language of gifts that is spelled in the alphabet of abundance. Each harvested tomato yields dozens of usable seeds; each harvested ear of wheat yields dozens of grains. Each seed gives life to dozens of seeds, so why skimp? The less you sow the fewer seeds you'll get.

Common goods

The market has been attempting for a long time to appropriate common goods, but it can only do so by ignoring their essence and forcing them into rules that cannot fit – the rules of scarcity, mentioned above.

Instead, common goods are characterized by abundance and for this very reason they become *revolutionary* vis-à-vis the rules of economics. Seeds, as a generic way to mention all the plant genetic resources, are given to us in a regime of abundance: it is their *indispensability*, not their scarcity, that makes them a common good. If we reflect on the main characteristic of common goods (water, air, creativity), we see how their quantity is always indeterminate, whereas their core feature is that we cannot do without them.

So, what does it mean that we have to manage genetic resources in a sustainable way? It means that we have to manage them remembering that they are indispensable for us, for the rest of the living beings, and for future generations. Again: the 'managers' of genetic resources are not only farmers and government, but also consumers. Because if it is true that in the market mechanism lies one of the main causes of the biodiversity erosion, then we have to admit that consumers are among the protagonists and they can play a heavy role in making it worse but also in radically counteracting the whole system.

How much does food cost?

If what we know as a cheap hamburger of any fast food outlet were really to cover all its production costs, we would have to pay tens of euros for it. Because its price should comprise not only the beef itself, the bread, the vegetables and the sauces (not to mention the sugars, unsaturated fats, colouring agents and chemical flavourings) the hamburgers contains, but also the environmental costs of deforestation to make room for intensive livestock breeding in the southern hemisphere, the health costs of the increase in cardiovascular diseases, diabetes and obesity resulting from a diet too rich in sugar and too poor in fibres and vitamins, the social cost of low salaries of workers and consequently fragile trade union relations and last but not least, the ecological costs generated by the incredible amount of energy needed to produce and sell a single kilocalorie (a ratio of around 1:150). All this considered, how much should our hamburger really cost? Probably the same price as a course in a 3-star Michelin restaurant.

There are many factors to bear in mind about food in each phase of the process that goes from production to consumption. What is more, such factors interact, and this makes things even more complex.

Let us start from what we call production. As happens in any kind of production, that of food relies on natural resources, some of them renewable, others not.

First questions: Leaving mere proprietary rights issues aside, non-renewable resources, such as fossil fuels (coal and oil, for example) – who do they belong to? Leaving stock exchange values aside, how much do they cost?

Moreover, when we talk about resources, sometimes we refer to energy, but other times we refer just to produce. Think of fish, for example. It is wrong here to consider fish as a 'product', insofar as we are speaking in terms of a withdrawal, the direct use of a resource (which is renewable following the natural rhythms of the sea, not the food production schedules we draw up on dry land). If the market we refer to is the fish market, we have to consider natural resources that 'support' fish production (fossil fuels for boat, for example), but there is also a natural resource that constitutes the basis for the withdrawal: namely, the sea itself.

In the same way, biodiversity, meaning also genetic plant resources, is part of the 'natural capital' that we have to keep in mind when we consider production in the classical economic way. We are used to considering two pillars: capital and labour. We need to learn that another pillar is involved: the natural capital made of all the resources we use, directly or indirectly for our production (Tiezzi, 1997).

That is not all; other factors need to be added. One such is 'social justice', meaning consideration of the living and workplace conditions of the people who contribute to the productive process. The second is animal welfare, where production presupposes the breeding, catching and involvement of animals. The third is, as mentioned above, the fact that we all live together on one planet, evident to us now thanks to globalization, but a fact since the dawn of time.

Arguably the most important and positive result of globalization is that it has given us the perception of being part of a planet, certainty that our actions have consequences not only on our own lives and those of people around us, but also on the lives of the rest of humanity, even far away from us.

Globalization helps us to understand that even when the price of our food respects the parameters of ecological economics and the social issues we have mentioned, it still has not done all its job: in this globalized world, consumption and production have consequences that also need to be taken into account.

The role of informed consumers in changing the rules

Citizen-consumers can work towards an ethical market, assuming that 'ethics' deal with the individual behaviours and their consequences on the community. Philosopher Emmanuel Kant, in his fundamental law of pure practical reason stated: *act only according to that maxim whereby you can at the same time will that it should become a universal law.* Which means: if the way you are going to behave could be bearable if everyone in the world did the same, than it is an ethical action. If it can be done only by one person it is not. The western countries are scared at the idea that China, India, Africa start consuming as much fuel, meat, water, as Europe, the USA and Japan are doing. 'It would be unsustainable!' they say. But they pretend not to see that the 'unsustainability' is already in their behaviours without considering developing countries.

Through their choices, consumers can orient production, which follows their lead on the basis of pragmatic considerations of customer satisfaction, not necessarily of ethical correctness. Of course a condition exists to make the consumer fully able to modify the market in an ethical sense. That condition, as said, is information. 'Good clean and fair food' (Petrini, 2007) must be recognized by the consumer, hence information must be available and reliable, and the consumer must have enough food culture to decide which is the best food for the common good. The first piece of information that the consumer must have is that food cannot be cheap, because when the price of a food is too low, someone or something is being damaged:

- It can be cheap because it is of poor quality, hence harmful for the health of the consumer.
- It can be cheap because not all the production costs, such as social or environmental costs, have been considered and have remained hidden. Sooner or later someone (or all of us) will have to pay for this, maybe in a multiplied amount.
- It can be cheap because it is the product of subsidized agriculture, meaning that it has damaged other (far away) agricultures and economies.

Going back to the example of the hamburger, when we buy it, it ends up being a form of collaboration with a production and distribution system that insists on damaging weak economies, weak workers and weak consumers. It damages health, it creates injustices. It feeds: but a food that just feeds – regardless to all that happens before and after, is a very bad food.

How pleasure can defend biodiversity

The Slow Food Movement was born in Italy in 1986 as an international association concerned with traditional food, good wine and small-scale tourism. Born as a movement for the 'defence of the right to pleasure', it began to consider all the implications of this concept. Pleasure means of course eating good and well identified food, whose origins and processing are known. It also means having a glass of top quality wine or beer or whatever is the traditional drink in the place we are in. Pleasure also means visiting areas whose rural landscapes can tell us their story and their habits, connected to the climate, the religion and the events of the people living there. Also pleasure is much more than that. More importantly, pleasure has to be taken into consideration as everybody's pleasure, and what is served at the table to be eaten is just the tip of the iceberg. Food comes from the land and those who eat must know that the action has been made possible by those who produced the food: farmers, producers, cooks, researchers. Nobody can enjoy their food without thinking that this is a universal right, and that every kind of food, even the simplest, has a story to tell: the story of a place, a population, an identity. Conserving the biodiversity of our crops and animals breeds, in order to save the great diversity of our traditional foods means, among other things:

- conserving regional traditions;
- encouraging young people to be interested in food and agriculture;
- working to avoid or at least curb the homogenization of food culture.

We all know the reasons why we have to protect and defend biodiversity: there are *agricultural reasons* (maintaining resources for disease resistance; the vulnerability of a monocultural rural system); *cultural reasons* (loss of knowledge, of memory, of culture; the higher adaptation of the traditional food to the needs of a certain population; the central part that farming, food and eating have in the definition of an identity); and *economic reasons* (small and medium farming has in many countries a big role and it allows the majority of populations to survive. This kind of agriculture is a mixed and traditional one: it has to be like that, for reasons of space, safety and … pleasure, because small farmers grow what their families love to eat).

The consumers, in the widest meaning of this word (the people who go shopping for food, but also the restaurant owners who buy the ingredients for their cooking …) are an important part of all this. They have to be informed and they want to be informed.

What has happened in a certain part of our world during the last 30 years is that the source of information about food has been lost by the majority of the people. The gap between who produces and who eats has grown bigger and bigger, and very often the young generations are completely unaware of the origin of what they are eating.

Again we come back to the theme of information: how important are information and education in the protection of biodiversity? In the last 50 years, food has been treated mainly as a problem of quantity. Starting with the Green Revolution, international attention has been focused on production per hectares, price and nutritional values. This is the quickest and most efficient way to lose biodiversity. And losing biodiversity is the best way to increase the quantity of people starving, making the planet more and more poor and vulnerable.

Today the poorest countries are those where the Green Revolution had its experimental and productive bases. The production problem has been virtually and factually solved, if it's true – and it is true – what Kofi Annan said on behalf of the United Nations: the planet is producing enough food to feed 12 billion people, which means almost twice the number currently living on Earth. So where is this food? Where does it go? Who does it belong to? Above all, what kind of food is it?

And we need new consumers and new professionals thinking and working with food, people, the environment and sustainability.

Conclusion

What the Treaty has successfully helped to understand is that our planet is a solid mechanism held together by thousands of fragile balances, that is protecting biodiversity gently, slowly, respectfully and in a very effective way to save these delicate balances. A big part of it has already been lost, forever. But it is still possible to save an important part of it and we cannot count only on individual wisdoms or commitment: we need laws, policies, and the Treaty is a crucial step in this direction. We have to gauge interventions every time in a different way because every time there is a different balance to save. But *accepting complexity* is the first step towards understanding this.

What is more, a new awareness is growing in the world of food production. Thousands of farmers, producers and even retailers and cooks are working wonderfully to rebuild or protect those products and traditions, and processes whose loss would make all of us poorer. The role consumers can play – together with institutions, researchers, politics, associations – is to help them in working better and better, sharing their experiences and their solutions, their 'seeds' for the future of food production. This is the main aim of *Terra Madre*, World Meeting of the Food Communities, held in Turin every other year since 2004.

This huge meeting, involving around 7000 people working together for 4 days in several different seminars – has been another tool to fortify a new, different way of thinking about food and agriculture, but also about progress and development. Who must help who? Who can teach what? And how everybody can help protect biodiversity?

References

Anderson, C. (2006) *The Long Tail: Why the Future of Business Is Selling Less of More*, Hyperion, New York

Greco, S. and Scaffidi, C. (2007) *Guarda che Mare*, Slow Food Editore, Bra, Piemonte, Italia, pp 9 and 19–20

Masini, S. and Scaffidi, C. (2008) *Sementi e Diritti*, Slow Food Editore, Bra, Piemonte, Italia

Petrini, C. (2007) *Slow Food Nation: Why Our Food Should Be Good, Clean, and Fair*, Rizzoli International Publications, New York

Petrini, C. (2009) *Terra Madre: Forging a Global Network of Sustainable Food Communities*, Chelsea Green Publishing, Claremont, NH

Tiezzi, E. Oikos: "*l'ambiente come laboratorio di creatività e di sostenibilità*" [Oikos: The Environment as a Laboratory of Creativity and Sustainability], in "Oikos, Rivista quadrimestrale per una ecologia delle idee", edited by Tiezzi, E. 1997

Part III

Experts' Views on Future Challenges in Implementing the Treaty: Trust and Benefit-sharing as the Key

Chapter 18

Our Heritage Is Our Future

Humankind's Responsibility for Food Security

Cosima Hufler and René Lefeber

The roots of the multilateral approach of the International Treaty on Plant Genetic Resources for Food and Agriculture

History explains global interdependence on plant genetic resources

As a multitude of studies have shown in the course of the past 30 years, global interdependence on plant genetic resources for food and agriculture (PGRFA) is nothing new, but merely a statement of fact. An often quoted FAO study dating from the year 1998 revealed the knowledge that only four crops (rice, wheat, sugar and maize) account for 65 per cent of the dietary intake worldwide (Palacios, 1998).

This is the result of a lively system of global exchange and movements of crops over hundreds of years, paired with the fact that crop varieties, if they are not nurtured through human care, will be neglected and are eventually endangered in their existence.

As a consequence of these processes of genetic uniformity and genetic erosion, the food base of humankind is already limited and even threatened to being reduced further through newly arising challenges, most prominently of all through climate change. Global interdependence results from these processes and is likely to increase further in the years to come.

Global interdependence requires global action

The organization of groups of persons in states emerged at a time human needs could be satisfied through either direct access to resources or trade to balance any deficits and surpluses in the domestic supply of such resources. Resource depletion resulting from continued and increasing demand has heightened awareness that the supply of the world's resources is finite. This does not only hold true for non-renewable resources, but also for renewable resources if the use of such resources is not sustainable. Competition among states for such finite resources has prompted the need for international regulation of their exploitation in order to secure their equitable use by present and future generations (Brundtland Report, 1987). Such need was especially felt as regards resources that are not subject to state sovereignty.

The prospect of benefits arising from the exploitation of mineral resources that are not subject to state sovereignty has led to the development in the second half of the 20th century of international frameworks for their legal status and use. Such resources can be found in common areas: the oceans, outer space and Antarctica. International agreements have designated the mineral resources of the deep seabed and celestial bodies within the solar system, other than the Earth, to be the 'common heritage of mankind'.[1] This means that there is common ownership over these resources and that their use is no longer free, but subject to international administration. Upon recovery, title to the resources can pass from mankind to third parties, but only in accordance with the applicable international framework. The international administration must secure that their use will be equitable. With respect to non-renewable resources, such as mineral resources, the principle governing their exploitation is the long-term maximization of benefits from the use of such resources. Implementing this principle is not without difficulty due to uncertainty regarding variables, such as the number of future generations and technological innovation that may impact on the use of resources for future generations.

In contrast to mineral resources, living resources in common areas have been exploited for centuries and the freedom of their use had long been established. However, the depletion of living resources that are not subject to state sovereignty, such as fish stocks, have led to the development of international frameworks governing their use that are based on different principles as regards their status and use. These resources are not subject to common ownership and the use of these renewable resources has not been subjected to international administration. Title to these resources is acquired through appropriation. However, international agreements limit the right of states, and their nationals, to freely appropriate and use these resources. Equitable use by present and future generations requires the conservation of living resources. The overarching principle guiding their exploitation is sustainable use; and the precautionary approach and the ecosystem approach must be taken into account to determine what use is sustainable.[2]

Similarly, concerns over the depletion of renewable resources that are shared by states, such as international watercourses, the ozone layer and the atmosphere, have triggered the development of international frameworks to secure their conservation and sustainable use. At the origin of this development is the recognition of a common interest of states in the conservation of these resources. International agreements related to the navigational and non-navigational uses of international watercourses are founded on the recognition of the community of interest of riparian states in the use of an international watercourse.[3] Similarly, the preamble of the United Nations Framework Convention on Climate Change acknowledges that climate change and its adverse effects are 'a common concern of humankind'. The concern over a common interest forms the basis for the concerned community to act and underlies the introduction of policies and measures by these international agreements to secure the equitable use of these resources by present and future generations (Shelton, 2009, p85).

A common interest in the conservation and sustainable use of natural resources is not necessarily limited to resources found in common areas or shared by states. This is recognized in the preamble of the Convention on Biological Diversity (CBD), which affirms that the conservation of biological diversity is 'a common concern of humankind'. This recognition is irrespective of the location of such resources within or beyond the limits of a state's jurisdiction. Accordingly, the location of a component of biological diversity within a state's jurisdiction does not prevent the introduction of internationally agreed policies and measures to control its use. Clearly, any such policies and measures cannot be imposed and must be based on respect for the sovereignty of states over their natural resources. The acceptance of internationally agreed policies and measures to control the use of resources within a state's jurisdiction reflects the exercise of sovereignty. The prevention of genetic erosion and genetic uniformity of plant genetic resources provides an example. This is a common interest and it has been recognized as such by the FAO.

Global action on plant genetic resources

The engagement of FAO in plant genetic resources dates well back into the 1960s. The year 1983 saw the adoption of the 'International Undertaking on Plant Genetic Resources' (IU), a voluntary instrument which has remained operational after the adoption and entry into force of the International Treaty on Plant Genetic Resources for Food and Agriculture (ITPGRFA or the Treaty) (see Annex 1 of this volume for the list of all Commission and Treaty meetings). The IU generally aims at the conservation and sustainable use of plant genetic resources. The objective contained in Article 1 states the main underlying principle of resource exchange: 'This Undertaking is based on the universally accepted principle that plant genetic resources are a heritage of mankind and consequently should be available without restriction.'

This principle is an expression of the interdependency of all countries with regards to PGRFA. If one looks at the four crops that account for 65 per cent of global energy intake and their centres of origin, it becomes evident how those have moved outside of these centres over the years and have been improved by farmers all over the world throughout the centuries. The diversity and variety of crops available to us nowadays is a result of the joint efforts of farmers and breeders all across the globe and cannot be accounted to one place of origin or one actor/stakeholder alone.

Therefore, it is easily understood why this principle found its entry into Article 1 of the IU and was repeatedly reaffirmed in the years to follow its adoption. The meaning of the principle is, however, less clear. The reference to 'mankind' points to the existence of a common interest in the conservation and sustainable use of plant genetic resources. The word 'heritage' connotes a temporal dimension and suggests that the use of the resources concerned should take into account the principle of intergenerational equity. However, the IU does not designate the plant genetic resources to be a 'common heritage'. The adjective 'common' associates the heritage with common ownership. The absence of that adjective in the IU allows for national ownership over the resources that fall within the scope of the IU. Since the IU does not provide for common ownership over plant genetic resources, it is not necessary to provide international administration of the use of such resources. Such international administration would also not seem to be compatible with the provision of the IU of which plant genetic resources should be available without restriction. The recognition of a common interest in the conservation and sustainable use of plant genetic resources, as evidenced by their designation as a heritage of mankind, is nevertheless significant as it provides the basis for the development of internationally agreed policies and measures to secure the equitable use of plant genetic resources by present and future generations. Facilitated access to plant genetic resources and equitable sharing of benefits arising out of the utilization of such resources are policies and measures that contribute to the achievement of this objective.

Nevertheless, the acceptance of the principle that plant genetic resources are the heritage of mankind was clearly not that romantic, as it may seem from a distance. Quite a number of developed countries held reservations to adhering to the IU, in particular, as related to plant breeders' rights and Farmers' Rights that might be affected by the application of the heritage-of-mankind principle. They feared that the implementation of the IU might still result in an international administration of resources, which would encroach upon their control over such resources.[4] The romance lasted until 1991, when the FAO Conference gave in to demands to clarify the principle further in its last Agreed Interpretation of the IU:

> *(a) the concept of humankind's heritage, as applied in the International Undertaking on Plant Genetic Resources, is subject to the sovereignty of the states over their plant genetic resources, [...]*
> *(d) conditions of access to plant genetic resources need further clarification.*[5]

What happened at that time? It was the point in time when, under the auspices of the United Nations Environment Programme (UNEP), the CBD was being negotiated and was soon to be adopted. This was the time when awareness grew significantly about the value of biological diversity for economic development and also the fear of biological diversity being exploited and degraded by multinational companies for purely monetary gains. Developing countries sought to prevent external interference with their domestic policies and measures to use natural resources under their jurisdiction. The negotiations resulted in the rejection of a multilateral approach to access and benefit-sharing within the framework of the CBD. The recognition of the sovereign rights of states over their natural resources was linked to the authority of national governments to determine access to its genetic resources. This provision reflects a complete 180 degree u-turn to the approach originally embarked upon by the FAO with the heritage-of-mankind principle and unrestricted availability of plant genetic resources. The only provision supporting the free availability of genetic resources is the call upon parties to endeavour to create conditions to facilitate access to genetic resources for environmentally sound uses. Access, where granted, is nevertheless subject to prior informed consent of the party providing such resources, unless otherwise determined by that party, and to mutually agreed terms. However, the sovereign rights based approach does not exclude a multilateral approach to access and benefit-sharing altogether.

The Conference for the Adoption of the Agreed Text of the Convention on Biological Diversity agreed that solutions need to be found with regard to access to ex situ collections of PGRFA not acquired in accordance with the CBD and the question of Farmers' Rights.[6] In 1993, the FAO Conference took on those outstanding matters and embarked on adapting the IU to the conditions created by the CBD. A mandate for negotiations was adopted for:

- the adaptation of the International Undertaking on Plant Genetic Resources, in harmony with the Convention on Biological Diversity;
- consideration of the issue of access on mutually agreed terms to plant genetic resources, including ex situ collections not addressed by the Convention; as well as
- the issue of realization of Farmers' Rights.[7]

Seven years later, in 2001, this resulted in the adoption of the ITPGRFA which is based on a multilateral approach to access and benefit-sharing (see Annex 3 of this book for details on the main provisions of the Treaty).

Common responsibility for access equals common responsibility for benefit-sharing

The Treaty's multilateral approach to access and benefit-sharing – A perfect circle

The Treaty's multilateral approach can be depicted as a circular system between access, benefit-sharing and the conservation and sustainable use of PGRFA. This approach gives recognition to the great level of interdependency in the food and agriculture sector as described earlier. The fundamental objective of the Treaty is the prevention of genetic uniformity and genetic erosion and hence, the maximum diversity of PGRFA. Although the Treaty does not designate plant genetic resources as a heritage of mankind in so many words, it appears from the preamble that the communal and temporal aspects of this notion are cornerstones of the Treaty. According to the preamble, the parties are '[*c*]*ognizant* plant genetic resources for food and agriculture are a common concern of all countries, in that all countries depend very largely on plant genetic resources for food and agriculture that originated elsewhere' and '[*a*]*ware* of their responsibility to past and future generations to conserve the World's diversity of plant genetic resources for food and agriculture'.

The Treaty recognizes the sovereign rights of states over their plant genetic resources, including their national government's authority to determine access to those resources (Article 10). In the exercise of its sovereign right over its genetic resources, a party – through the multilateral system of access and benefit-sharing of the Treaty – offers facilitated access to other parties as well as legal and natural persons therein of its plant genetic resources under state control for the purposes of research, breeding and training. This is a formalization of the practices that were in place already for hundreds of years among farmers worldwide. However, this is a system that is adapted now to the new advances in a globalized world and turned towards greater efficiency by minimizing transaction costs. For instance, the exchange of PGRFA takes place based on one standardized material transfer agreement, the SMTA, which lays down the terms and conditions of access to the resource(s) and benefit-sharing from the utilization of the accessed PGRFA.

The multilateral system allows any party to tap into the joint pool of PGRFA listed in Annex I of the Treaty and in return maximizes benefit-sharing again through a multilateral approach. For this purpose, the heart of the benefit-sharing approach is a multilateral fund that is, in principle and among other sources, being nurtured by an equitable share of the benefits arising out of the commercialization of a product based on the material derived from the Treaty.

The Treaty has one distinct feature that is unique – its multilateral system has created a plant genetic resources pool that resembles a global public good. Public goods are usually described by contrasting them to private goods which can be made excludable and exclusive in consumption. An example of a private good would be a car whose use (or 'consumption') is controlled by the owner in possession of the car keys. By contrast, the air we breathe would be denoted as a public good, as one person is in general not capable of reducing the amount of air available or controlling access to the air. In 1954, the economist Paul Samuelson was the first to describe public goods as '[goods] which all enjoy in common in the sense that each individual's consumption of such a good leads to no subtractions from any other individual's consumption of that good' (Samuelson, 1954, p387). Consequently, *global* public goods have been described as 'public goods with benefits ... that extend across countries and regions, across rich and poor population groups, and even across generations' (Kaul et al, 2003, p3).

Plant genetic resources as such would not feature as public goods and the multilateral system certainly is not comparable to the example of the air used above. However, the multilateral system does resemble the characteristics of a global public good, in so far as Article 12 of the ITPGRFA establishes that parties to the Treaty are to provide access – obviously under certain conditions further detailed within the Treaty – to those PGRFA held within the joint pool to other parties and to legal and natural persons under the jurisdiction of any party through the multilateral system. Hence, by the terms of the Treaty, those PGRFA contained in the multilateral system are available to all parties to the Treaty and one party in principle cannot prevent another party from accessing ('consuming') PGRFA held within the multilateral system. Furthermore, the Treaty regards such access as a benefit for all parties: 'The Contracting Parties recognize that facilitated access to Plant Genetic Resource for Food and Agriculture which are included in the Multilateral System constitutes itself a major benefit of the Multilateral System' (Article 13.1).

With the creation of the multilateral system, the same problems appear pertinent that are commonly known in relation to public goods, such as: Who feels responsible for maintaining what is contained in the multilateral system and who pays? Is there a free-rider problem and how is this to be addressed? Is the multilateral system in its current form sufficient or does it need to be expanded? The first question on responsibility is the most significant for the purposes of this article.

Public goods 'simply put' suffer from the fact that they are being taken for granted. Biodiversity is a shining example of this. However, the most recent 30 years or so have seen a greater consciousness that the loss of biodiversity constitutes a significant cost that only comes to bear over time. However, only from that moment onwards, where this loss of a public good has a tangible impact on the individual, is the individual willing to take a share in the responsibility to address this loss. The Treaty's multilateral system thus also serves the purpose to make a potential loss tangible, palpable. While every party benefits from the access to the plant genetic resources contained in the multilateral system, every party will also lose out when the system is compromised including through genetic erosion that would reduce the availability of plant genetic resources accessible through the system.

So, who feels responsible and who pays? The answer to this question is that the Treaty itself foresees an in-built mechanism that allows for the Treaty community as a whole to take responsibility for the maintenance of the core ingredients of the Treaty's multilateral system, that is the PGRFA. That mechanism is twofold: on the one hand, it is the financial support provided by the parties to the administration of the Treaty, as is usual practice in multilateral agreements and, on the other hand it is the benefit-sharing fund as an in-built mechanism of the Treaty, and more widely so, its funding strategy.

The funding strategy should close in on the other half of the Treaty's circular system: There are certain limitations to the system and the most important one is that PGRFA constitute a resource that could potentially become extinct. The other limitation is the same as with any other multilateral agreement, its effective implementation depends on a level playing field for all parties in terms of their capacity and ability to implement the system. This is the reason why the funding strategy of the Treaty lays down the three priority areas, namely conservation, sustainable use and assistance (capacity-building and technology-transfer), towards which funding for plant genetic resources for food and agriculture should be directed. For the area of PGRFA, sustainable use is of greatest importance – diversity can not only stem from conserving what is already in major use, but also by making sustainable use of neglected or underutilized crops so as to create incentives for their conservation as well as eventually increasing genetic variety.

The funding strategy recognizes the number of finance streams directed towards plant genetic resources and aims at a comprehensive strategy in the best interests of the parties to the Treaty. Its heart is formed through its benefit-sharing fund which holds the financial resources that are within the direct control of the Treaty's Governing Body.

The benefit-sharing fund was mainly conceived as the fund that would be nurtured through the monetary benefits derived directly from the utilization of PGRFA. This fund would finance projects in the three priority areas, targeted towards in situ conservation and on-farm management to be able to make full use of the mechanisms of the Treaty. In this conceptualization it would form one puzzle piece in the entire funding landscape for biodiversity, including for PGRFA. As the Fund's resources would be under the direct control of the Governing Body, parties collectively would be able to select projects that would fill urgent implementation gaps and would allow for quick responses.

In short: The circular system of the Treaty foresees facilitated access to a joint pool of resources that is being commonly cultivated by all parties and accessible on the basis of a standardized benefit-sharing arrangement. Contributions based on the benefit-sharing arrangement would in return flow back into a multilateral benefit-sharing fund of the Treaty. Apart from capacity-building projects and programmes, this fund should contribute to the conservation and sustainable use of PGRFA, so as to achieve the objectives of the Treaty and maintain maximum diversity of plant genetic resources. In theory, the circle seems complete.

Developments in the benefit-sharing fund – going round in circles

Reality could appear different. The entry into force of the Treaty only dates back six years. Therefore, it is still partly adjusting to get into the flow of things. Looking at breeding cycles, direct contributions from commercialization can realistically be expected only in several years time from now. Urgently required capacity-building programmes have therefore been facilitated through a newly created Joint Programme on Capacity-Building of the FAO, the Treaty Secretariat and Bioversity International, which not only allows advances towards the required level playing field among the parties but also allows more countries to become parties to the Treaty. Considerable funding flows are taking place towards the conservation and sustainable use of PGRFA (e.g. including through the Global Environment Facility (GEF), or in terms of in situ conservation through the Global Crop Diversity Trust), but they constitute financial resources that are not under the direct control of the Governing Body.

The crux: While a common effort was made to building the multilateral system, the in-built mechanism described above had difficulties delivering a tangible perception that the common responsibility for the sharing of benefits was taken seriously.

It was only after the Second Session of the Governing Body in 2007 that some parties (Spain, Italy, Norway and Switzerland) committed voluntary contributions to the benefit-sharing fund in order to facilitate the execution of a number of projects selected and approved by the Governing Body. This received great appreciation of all parties at the Third Session of the Governing Body in 2009. The political response to this was twofold. The Governing Body decided, on the one hand, that PGRFA resulting from projects funded by the benefit-sharing fund and listed in Annex I be placed under the multilateral system. On the other hand, the Governing Body has established a target of US$116 million to be reached over the next five years. This constitutes an acknowledgement of the time-lag occurring before the monetary benefits arising from the utilization of PGRFA accessed through the multilateral system will be committed to the benefit-sharing fund. During this time, more intensified provision of voluntary contributions will be required to address conservation, sustainable use and capacity-building needs.

Clearly, the benefit-sharing fund cannot assume the role of a financial mechanism of the magnitude of the GEF, for example, and it was not created as such. Yet, the benefit-sharing fund, and more broadly, the funding strategy have two very significant purposes: First, they address needs that are directly related to PGRFA and the implementation of the Treaty, and second, they add greater coherence in the wide and broad funding landscape for plant genetic resources by setting clear priorities and directions.

This is the direction towards which the Treaty is currently heading. The fundamental underlying motivation of the global exchange of PGRFA even centuries before its formalization through the Treaty has always been the provision of food crops that fit the climatic and socio-economic environment of a region or country. Food security is the overarching expression for this – it is no coincidence that the first Millennium Development Goal of the United Nations ('Eradicate extreme poverty and hunger') finds its origins in the policies and activities of FAO. In addition to all the obstacles towards achieving this goal, food security is confronted with another threat: climate change.

At the 12th session of the Commission on Genetic Resources for Food and Agriculture in October 2009, a study on 'The Impact of Climate Change on the Interdependence of Countries and the Genetic Resources for Food and Agriculture' (Fujisaka et al, 2009) was submitted. This study shows that levels of interdependence of countries on genetic resources for food and agriculture will grow even further through the results of climate change, in particular, for plant genetic resources. Climate change will impact on the suitability of currently adapted landraces and varieties for various regions and increase the demand in general for PGRFA globally. One of the main findings is that '[i]nternational cooperation/coordination between farmers, government institutions, and research agencies will be critical in order to support the moving production system of germplasm from present locations that become unsuitable to future suitable areas as well as to support continued agricultural production in areas that will experience unprecedented climate-related stresses'. This shows that self-sufficiency will also not be possible in the future. The Treaty holds the key for early preventive and precautionary measures to assist farmers to adapt to climate change before the effects of climate change will affect food security, which is expected to take place in the next 30–50 years.

Common and joint responsibility – closing the circle

Facts show that from one biennium to the next, voluntary funding for the Treaty's benefit-sharing fund increased from a sum of approximately US$600,000 to a sum currently approaching US$13 million and projected to rise further. This is particularly remarkable against the backdrop of the current global recession and economic crisis and a general serious pressure on public spending.

While the success of this is surely a combination of factors, it should be taken for what it is in the first place: a conviction of the international community that the multilateral system of the Treaty is to be maintained if we want to secure the conservation and sustainable use of PGRFA. The parties have collectively taken common and joint responsibility for those PGRFA that they have placed in the multilateral system – both in terms of maintenance of the system as well as for maintaining maximum diversity of plant genetic resources. Diversity of PGRFA is of course not limited to those that are currently contained in the multilateral system, even if those selected 64 crops reflect the criteria of food security and interdependence. Growing interdependence and climate change as a serious factor in genetic uniformity and genetic erosion are important arguments that underline the need for a comprehensive approach to food security. In the acknowledgement that the multilateral system is an expression of joint responsibility and joint custodianship, a comprehensive approach to food security could entail that those PGRFA that are currently not covered by the multilateral system might become subject to the system in the future.

Conclusion

The ITPGRFA demonstrates that the permanent sovereignty of states over their natural resources does not preclude a multilateral approach to the use of such resources. The common interest in the prevention of genetic uniformity and genetic erosion has induced states to design a system that facilitates international access to PGRFA in return for an equitable share of the benefits arising out of their utilization. This multilateral approach respects sovereign rights over plant genetic resources and is designed to secure their equitable use by present and future generations.

There is a delicate balance between access to and use of resources, on the one hand, and the sharing of benefits arising from such use, on the other. There will be no benefits to share without the use of resources, but a system that allows for use without the return of benefits is not sustainable. Since there may be a considerable lag between access and the return of benefits, the survival and further development of the multilateral system of the ITPGRFA became critically dependent on the availability of funds to span the time between the use of plant genetic resources and the return of benefits arising from such use. The development of a funding strategy and its successful implementation are thus essential to come to a full circle and to preserve the circle of life.

Notes

1 See 1982 United Convention on the Law of the Sea (Article 136); 1979 Agreement Governing the Activities of States on the Moon and Other Celestial Bodies (Article 11).
2 See, for example, Practical Principle 5 of the Addis Ababa Principles and Guidelines for the Sustainable Use of Biological Diversity, Annex II of Sustainable Use (Article 10), CBD Decision VII/12 (2004), calling for the application of the precautionary approach in accordance with Principle 15 of the 1992 Rio Declaration on Environment and Development and the ecosystem approach in accordance with Principles 3, 5 and 6 of the Ecosystem Approach, CBD Decision V/6 (2000).
3 See International Court of Justice, Case Concerning Gabčíkovo-Nagymaros Project (Hungary/Slovakia), Judgment of 25 September 1997, *1997 ICJ Reports*, p7, para 85.
4 These countries were Canada, France, the Federal Republic of Germany, Japan, New Zealand, Switzerland, the United Kingdom and the United States. See H. J. Bordwin (1985) 'The Legal and Political Implications of the International Undertaking on Plant Genetic Resources', *Ecology Law Quarterly*, vol 12, pp1053–1069.
5 See Resolution 3/91, 'Annex 3 to the International Undertaking on Plant Genetic Resources', of the 26th session of the FAO Conference.
6 See Resolution 3, 'The Interrelationship between the Convention on Biological Diversity and the Promotion of Sustainable Agriculture', Nairobi Final Act of the Conference for the Adoption of the Agreed Text of the Convention on Biological Diversity, 22 May 1992.
7 See Resolution 7/93, 'Revision of the International Undertaking on Plant Genetic Resources', of the 27th session of the FAO Conference.

References

Brundtland Report (1987) *Our Common Future*, available at www.ace.mmu.ac.uk/eae/sustainability/Older/Brundtland_Report.html

Fujisaka, S., Williams, D. and Halewood, M. (2009) 'The impact of climate change on countries' interdependence on genetic resources for food and agriculture', CGRFA Background Study Paper No. 48

Kaul, I., Conceicao, P., Le Goulven, K. and Mendoza, R. U. (eds) (2003) *Providing Global Public Goods: Managing Globalisation*, Oxford University Press, New York

Palacios, X. F. (1998) 'Contribution to the estimation of countries' interdependence in the area of plant genetic resources', CGRFA Background Study Paper No. 7, Rev.1

Samuelson, P. A. (1954) 'The pure theory of public expenditure', *Review of Economics and Statistics* (The MIT Press), vol 36, no 4, pp387–389

Shelton, D. (2009) 'Common concern of humanity', *Environmental Policy and Law*, vol 39, pp83–90

Chapter 19

An Innovative Option for Benefit-sharing Payment under the International Treaty on Plant Genetic Resources for Food and Agriculture

Implementing Article 6.11 Crop-related Modality of the Standard Material Transfer Agreement

Carlos M. Correa

Introduction

This chapter discusses the crop-related payment established by Article 6.11 of the Standard Material Transfer Agreement (SMTA), as adopted by the Governing Body of the International Treaty on Plant Genetic Resources for Food and Agriculture (ITPGRFA or the Treaty). It argues that this option, based on a proposal by the African group, might be attractive for recipients and important to generate funding for benefit sharing under the Treaty. The ITPGRFA has been developed 'in harmony with the Convention on Biological Diversity' (CBD)[1] (see Annex 1 of this volume for the list of all Commission and Treaty negotiating meetings). Accordingly, the Treaty recognizes 'the sovereign rights of States over their own plant genetic resources for food and agriculture', and that the 'authority to determine access to those resources rests with national governments and is subject to national legislation'.[2] The ITPGRFA sets out as one of its principal objectives to ensure 'the fair and equitable sharing of the benefits' arising out of the use of plant genetic resources for food and agriculture'[3] (see Annex 3 of this book for details on the main provisions of the Treaty).

Despite aiming at the same objective of the CBD as regards benefit sharing, the system established for this purpose under the Treaty is significantly different from the CBD's mechanism. While the latter is essentially conceived as resulting from a bilateral relationship between the country providing genetic resources and the recipient thereof, benefit sharing under the ITPGRFA is of a multilateral nature.

In addition to the facilitated access to plant genetic resources for food and agriculture (PGRFA) which are included in the 'multilateral system',[4] contracting parties to the ITPGRFA (see Annex 2 of this volume for the list of contracting parties to the Treaty) may benefit from 'the exchange of information, access to and transfer of technology, capacity-building, and the sharing of the benefits arising from commercialization'.[5] Importantly, these benefits are not intended to accrue to individual countries, but to be distributed 'fairly and equitably'[6] among contracting parties 'taking into account the priority activity areas in the rolling Global Plan of Action'.[7]

Obviously, the capacity to fulfill the benefit sharing objectives of the ITPGRFA will depend on the funding available for this purpose. The benefit-sharing fund established in pursuance to the ITPGRFA has already funded more than a half-million US dollars in awards aimed at supporting 11 developing countries for the protection of existing collections of seeds and other genetic resources.[8] Funding has relied so far on voluntary contributions from Norway, Italy, Spain and Switzerland, which have contributed seed money for the benefit-sharing scheme.[9]

The benefit-sharing fund should also receive, under certain circumstances, contributions from the recipients of materials in the multilateral system. One of the components of the benefits to be shared under the Treaty is to be derived, in effect, from the obligation imposed by Article 13.2 (d)(ii) of the Treaty: a recipient who commercializes a product that is a PGRFA and that incorporates material accessed from the multilateral system, must pay to the international fund set up by the Treaty, 'an equitable share of the benefits arising from the commercialization of that product, except whenever such a product is available without restriction to others for further research and breeding, in which case the recipient who commercializes shall be encouraged to make such payment'.[10]

Product-related payment under the SMTA

Consistently with the principle of facilitated access that underpins the multilateral system created by the Treaty, Article 13.2 (d)(ii) of the Treaty only requires a payment to be made when the PGRFA that incorporates a material obtained from that system is commercialized and subject to 'restriction'. In accordance with the SMTA approved by the Governing Body, the recipient shall pay a fixed percentage of the sales of the commercialized product, which has been set at 0.77 per cent (1.1 per cent less 30 per cent) of the sales value.[11]

The Treaty does not define the type of restriction that would trigger payment; as defined by the SMTA (see below), such a restriction might be of legal, technological or contractual nature. Despite the broad range of possible measures that may restrict access for further research and breeding, the likelihood of immediate and substantial payments under the referred Article 13.2 (d)(ii) is low. During the negotiations of the Treaty, there were significant expectations about the funding that the implementation of this obligation could generate; however, its actual potential is probably rather limited.

There are two main reasons for this hypothesis:

1 Developing a new variety by conventional breeding methods may take several years and, hence, payments by potential recipients may not be received soon. The payment obligation is triggered when a product is 'commercialized'. This means that the product must be actually introduced into commerce. The logical linkage to commercialization rather than access delays the possible generation of income for benefit sharing. In accordance with the Secretariat of the Treaty, 'plant breeding is a slow process and it can take ten years or more for a patented product to emerge from the time the genetic transfer took place which is why the aforementioned governments have backed the scheme'.[12]
2 Legal restrictions are likely to arise out only in those few countries where plant varieties are patentable per se. Most countries have implemented the exception specifically allowed by the Agreement on Trade-Related Aspects of Intellectual Property Rights (TRIPS Agreement is Annex 1C of the Marrakesh Agreement Establishing the World Trade Organization, signed in Marrakesh, Morocco on 15 April 1994) and do not allow such patents.[13] As a result, the payment obligation may arise in a relatively small number of countries.[14]

The 'African proposal'

During the negotiations regarding the text of SMTA, the African group proposed an alternative to the obligation to pay a royalty on *each* product that incorporates material received from the multilateral system, as described above. The proposal, with some amendments, was finally incorporated in Article 6.11 of the SMTA.

The main reason underpinning this proposal was the African group's concern about the long period that would normally be necessary to develop new varieties which would eventually incorporate materials from the multilateral system and the limited circumstances in which the obligation to pay might arise. The proposal emerged from discussions between the African group members and Mr José Esquinas-Alcázar, Secretary of the Commission on Genetic Resources for Food and Agriculture. It essentially aimed at providing a simple method of payment that could reduce transaction costs for recipients in obtaining materials from the multilateral system. At the same time, it would accelerate and increase the generation of income to support the various types of benefit sharing activities contemplated in the Treaty.

The proposal suggested another option to the product-related payment obligation. Choosing this alternative was left to the discretion of the recipient because the proposed royalty (as finally established by the SMTA) would be applicable not only to the sales of the product that incorporated the material received from the multilateral system, but to *any* products that are PGRFA belonging to the same crop to which the material received from the multilateral system belongs. This means that, by selecting this option, the recipient would pay a royalty on all products of a certain crop regardless of whether they incorporate the material received from the multilateral system or whether the further use of the material by third parties for research and breeding is limited. A clear advantage of this option from the perspective of contracting parties is that the payment obligation would be triggered as soon as the recipient sells any product of the respective crop.

An important feature of this option is that, once the choice is made, it becomes the *mandatory* form of payment applicable to the recipient. This means the recipient is free to choose but, after selecting his preferred option, he is bound by the respective terms and conditions of the SMTA.

The African proposal was received with some scepticism by some of the negotiating parties. Doubts were raised about the compatibility with the IPGRFA given that payment is not linked to the effective commercialization of a product incorporating material received from the multilateral system. It might even happen that such a product was never developed; despite this, the recipient would be obliged to pay the established royalty. Strictly speaking, there would be no 'benefit sharing' since no such benefit would have been created at that stage. This observation, however, can be dismissed on the argument that the Treaty provides for mandatory and voluntary payments. The African proposal introduced a hybrid solution: it is voluntary to opt for it but, as noted, payment becomes mandatory when the recipient has exercised his right to choose.

It was also argued that the Treaty required the establishment of a single level of payment. The commented proposal introduced, in fact, a different (discounted) royalty rate. But Article 13.2(d)(ii) of the Treaty provides that '(t)he Governing Body may decide to establish different levels of payment for various categories of recipients who commercialize such products ...'. Recipients that accept to pay a royalty over all the products belonging to a crop may be considered a different 'category' of recipients.

The duration of the obligation to pay was also questioned. The mandatory product-related payment under the SMTA has no definite term. It will be enforceable as long as the conditions that trigger the payment obligation continue to exist. Since the African proposal delinked payments from the presence of the received material in the products sold, the determination of a term was necessary and introduced in the adopted SMTA.

Finally, doubts were raised about the potential interest of seed companies and other recipients to subscribe to an option that might create a financial burden higher than that emerging from the mandatory payment. However, the African proposal, received support from the representative of the seed industry (for details on the seed industry, see Chapter 12). In fact, it may be particularly suitable to companies that are unwilling or unable to track the presence of a received material in its breeding lines and could eventually become the preferred option for some companies in the seed industry.

Crop-related payment under the SMTA

The African proposal, as implemented in the SMTA, allows for a simplification of the procedures to receive materials in the multilateral system. While subscribing to an SMTA, the recipient must notify the Governing Body that he has opted for this modality of payment.[15] The recipient is relieved from 'any obligation to make payments under Article 6.7 of this Agreement or any previous or subsequent Standard Material Transfer Agreements entered into in respect of the same crop'.[16] This means that once a recipient has opted for this alternative to receive a particular material or set of materials, he may receive other materials *belonging to the same crop* by signing the respective SMTA. The recipient will be waived from complying with the product-related payment obligation under these new SMTAs, but he will be subject to the other obligations established by the agreements, including not to seek intellectual property protection over the received materials.

In implementing the proposal, the SMTA took into account the referred concern raised during the negotiations about the period of validity of the option. In accordance with the SMTA, the payment clause will be valid for ten years and would be renewed for additional periods of five years unless the recipient notifies his intention to opt out.[17] If the application of the clause were terminated, the recipient would be bound to make payments only on the products that incorporate material received during the period in which the clause was in force, and only in cases where such products are not available without restriction.[18] These payments would be calculated at the same rate as that applicable during the period in which the clause was in force.

In order to make the crop-related payment attractive to potential recipients, as mentioned, a discounted royalty rate was set out by the SMTA. It was set at 0.5 per cent of the sales of any products that incorporate the received material and of the sales of any other products that are PGRFA belonging to the same crop to which the material received under the SMTA belonged.

Given that the product-related payment under the SMTA is, as noted, 0.77 per cent, the discounted rate means a saving of 0.22 per cent, but the crop-based rate applies to all products of the recipient for the relevant crop. This modality of payment, hence, would generate considerably more income from individual recipients than the product-related modality. It is even possible to speculate that a greater discount could attract more recipients and increase the funds available for benefit sharing under the Treaty. The Governing Body might consider reviewing the royalty rate applicable to this option in the future, in order to expand the difference with the ordinary rate.

Complying with the payment obligation: A comparative analysis of the two options

Compliance with the product-related payment obligation requires the recipient to submit to the Governing Body an annual report setting forth the sales of the product that incorporates the material received from the multilateral system, including the amount of the payment due, and information that would allow for the identification of any restrictions that have given rise to the benefit-sharing payment.[19]

Since the product-related payment under the SMTA would be mandatory only when a restriction is imposed by the recipient for further research and breeding by third parties, the recipient would have to assess whether a 'restriction' encumbers a product that incorporates the received material in a way that limits research and breeding by others.

The SMTA defines in Article 2 'available without restriction' as follows: 'A Product is considered to be available without restriction to others for further research and breeding when it is available for research and breeding without any legal or contractual obligations, or technological restrictions, that would preclude using it in the manner specified in the Treaty'. This definition suggests that a 'restriction' would exist when the owner of the product is able 'to exclude, prevent, make impracticable'[20] access for research and breeding. This interpretation raises, among others, the question whether the establishment of certain *conditions* (for instance, payment of a predetermined royalty) to get access to a product would be sufficient or not to consider that a 'restriction' exists. The recipient may have reasonable doubts in these cases about the need or not to effect the payment provided for, and should eventually seek clarification from the Governing Body or any subsidiary body dealing at that time with this issue through the Secretariat.[21]

In addition, the implementation of the product-related payment obligation imposes on the recipient the burden of tracking the use of the material received from the multilateral system, keeping separate records of the products that incorporate such material, calculating and paying the established royalty on each of the products in this situation. Further, the recipient would be responsible not only for payment of the royalties calculated on the sales of his own products, but also on the sales made by its affiliates, contractors, licensees and lessees. This might create a significant additional burden on the recipient.[22]

Opting for the crop-related payment obligation would not mean that the recipient would be relieved from signing new SMTAs to obtain other materials from the multilateral system, even if they belonged to the same crop as the material obtained under the first SMTA. Likewise, if he had previously signed other SMTAs, he must comply with them, except with regard to the product-related payment obligation. The integrity of the system, hence, is not affected in any way by the implementation of the crop-related payment option.

The crop-related payment option presents some advantages for the operation of the multilateral system set out by the Treaty. They include the possibility of generating income faster than under the product-based modality, as well as of reducing the monitoring costs. In effect, there would neither be a need to verify whether a material received from the multilateral system has been incorporated by a recipient into a commercialized product, nor to establish whether further access for research and breeding is restricted. This would reduce the burden of the third party beneficiary and, possibly, avoid litigation. In addition, as noted, the income generated by individual recipients may be much greater than under the product-based payment modality, since the 0.5 rate would be applicable to all the recipient's sales of products belonging to the same crop.

On the other hand, the modality of crop-related payment may have a number of distinct advantages for recipients as compared to the product-based payment, namely:

- No need to track the incorporation of the material received from the multilateral system.
- No obligation to provide the Governing Body with information about restrictions for further use.
- Straight and simple annual calculation of the royalty payments to be made.
- Disputes about compliance with the SMTA are less likely to arise.
- Opting for the crop-related modality may be positive in terms of public relations for the image of seed companies (as supporters of the implementation of the Treaty).

In sum, this option may be far less bureaucratic and much easier to administer and enforce by recipients than the product-related alternative. There are, in fact, indications that some seed industry circles are interested in investigating more deeply the potential advantages of the crop-related modality as the preferred alternative.

Conclusion

The crop-related modality of royalty payment represented an innovative way of looking at the implementation of the obligation established by Article 13.2 (d)(ii) of the Treaty. Through this hybrid (mandatory/voluntary) option, transaction costs may be reduced for both the Governing Body (and FAO as third beneficiary) and the recipients that choose to apply it. The benefit-sharing fund created in pursuance of the Treaty might receive royalty payments earlier than under the product-based modality, given that there will be no need to wait until a product incorporating material from the multilateral system is developed and commercialized. Moreover, since payment is to be made independently of the existence of any restriction for the further use of the improved material, if that option were chosen by a large number of recipients and/or by companies with significant seed sales, it might possibly generate more funds than its contractual alternative. Although the proposal by the African group was essentially aimed at speeding up and improving funding for benefit sharing, the optional mode of payment incorporated into the SMTA may, due to its lower cost and greater simplicity, serve well the immediate and long term interests of a wide range of recipients.

Notes

1 Article 1.1 of the ITPGRFA.
2 Article 10.1 of the ITPGRFA.
3 Article 1.1 of the ITPGRFA.
4 Article 13.1 of the ITPGRFA.
5 Article 13.2 of the ITPGRFA.
6 Article 13.1 of the ITPGRFA.
7 Article 13.2 of the ITPGRFA.
8 See ftp://ftp.fao.org/ag/agp/planttreaty/news/news0009_en.pdf (accessed 2 September 2010). See also document IT/GB-1/06/Report, Appendix F establishing the 'funding strategy' for the implementation of the ITPGRFA, available at ftp://ftp.fao.org/ag/agp/planttreaty/funding/fundings1_en.pdf (accessed 2 September 2010).
9 Norway introduced a small tax on the sale of seeds on its domestic market to fund its donation. See www.itpgrfa.net/International/content/delegates-120-nations-tunis-share-benefits-treaty-food-plant-genes (accessed 2 September 2010).
10 Article 13.2 (d)(ii) of the ITPGRFA.
11 Annex 2, Article 1 of the SMTA.
12 See www.itpgrfa.net/International/content/delegates-120-nations-tunis-share-benefits-treaty-food-plant-genes (accessed 2 September 2010).
13 See Article 27.3(b), which permits WTO members to protect plant varieties under a *sui generis* regime, patents or a combination of both.
14 The USA is one of the few countries where patents may be granted over plant varieties. Although the USA has signed the ITPGRFA, it has not ratified it yet.
15 SMTA, Article 6.11(h).
16 SMTA, Article 6.11(f).
17 SMTA, Article 6.11(h). In accordance with Annex 3, Article 4 '[A]t least six months before the expiry of a period of ten years counted from the date of signature of this Agreement and, thereafter, six months before the expiry of subsequent periods of five years, the Recipient may notify the Governing Body of his decision to opt out from the application of this Article as of the end of any of those periods. In the case the Recipient has entered into other Standard Material Transfer Agreements, the ten years period will commence on the date of signature'.
18 SMTA, Article 6.11(h).
19 SMTA, Appendix 2, Article 3.
20 The *Concise Oxford Dictionary*, p808.
21 SMTA, Article 8.3.
22 SMTA, Annex 2, Article 3(a).

Chapter 20

General Conclusions

Summary of Stakeholders' Views and Suggestions to Cope with the Challenges in the Implementation of the International Treaty on Plant Genetic Resources for Food and Agriculture

*Christine Frison, Francisco López and José T. Esquinas-Alcázar**

The purpose of this chapter is to provide a general analysis of the comments made by the authors of the book chapters in the ongoing implementation of the Treaty. The reader will have noticed that, on the one hand, many authors remain fairly optimistic about the Treaty and note that considerable progress has been achieved in a very short period of time, even beyond their initial expectation. On the other hand, some authors, while recognizing that the Treaty is a useful and flexible instrument, point at the risk that the lack of appropriate and quick decisions and actions to speed up the implementation process may lead to a decreased level of confidence in the general framework set up by the Treaty. Most of them recognize that it is now the moment to advance on its implementation.

In order to analyse most of the appraisals and concerns provided by authors on the implementation of the Treaty in a systematic way, these concerns have been grouped in several sections in line with the structure of the Treaty (Part I of this chapter). The editors, have tried to go one step further by sharing thoughts on possible ways and means to address these concerns (Part II).

Part I – Appraisals and concerns raised by the authors on the Treaty and on the implementation of its provisions

Following as much as possible the structure of the Treaty, this first part is divided into eight sections: General Considerations, Conservation and Sustainable Use, Farmers' Rights, the Multilateral System of Access and Benefit-sharing, Instruments for International Cooperation, and Financial Provisions. For the benefit of those readers not familiar with all the details of each Treaty provision, the editors decided to add a short explanatory note under the title 'Thematic content' for the most complex issues addressed. Then, appraisals and concerns made by the authors are summarized.

General considerations: Public awareness, policy coherence, legal certainty and trust created by the Treaty

Many authors have made general comments on the Treaty and many of them have said that public awareness and policy coherence as well as legal certainty and trust are important factors contributing to an efficient implementation of the Treaty.

Regarding public awareness and policy coherence, there is a belief that the mere existence of the Treaty as a legally binding instrument is crucial for three reasons. First, the recognition of the importance of plant genetic resources in the national political arena has significant value in itself as it puts agriculture at the forefront (e.g. Chapters 5, 8, 9 and 18). Second, authors acknowledge that the Treaty provides for a renewed belief that protecting PGRFA is an urgent matter (e.g. Chapters 6 Appendix, 7, 13 and 16). Finally, authors also note that the adoption of the Treaty, its rapid entry into force and implementation have significantly contributed to put the agricultural sector in the limelight within the constellation of international fora and associated UN institutions (e.g. Chapters 6, 7, 10, 12 and 14). However, some authors also point to a number of general shortcomings, in particular, the insufficient coordination and coherence at different levels (e.g. Chapters 3 and 7). Some authors pledge in favour of stronger coordination and synergy in the development of policies, legislation and regulations among related international instruments (such as the Treaty, the CBD and TRIPs), and among the various competent ministries, governments and other institutions with responsibility for different aspects of PGRFA (e.g. Chapters 3 and 4). Some authors also feel that awareness of the Treaty at the national level is too low (e.g. Chapters 3, 7 and 17). In certain cases, the limited capacity, funds and training make it hard to organize wide national consultations (e.g. Chapters 9 and 13).

Regarding legal certainty and trust, many authors tend to agree that the Treaty constitutes a framework providing legal certainty and clarity in the exchange of PGRFA. They recognize that the Treaty has fostered trust among stakeholders and also between developed and developing countries (e.g. Chapters 2, 5, 6 and 8). Nevertheless, many authors point to the lack of clarity of some specific Treaty provisions (e.g. Chapters 2, 4, 5, 6 Appendix and 15), such as Article 12.3(d), or Article 11.2 (see below section 4). Some authors have signalled a possible decrease in the trust they confer to the instrument since its entry into force, as a result of these ambiguities (e.g. Chapter 3).

Conservation and sustainable use (Articles 5 and 6 of the Treaty)

> **Thematic content**
>
> Article 5 and 6 of the Treaty deal with 'Conservation, Exploration, Collection, Characterization, Evaluation and Documentation of Plant Genetic Resources for Food and Agriculture' and with 'Sustainable Use of Plant Genetic Resources'. These provisions constitute obligations of contracting parties.

Some authors recognize that in their national implementation the conservation obligations have attracted more attention than the sustainable utilization ones (e.g. Chapters 6 and 7). Other authors request more clarity as to what should be understood under the obligation of sustainable use of PGRFA (e.g. Chapter 17). Many authors consider that these two articles are non-separable and argue for a stronger and faster implementation of both conservation and sustainable use provisions at the domestic level (e.g. Chapters 6 Appendix and 7). Authors stress that the main constraints for the implementation of these provisions relate to technical and scientific limitations regarding the maintenance and management of genetic diversity and the sustainable use of genetic resources, such as poor safety and storage conditions, lack of financial and human resources (e.g. Chapters 7 and 16).
Some authors favour the promotion of in situ conservation and use of wild crop relatives and wild plants, going as far as proposing the establishment of regional research sites (e.g. Chapter 7). Moreover, some authors stress that more information on genetic erosion should be generated (e.g. Chapter 7, 8).

Farmers' Rights (Article 9 of the Treaty)

> **Thematic content**
>
> Part III of the Treaty is entirely devoted to Farmers' Rights and to contracting parties' responsibility for their realization. Article 9 reiterates the broad rationale for Farmers' Rights in the first paragraph (Art. 9.1). In the second paragraph, specific rights are identified (Art. 9.2). However the weight of this second paragraph in terms of an international obligation is limited since the main responsibility for their realization rests with national governments. The Preamble also refers to Farmers' Rights and to the importance to promote them at both national and international levels (Treaty Preamble § 7 and 8). The last part of the article (Art. 9.3) deals specifically with the rights of farmers to save, sell and exchange seeds, and remains neutral on the status of these rights.

Many authors consider that the Treaty provides incentives for stakeholders, in particular, farmers, to conserve and use PGRFA in a sustainable manner, through the recognition of Farmers' Rights and the benefit-sharing mechanism of the MLS (e.g. Chapters 2, 3, 7 and 17). Nevertheless, some authors point out that for them Farmers' Rights was not a primary concept during the negotiations of the Treaty (e.g. Chapters 5 and 8). For others, the inclusion of a provision for Farmers' Rights in the Treaty is seen as an important first step (e.g. Chapters 3 and 13). They state that the Treaty has allowed for the recognition of farmers' movement and for more coherent and larger organization of farmers' communities (e.g. Chapters 10 and 13). The establishment of a Global Fund for Farmers, similarly as the Global Crop Diversity Trust, to support farmers' work on on-farm conservation and crop development is also proposed (e.g. Chapter 13).

Several authors sustain that there is a lack of sufficient international recognition[1] and national implementation of Farmers' Rights (e.g. Chapters 3 and 13). Difficulties in implementing Farmers' Rights at the national level are attributed to a lack of legal expertise and prior experience in the field (e.g. Chapter 3). Moreover, the authors from the chapter on farmers' communities consider that Farmers' Rights should encompass many other rights (e.g. Chapter 13). They argue that Farmers' Rights are a bundle of rights, which should characterize the interrelationships of seeds with land, water, energy, culture, social fabric, household and individual well-being. Authors from the seed industry chapter believe that Article 9.3 does not provide any legitimacy to save, use and sell farm saved seed, and they interpret Article 9 in a limited manner (e.g. Chapter 12). Finally, some authors have stressed the fact that the provisions of the Treaty and their implications for smallholder farmers are yet to be substantially 'processed' by farmers and their communities. A large number of farmers and their organizations have yet to identify themselves within this 'social construct' (e.g. Chapter 13). Still, while some authors stress that it is strictly a domestic issue (e.g. Chapter 8), other authors contend that there is some lack of clarity and no common interpretation of this Treaty provision, which contributes to misunderstandings and requires clarifications (e.g. Chapter 3). Up to now, only a few countries, such as for instance India, have implemented legislation on Farmers' Rights (e.g. Chapter 14). [2]

The multilateral system of access and benefit-sharing (Articles 10–13 of the Treaty)

Thematic content

Part IV of the Treaty is devoted to the Multilateral System (MLS) of Access and Benefit-Sharing. Article 10 recognizes the sovereign rights of countries over their own PGRFA. It states that Contracting Parties establish the MLS in the exercise of these sovereign rights. Article 11 limits the scope of the MLS to PGRFAs listed in Annex I to the Treaty, while Article 12 defines its facilitated access mechanism and Article 13 deals with benefit-sharing.

Many authors consider the MLS as the core of the Treaty. The MLS is generally regarded as a unique instrument because it ensures multilateralism (e.g. Chapters 2, 5 and 6). Most authors recognize that the success of the Treaty will depend on the effective implementation of its MLS (e.g. Chapters 3, 6 and 19). Several authors recall that the MLS creates a balance between access and benefit-sharing (e.g. Chapters 5, 18, and Annex 4). Some of them express their conviction that a balanced implementation of the MLS, equally fostering access and benefit-sharing, is the only manner in which to implement the Treaty in a sustainable way (e.g. Chapter 18).

Coverage of the MLS

Thematic content

The negotiations of the coverage of the MLS were difficult and often caused considerable tensions. While some Parties initially wanted to apply the MLS to all PGRFA (similar to the other Treaty provisions), others strongly opposed this wide scope of application. Negotiators used the criteria of 'interdependency' and 'food security' to determine which crop should be covered by the MLS. The compromise resulted in the Annex I list of 64 crops and forages.

While stories and strategies regarding the design of the scope of application of the MLS differ significantly, most authors contend that the criteria of 'interdependency' and 'food security' led to the designation of a fairly wide list of crops and forages in Annex I to the Treaty (e.g. Chapters 5, 6 and 8). While a few authors indicate that in their opinion the list is too broad (e.g. Chapter 4), a few others propose that the MLS should apply to all PGRFA (e.g. Chapter 12). However, many authors suggest that Annex I should eventually be modified, especially because very important crops, such as tomatoes, soybeans or peanuts are not included in the Annex, and because climate change impacts on the interdependency and relative importance of the crop for food security (e.g. Chapters 5, 7, 8, 15, and Annex 4). Other authors, do not reject the idea of a modification of the list, or even its expansion to all PGRFAs, but they do not support such a development before it is clear that the MLS functions efficiently, in particular, with respect to its benefit-sharing provisions (e.g. Chapters 3 and 6). Some authors, wishing a much broader coverage, emphasize that several countries as well as the CGIAR already use the SMTA to distribute both Annex I and non-Annex I materials acquired before the Treaty came into force, thereby *de facto* widening the scope of the MLS (e.g. Chapters 5 and 11). Several authors claim that the identification of the material covered by Annex I of the Treaty, which should be included in the MLS (as per Article 11.2), is difficult and remains a challenge (e.g. Chapters 7 and 14).

Facilitated access to genetic resources (Article 12)

Authors welcome the adoption of the SMTA at the first meeting of the Governing Body and many consider it as an essential element to implement the MLS. Some authors recognize its facilitative purpose in accessing PGRFA through a standard contract (e.g. Chapters 5 and 12). Only a few authors actually report on the use of the SMTA (e.g. Chapters 8, 11), because of the early stage of the implementation process at the domestic level. Yet, a faster implementation of the SMTA is recommended, urging governments to take the necessary policy and regulatory measures to this end (e.g. Chapter 5). The CGIAR Centres notice that they have been the main providers of materials using the SMTA in the first years of operation of the Treaty's multilateral system (e.g. Chapter 11).

As for non-Annex I material, few authors recognize the importance of the decision taken by the Governing Body for the CGIAR Centres to use the same SMTA for both Annex I and non-Annex I material (e.g. Chapter 11). Some authors mention that a few countries, such as The Netherlands and Germany, de facto expand the use of the SMTA to non-Annex I crops and forages under their management and control (e.g. Chapter 5). The seed industry considers that the breeders' exemption embedded in the SMTA is positive and strongly defends a wide use of the SMTA for non-Annex I material (e.g. Chapter 12). Moreover, plant breeders stress for instance that they encounter difficulties in gaining access to genetic resources that are not part of the list in the daily breeding practice (e.g. Chapter 15).

Some authors express concern about the difficulties in accessing material because of restrictions due to Intellectual Property Rights (IPRs) (e.g. Chapters 3 and 15). The lack of clarity, especially regarding Article 12.3(d) on the interpretation of the terms 'parts and components' and 'in the form received from the Multilateral System', (e.g. Chapters 2, 3, 4 and 15), and the lack of guidance as to the practical use of the SMTA (e.g. Chapters 12 and 14) are identified as a significant constraint in the implementation of the facilitated access obligation. Other authors mention that some countries have delayed the implementation of the MLS due to the negotiations on the International Regime on ABS under the CBD (Nagoya Protocol), in order to implement both instruments in a coherent way (e.g. Chapters 4, 6, 7 and 14).

Finally, some authors mention that traceability and control of the transfer of MLS material remains a challenge and that the increasing number of SMTAs might create an administrative burden in particular for providers (e.g. information that needs to be provided to the Third Party Beneficiary) (e.g. Chapters 5 and 19). This may lead to a limited distribution of samples of genetic material within the scope of the MLS (e.g. Chapters 12 and 15). Finally, some authors claim that providing material to the recipient under prompt and free access conditions for all PGRFA might sometimes be dependent on multiplication or regeneration or genetic resources costs and time efforts in the gene bank (e.g. Chapters 11 and 14).

Benefit-sharing in the MLS (Article 13 of the Treaty)

Thematic content

The negotiations on the benefit-sharing provisions of the MLS were closely related to those on facilitated access. The Treaty provides that benefits should be fairly and equitably shared by way of the exchange of information (Article 13.2(a)); access to and transfer of technology (13.2(b)); capacity-building (13.2(c)); and the sharing of monetary and other benefits of commercialization (13.2(d)). Moreover, voluntary benefit-sharing strategies are also sought to be considered as a contribution from food-processing industries (Article 13.6).

It is generally agreed by authors that the MLS creates a unique and innovative benefit-sharing mechanism by sharing monetary and non-monetary benefits derived from the use of PGRFA (e.g. Chapters 2, 3, 5, 6, 8, 10 and 18). Some authors mention that the Benefit-sharing Fund is the most important instrument for benefit sharing (e.g. Chapters 3 and 6). Other authors also consider the Global Crop Diversity Trust (GCDT) an important instrument (e.g. Chapters 5 and 16). Many authors foresee the availability of genetic resources as a major benefit of the MLS in itself (e.g. Chapters 5, 7, 9, 11, 15, 16, and Annex 4), while others do not consider access to PGRFA as a major benefit of the MLS, particularly because some countries have limited financial and technological capacity to utilize PGRFA, either conserved in their own gene banks or accessed elsewhere (e.g. Chapter 3). What these authors consider more important is to ensure that benefits derived from the use of genetic resources reach those who need them most and that capacity-building and transfer of technology and information is effectively implemented as a benefit-sharing instrument. It is commonly acknowledged that the exchange of information and results of technical, scientific, and socio-economic research on PGRFA constitute important benefits which should be shared. The same can be said for the access to and transfer of technology related to PGRFA. An example of technology transfer and capacity building projects is provided in Annex 4 of this book. However, some authors emphasize that concrete realization of non monetary benefits such as information sharing, access to and transfer of technologies and capacity building has not occurred yet. Similarly, some authors think that so far the benefit sharing, both monetary and non-monetary has been too limited (e.g. Chapters 3, 6 and 7). For example, the authors of the African Regional Group (e.g. Chapter 3) consider that the apparent delays in expanding the Benefit-sharing Fund under the Funding Strategy create a major obstacle in the implementation of the ITPGRFA.

Instruments for international cooperation (Articles 14–17 of the Treaty)

> **Thematic content**
>
> The instruments for international cooperation (Part V of the Treaty), include the rolling GPA, *ex situ* Collections of PGRFA held by the IARCs of the CGIAR and other International Institutions, International Plant Genetic Resources Networks and the Global Information System on PGRFA.

Few comments on these instruments and processes were received. General comments are made by some authors as to the lack of sufficient means devoted to international cooperation, and the limited implementation and results of these Treaty provisions at the national level. Some authors state that a common implementation framework would be useful to help assist countries, especially developing countries, with the effective implementation of the Treaty (e.g. Chapter 7).

Global plan of action (Article 14 of the Treaty)

> **Thematic content**
>
> The development of the GPA is one of the two major outcomes of the FAO Global System on Plant Genetic Resources, which was initiated by the establishment of the Commission on Genetic Resources in 1983 (see the structure of the global system in appendix 1 of Annex 1 of this book). The GPA provides an operational framework for the development of national programmes on PGRFA, and for regional and international cooperation. The GPA contains a set of recommendations and priority activities as a response to the needs, gaps and challenges identified in the first report of *The State of the World's Plant Genetic Resources for Food and Agriculture* (SoW) in 1996. The 20 priority activity areas of the GPA were recognized by the Governing Body of the Treaty as the reference for the establishment of initial priorities of its Funding Strategy. Some of these were grouped in three sets of priorities to guide the first and second benefit-sharing project cycles (information exchange, technology transfer and capacity-building; managing and conserving plant genetic resources on farm; and the sustainable use of plant genetic resources).

Some authors mention that the rolling GPA has not received sufficient attention since the entry into force of the Treaty (e.g. Chapter 3) and that the review of the GPA could contribute significantly to implement Treaty provisions that are not implemented yet in an effective way (e.g. Chapters 7 and 8). They point that the publishing of the second report on the 2010 SoW should allow for the update of the rolling GPA (e.g. Chapter 7).

Ex situ collection held by the IARCs of the CGIAR and other international institutions (Article 15 of the Treaty)

> **Thematic content**
>
> Under Article 15, the IARCs and other international institutions holding PGRFA collections in trust and which signed an agreement with the GB distribute Annex I PGRFA following the MLS provisions (in particular, using the SMTA). The centres are subject to policy guidance of the Governing Body for the ex situ collections held by them.

Some authors mention that, on the whole, the experience of the centres with the implementation of the Treaty has been quite positive, as the Treaty considerably simplifies the task of the centres in making PGRFA available and notably reduces the administrative costs involved (e.g. Chapter 11). Even more so since the Governing Body at its second meeting took note of the preferences of the centres and endorsed the option to use the same SMTA for both Annex I and non-Annex I material with an interpretative footnote, thus avoiding the need for two versions of the SMTA.[3] Other authors recall that – up to now – the majority of the materials distributed through the MLS are those of the CGIAR centres (e.g. Chapter 5).

International plant genetic resources networks (Article 16 of the Treaty)

> **Thematic content**
>
> The Treaty provisions on International Plant Genetic Resources (PGR) Networks (Article 16) are grouped in three categories (i.e. crop-based networks, regional networks and thematic networks).

Some authors argue that networks are very important to the conservation, sustainable use and exchange of PGRFA, including through their role in raising public awareness, and that they should be developed and promoted in to the framework of Treaty initiatives (e.g. Chapters 2 and 9). They state that regional and national PGRFA networks are important instruments for the GPA implementation (e.g. Chapter 7), in particular, for PGRFA exchange, information sharing and technology transfer (e.g. Chapters 3, 4). Authors stress that the implementation of the GPA at the national level strengthens national networks which have also a direct positive effect on regional and global networks (e.g. Chapter 7). Some authors provide examples of national or regional networks active in their country (e.g. Chapters 9 and 14).

The Global Information System (GIS) on PGRFA (Article 17 of the Treaty)

Thematic content

Article 17 states that Contracting Parties shall cooperate to develop and strengthen a GIS to facilitate the exchange of information, based on existing information systems, on scientific, technical and environmental matters related to PGRFA. Several types of existing information systems could be relevant to its development, such as the World Information and Early Warning System on Plant Genetic Resources for Food and Agriculture (WIEWS) in FAO, the System-wide Information Network for Genetic Resources (SINGER) for the CGIAR, the National Plant Germplasm System and the European Plant Genetic Resources Search Catalogue (EURISCO), and more recently Genesys.[4]

Only a few references are made to information systems, with little specific comment: SINGER is mentioned a couple of times (e.g. Chapters 11, 14) and Genesys is named once (Annex 4 of this book), (see below Part II for more details on the GIS).

Financial provisions (Article 18 of the Treaty)

Thematic content

Through Article 18 of the Treaty the contracting parties undertake to implement a Funding Strategy for the implementation of the Treaty. The objectives of this Funding Strategy shall be to enhance the availability, transparency, efficiency and effectiveness of the provision of financial resources to implement activities under the Treaty. Several measures are listed, enabling contracting parties to reach these objectives and a funding target has to be periodically established. Funding for priority activities, plans and programmes are focused, in particular, on projects in developing countries and countries with economies in transition taking into account the GPA. Article 18.4(b) states that the extent to which the latter Contracting Parties will effectively implement their commitments under the Treaty will depend on the effective allocation, particularly by developed country Parties, of the resources referred to in this article.

Contracting Parties have developed this Funding Strategy between 2006 and 2009. In 2006, the GB adopted its relationship agreement with the Global Crop Diversity Trust (GCDT),[5] recognizing that 'FAO and the Future Harvest Centres of the CGIAR have promoted the establishment of a GCDT, in the form of an endowment with the objective of providing a permanent source of funds to support the long-term conservation of the ex situ germplasm on which the world depends for food security, to operate as an essential element of the Funding Strategy of the International Treaty, with overall policy guidance from the Governing Body of the International Treaty, and within the framework of the International Treaty' (ITPGRFA, 2006). Up to March 2010, the GCDT has raised US$136 million. In 2008, the Benefit-sharing Fund became operative with voluntary contributions from Contracting Parties (Spain, Italy, Norway and Switzerland), and in 2009 Contracting Parties established a target of US$116 million to be raised for the period 2009–2014.

Some authors believe that the Funding Strategy functions as a strategy to primarily mobilize money from existing sources and channels, including the monetary benefits of the MLS (e.g. Chapter 2). Many authors point to the lack of sufficient funds (e.g. Chapters 2, 3, 7, 13, 19). Several authors state that the ITPGRFA will have great difficulty in generating new and additional financial resources to support programmes to conserve and utilize PGRFA in a sustainable way at the regional, national and local community level. Some add that although at the third meeting of the Governing Body in 2009 a target of US$ 116 million was agreed to be raised within the next 5 years (implementation of Article 18.3), much of these funds are not available yet and might be difficult to obtain (e.g. Chapter 3). Authors from the European Regional Group Chapter recall that some European countries have contributed to the Benefit-sharing Fund as a response to queries from developing countries for more money. They also note the pledge for benefit-sharing from Norway, which adopted a national policy where the equivalent of 0.1 per cent of all annual seed sales is transferred to the Fund.

Besides the clear concerns about the lack of funds, authors explain that the GCDT, although an independent institution separate from the Treaty, operates within the framework of the Treaty as it constitutes an essential element of its Funding Strategy (e.g. Chapter 16) and receives policy guidance from the Governing Body (e.g. Chapters 5 and 16). Some authors consider that the GCDT also contributes to an efficient implementation of the Treaty by supporting the benefit-sharing and conservation provisions of the Treaty (e.g. Chapter 7). However, other authors consider the GCDT useful but not central to the efficient implementation of the Treaty provisions, since it focuses mainly on ex situ conservation activities (e.g. Chapter 8).

Part II – Editors' analysis of shortcomings and suggestions to cope with present and future challenges

This second Part of the conclusion aims at processing the appraisals and concerns regarding the implementation of the Treaty as described by the various authors. In doing so, the editors wish to share possible ways forward to facilitate the further implementation of the Treaty. The purpose is not to be exhaustive but to put forward a preliminary analysis of some of the major constraints on the implementation of the Treaty. While recognizing that many actions could be taken at the local, national or regional level, the editors, following the proposals made by the authors throughout the book, will stay within the limits of actions that can be taken at the Governing Body level. Part Two of this concluding chapter is divided into four sections, as summarized and illustrated in columns 1 to 4 of Table 20.1 below. Section 1 deals with the constraints identified by authors related to the implementation of the Treaty (Table 20.1, column 1). The editors have attempted in Section 2 to categorize these constraints into four types of needs (Table 20.1, column 2). Then, a selection of tools is discussed as possible ways to mitigate the identified constraints and needs (Section 3, Table 20.1, column 3). Finally, the editors point to some of the Treaty articles which have received little attention until now and which implementation or further implementation could contribute to mitigate some of these constraints (Section 4, Table 20.1, column 4). Throughout the analysis, concise and concrete examples are given.

In the following text, the editors do not try to be prescriptive neither comprehensive but just to provide an input to promote discussion.

Section 1: Constraints identified by authors of the book (cf. Table 20.1, column 1)

The editors have selected some of the major constraints raised by authors of this volume, which are mentioned in Part I of this chapter. The constraints are based on the authors' experience with the implementation of the Treaty. They are very diverse and deal with scientific and technical, legal, political, and/or economic aspects of the Treaty implementation. Many constraints cover several aspects at the same time. This makes it even harder to tackle them. The editors make an attempt to categorize the needs associated with these constraints in order to discuss potential solutions to address them in the following Sections 2 and 3.

Section 2: Categories of needs associated with the identified constraints (cf. Table 20.1, column 2)

In an attempt to facilitate the analysis of all the constraints identified, the editors propose to categorize and qualify these constraints into four types of needs: the need for more clarity; the need for review and update; the need for further development; and the need for more coherence and coordination. These four categories of constraints are not exhaustive, and respond to the following four questions.

Is there a need for clarification of Treaty provisions?

Many authors indicate the need for clarification of various Treaty provisions to guide the implementation, in particular, of Article 11.2 and Article 12.3 (d). A clear example of ambiguity concerns the scope of the following expression 'PGRFA [...] that are under the management and control of the Contracting Parties and in the public domain' (see Article 11.2). The need to provide guidance in the interpretation of this and other ambiguities related to the MLS has led to the establishment of an Ad Hoc Technical Advisory Committee by the Governing Body to provide inter alia some guidance regarding the identification of plant genetic resources for food and agriculture under the control and management of Contracting Parties, and in the public domain.[6] Another example relates to Article 12.3(d), where different interpretations can be given to the terms 'parts and components' and 'in the form received' and therefore to the definition of the material that can be protected by IPRs or not.

Is there a need for further development of Treaty mechanisms and strategies?
Authors plead in favour of rapid action by the Governing Body to develop further mechanisms and strategies in various aspects. A major example concerns the non-monetary benefit-sharing obligations (Article 13.2 (a), (b) and (c)), which is poorly taking place, according to many authors.

A second example relates to the need to further develop financial mechanisms helping countries to implement the GPA priority area activities, and especially the priorities that are not directly covered by funds already established (the GCDT or the Benefit-sharing Fund). Indeed, many priorities of the GPA do not foresee an appropriate and specific financing mechanism to implement them yet. This could perhaps be done taking advantage of the experience of other existing fundraising mechanisms and funding organizations active in the agricultural sector (such as the Global Environmental Facility (GEF), the United Nations Development Programme (UNDP) or the International Fund for Agricultural Development (IFAD)).

A third example where further development is needed relates to the Global Information System.

Is there a need for review or update of Treaty mechanisms and strategies?
The text of the Treaty and its implementation mechanisms and strategies request, in certain cases, such review and update processes. Several examples can be mentioned, such as the review of the levels of payment in the SMTA by the GB (Article 13.2(d)(ii); or the periodic establishment of a funding target (Article 18.3). Another example concerns Article 17.3 of the Treaty, which requires Contracting Parties to collaborate with the Commission on Genetic Resources for Food and Agriculture to periodic reassess the State of the World's plant genetic resources for food and agriculture in order to update the GPA A last example is constituted by the priorities set for the Benefit-sharing Fund, where Annex 1 of the Funding Strategy sets out eligibility, selection criteria and additional requirements, that can be updated regularly by the Governing Body.

In addition, the editors further consider the possibility to modify and review Treaty mechanisms and strategies in reaction to external circumstances, which were not foreseen at the moment of the Treaty negotiations and which may have a substantial impact on its implementation. A good example would be the updating of Annex I list as a consequence of external factors. Indeed, the identification of the list of crops and forages were negotiated according to the double criteria of interdependency and food security, which are currently being affected by climate change and technological developments.

Is there a need for a stronger coordination in order to facilitate the implementation of this Treaty provision?
Many authors have stressed the limited coordination and coherence at three levels, resulting sometimes in numerous competing and/or conflicting international obligations: (1) between governing bodies and secretariats of international institutions; (2) between national representatives attending different but co-related

international fora such as the WTO, the CBD and the ITPGRFA; and (3) between different sectors and people at the national level responsible for the implementation of these different international obligations. (1) At the secretariat and governing body level, periodic meetings between Secretaries and joint meetings between Governing Bodies of different international organizations could be two options leading to the development of common programmes and activities, mitigating the limited coordination and coherence problems. An example of successful inter-sectorial cooperation in the negotiating process is provided by the mutual recognition and support between the Treaty and the Nagoya Protocol on Access and Benefit-Sharing. The new Protocol expressly refers to the Treaty as a complementary instrument of the international ABS regime[7] (2) At the national delegation level, common preparatory meetings and inter-sectorial composition of delegations could be envisaged to prepare for international meetings. (3) At the national level, coordination by national inter-sectorial committees could contribute to favour coherence and coordination when implementing international obligations at the national level.

Section 3: Specific legal tools to improve the implementation of the Treaty (cf. Table 20.1, column 3)

The aforementioned needs for clarification, review, further development and coordination require Contracting Parties and the Governing Body to take action to further implement the Treaty. Following directions proposed by authors, the editors limit the discussion to possible actions to be taken at the Governing Body level. Article 19 of the Treaty and its Rules of Procedures empower the Governing Body to take actions to promote the full implementation of the Treaty. In the following section, four possible tools are suggested – each having a different level of obligation – to mitigate the identified constraints and needs. These tools are: to design common implementation frameworks, to develop soft law tools such as guidelines, to adopt agreed interpretations on specific Treaty articles, and to reopen the negotiation of some Treaty provisions. For each constraint, these tools have been classified using numbers from 1 to 3 according to their level of suitability. Where no numbers are mentioned, the tool is found not to be applicable to the specific constraint.

Should Contracting Parties design a Common Implementation Framework (CIF) in order to facilitate a harmonious and systematic implementation in all member countries?

Many countries, especially developing countries and countries with economies in transition, experience difficulties in implementing the Treaty due to the lack of legal, technical, economic or human resources. To mitigate this constraint, the Governing Body may establish comprehensive plans and programmes on specific subject matter (as per Article 19.3(b)) or by strengthening existing programmes, for scientific and technical education and training (Article 13.2(c)(i)). Furthermore, Article 14 promotes national actions and international cooperation to provide a coherent implementation framework for the rolling GPA. However, these efforts might appear scattered and disconnected from one another. Therefore, proposing a CIF for the implementation of the Treaty at the domestic level might be a useful tool, in particular, for developing countries. This is not to say that the Treaty would develop a 'one-size-fits-all' tool. On the contrary, such CIF should provide sufficient flexibility to countries to be able to fit their specificities and particularities. The commonality of the framework would lay in the common objectives and principles set out by the GB to help countries implementing the Treaty obligations, through a wide range of diverse information, administrative, legal, scientific and technical systems, instruments, toolboxes etc. This should be accompanied by a roadmap with specific and quantifiable targets (percentage), periodically reviewed, in order to facilitate the development and funding of national strategies in line with international priorities. This approach would avoid the omission of a priority from any funding mechanism. The GPA may partly cover such a CIF for the Treaty with respect to the conservation and sustainable use of PGRFA obligations falling under the 20 GPA activity area priorities. The editors believe there are two options at this point in time. Either the GPA is broadened to become the CIF for all Treaty obligations, or a new other CIF is developed to integrate and complement the GPA. In both cases, this CIF could be broader, taking into account informal networks, or even obligations deriving from other related international instruments. As a matter of fact, these instruments, such as the CBD, the Cartagena Biosafety Protocol and the Nagoya ABS Protocol, TRIPS, WIPO, or the Kyoto Protocol, may compete or overlap with some Treaty provisions. Further coherence and coordination could be reached by way of common programmes and activities aimed at ensuring cooperation and coordination, avoiding duplication, gaining synergies and effectiveness in the use of limited funds, within the ambit of a CIF.

Should Member States develop soft law tools such as codes of conduct and guidelines to lead Parties in their implementation efforts?

It is unlikely that in the short-term, new binding text will be negotiated as a solution for Treaty provisions that need further clarification or development in view of the need for rapid and effective implementation of the Treaty. However, soft law tools, such as guidelines, codes of conducts or standards, might be more adequate to provide prompt guidance. According to Article 19.3(a) the Governing Body as well as the subsidiary bodies established by the Governing Body should provide policy directions. This could be done through GB resolutions, but also through the design of guidelines, standards or codes of conduct to facilitate countries' implementation of the Treaty. Such tools have been used widely in the past and have proven to be effective. The Treaty specifically states that, in the absence of national legislation dealing with access to in situ PGRFA, the Governing Body may set standards (Art. 12.3(h)). Guidelines are a commonly used tool in many different fora. Without entering into the details of the different types of guidelines, the editors would like to stress the fact that there are very useful instruments to raise awareness, promote public participation and training and allow for the necessary flexibility Contracting Parties often need in order to implement international obligations at the domestic level according to their specificities and needs. A successful example is provided with the Voluntary Guidelines for the Implementation of the Right to Food. These Voluntary Guidelines were developed in the framework of the Committee on World Food Security (CFS)[8]. However, collaboration and coordination between existing national or regional guidelines and the Treaty is important.

Such tools might be useful for several constraints identified by authors. The first one to be highlighted regards the national implementation of Farmers' Rights. In this case, voluntary guidelines might be designed. A second example could be the development of guidelines or a code of conduct to help countries and their gene banks in the identification process of designing the material covered by the MLS.

Should Contracting Parties seek to adopt agreed interpretations on specific articles?

In 1983, the FAO Conference adopted the non-binding International Undertaking (IU) by Resolution 8/83. To overcome the reservation of certain countries on the IU, the FAO Conference later adopted further resolutions, which were annexed to the IU as agreed interpretations. Although the Treaty is a binding agreement, and therefore different from the IU, Contracting Parties might envisage developing agreed interpretations through the adoption of a GB decision (on Articles 11.2, or 12.3(d), for example) in order to provide clarity on or review specific obligations. This is consistent with the mandate of the Governing Body, which may take all decisions, by consensus, in order to promote the full implementation of the Treaty. It would be less burdensome to adopt a resolution which would be annexed to the Treaty than to reopen negotiations, however, it might be a risky tool, as it might facilitate attempts to rewrite the Treaty through interpretations. Therefore, this option should very cautiously be envisaged, and surely only to very well defined and limited provisions, with very strict rules of procedures to be applied.

Should there be a reopening of the negotiation to modify the text of the Treaty to address the limitations in its implementation?

Although is not recommended as primary solution for any of the constraints identified by authors, clarification and further development needs could eventually come about through new negotiations aimed at 'improving' or 'complementing' a very specific provision of the Treaty. Indeed, the Treaty provides for the possibility to amend its text (Articles 19.3 (h, i); 23; 24). Article 23 states that 'Amendments to this Treaty may be proposed by any Contracting Party [and] shall be adopted at a session of the Governing Body'. Amendment can be made only by consensus of the Contracting Parties present at the Governing Body, and will enter into force following the same procedure used for the Treaty. However, unless this tool is used under very strict conditions, with the understanding that a failure to adopt such amendment would automatically bring Contracting Parties back to the *status quo ante*, this approach might put at risk the climate of cooperation that exists today, and might facilitate attempts to rewrite the Treaty through interpretations.

Moreover, this approach would be quite costly, time-consuming and complex, as the whole national process of ratification, acceptance or approval should again be pursued by each Contracting Party (i.e. be discussed and adopted at National Parliaments).

Section 4: Treaty Articles, which further implementation could contribute to mitigate some identified constraints (cf. Table 20.1, column 4)

Finally, the editors believe that implementing further Treaty and SMTA obligations that have received little or no attention in the implementation process until now, could actually significantly contribute to solving some constraints identified by authors. The editors will concentrate on three provisions: Article 6.11 of the SMTA; Article 13.6 of the Treaty; and Article 17 of the Treaty. However, other provisions should require more attention some of which are currently under developed, such as inter alia PGR Networks (Article 16), compliance (Article 21) or sustainable use of PGRFA (Article 6).

The crop-based alternative payment scheme (Article 6.11 of the SMTA)

Some authors stress that promoting the use of the crop-based alternative payment scheme could at least partially mitigate some of the identified constraints (see Chapter 19 for details). Until now, it has received little attention, but it could provide very practical solutions for the funding of the Benefit-sharing Fund within the MLS. In fact, this scheme offers a more general and less bureaucratic approach for dealing with SMTAs, thereby decreasing significantly the administrative burden and increasing transparency. Mandatory monetary benefits would be immediate, thereby providing funds quickly to the Benefit-sharing Fund. The provision allowing the Governing Body to predict contributions and to review periodically the levels of payment in order to achieve fair and equitable sharing of benefits (Article 13.2) could be an opportunity to match the priorities with the monetary benefits to be transferred to the Benefit-sharing Fund. Chapter 5 explains why this alternative payment scheme has not received more attention from users of PGRFA up to now. One of the reasons put forward is that the discounted rate of 0.5 per cent of the sales is too close to the rate in the Article 6.7 (SMTA) payment scheme (of 1.1 per cent less 30 per cent). Perhaps, if the difference between the two rates was bigger, it would render the alternative payment scheme more attractive, thereby answering several of the constraints identified by authors. Implementing further this obligation could therefore mitigate at least two major constraints: the technical and administrative constraints related to the daily use of the SMTA, and the lack of predictability of funds for the Benefit-sharing Fund.

Contribution by the food-processing industries to the MLS (Article 13.6 of the Treaty)

Another provision, which has barely received any attention up to now, but which further development and implementation could lead to new and additional monetary benefit-sharing, as well as raising awareness to the wider public, can be found in Article 13.6 of the Treaty. This Article deals with voluntary contributions of the food-processing industries to the MLS. Process and commercialization of wider diversity of crops and crop varieties increases the number of options for consumers and food industry. However, up to now, neither consumers nor the food industry have been much included in the international discussions between the various stakeholders. This is surprising if one considers that we are all consumers, whereas, in developed countries, for instance, farmers represent only 3 per cent of the population (see Chapter 17). Therefore, it is vital to raise awareness amongst consumers and consumer organizations to identify and motivate the food industry to contribute to the MLS and to design mechanisms for this purpose. Strong incentives for the food industry to contribute to the Benefit-sharing Fund are required. An example could be to create a 'green tag' for products coming from these industries contributing to the Fund or for industries agreeing to contribute to the MLS. With this green tag label consumers would be able to decide to buy products that contribute to the conservation and sustainable use of PGRFA. But for this to happen, consumers should be conscious that their choices regarding food products provide them with considerable leverage to influence the food industry's economic and policy decisions. Contracting Parties should therefore target consumers as well as farmers' organizations in their public awareness programmes.

The Global Information System on PGRFA (Article 17 of the Treaty)

The GIS is still at a very early stage of implementation. Contracting Parties requested the Secretariat to develop a vision paper presented at the fourth meeting of the Governing Body (ITPGRFA, 2011). This vision paper takes stock of existing information systems and outlines a process for the development of the GIS. The galaxy of information systems makes it difficult to have a clear vision of the current situation.[9] It is believed that a mere catalogue of existing databases is not enough. It is important to identify the gaps in current information systems and the needs for information of providers and users. The editors are convinced that the GIS should constitute the general and interactive database for all Treaty information, facilitating the implementation of all its provisions, including an information Clearinghouse.[10] The GIS should include online updated information relevant to every Treaty provisions and its implementation, including inter alia scientific and technical information (e.g. genetic diversity, erosion and vulnerability; scientific and technical developments; PGRFA conservation and use; and other areas covered by the GPA), legal and policy information (e.g. policies, laws and regulation relevant to PGRFA and related technologies, including on IPR regulation and traditional knowledge protection; disputes under the Treaty); financial information (e.g. financial contributions, financial disbursement; projects financed through the Treaty; funds availability), as well as the state of implementation the Treaty (e.g. Farmers' Rights; MLS and its SMTA; national and regional reports and inputs being received for the updating process of the SoW and GPA).

Table 20.1 *Constraints, needs and implementation tools*

To facilitate understanding, the following table (on page 276) summarizing and illustrating the findings of the analysis is provided. It contains four columns: 1) Specific implementation constraints identified by authors of the book; 2) Categories of needs associated with the identified constraints; 3) Specific legal tools to improve the implementation of the Treaty; 4) Treaty Articles weakly implemented up to now, which implementation could mitigate some identified constraints. Under the specific tools column, the tools are classified using numbers from 1 to 3 according to their level of suitability; 1 is the option the editors find most appropriate to deal with the concern, 3 is the option found to be the least appropriate to deal with the constraint. Where no numbers are mentioned, the tool is not applicable to the constraint.

Treaty Part	*Column 1* *Specific implementation challenges and constraints identified by authors*	*Column 2* *Needs related to the constraint* *Clarification*	 *Further development*	 *Review/ update*	 *Coordination/ coherence*
Part I and General constraints					
	Policy coherence between the ITPGRFA and other international instruments (CBD, TRIPS, UPOV)		X	X	X
	Public awareness & capacities at the national level		X	X	X
	Trust between Contracting Parties	X	X		X
	Clarity of Treaty provisions (e.g. Art. 6, 9, 11.2, and 12.3(d))	X	X		X
Part II Conservation and sustainable use of PGRFA					
	Financial, technical & scientific constraints		X		X
	Weak implementation of in situ conservation obligation		X		
Part III Farmers' Rights					
	Recognition & national implementation	X	X		X
	Participation of farmers' organizations		X		X
Part IV The multilateral system of access and benefit-sharing					
	Modification of Annex I list to face new challenges		X	X	
	Limitations in access to PGRFA	X	X	X	
	Notification of inclusion of material in the multilateral system	X			X
	Limited realization of benefit sharing		X	X	X
	Limited realization of non-monetary benefit sharing		X		X
Part V Supporting components					
	Limited implementation of the GPA		X	X	X
	Little use of existing formal and informal networks		X		X
	Limited implementation of the GIS		X		X
Part VI Financial provisions					
	Limited and unpredictable funding		X	X	X

Column 3 *Specific tools to tackle the constraint*				*Column 4*
Common implementation framework	*Soft law (codes of conducts, guidelines)*	*Agreed interpret-ations*	*Reopen negotiations*	*Provisions, which implementation or further implementation could contribute to mitigate the identified constraint*
1	1			Cooperation with other international organizations (Article 19.3(g))
1	1			Non-monetary benefit-sharing (Art. 13.2 (a,b,c), 14)
1	2	3		
2	1	3		
1	1	2		Compliance (Article 21) & third party beneficiary (SMTA 4.4; 4.5; 8.3)
1	2			In situ conservation (Article 5.1(d))
1	1	3		
1	1			Articles 6, 9.1, 9.2(c), 13.2 (a, b, c) & 14
1	1	2	3	
1	1	3		Compliance & third party beneficiary
1	1	3		Compliance & third party beneficiary
1	1	3		Article 13.6, compliance & third party beneficiary
1	1			Articles 13.2 (a, b, c) & 14
1	1			Articles 14, & 17.3
1	1			Article 16
1	1			Article 17
1	2			Article 6.11 SMTA; Articles 13.6, 18 & 19.3(f)

Notes

* This chapter only represents the opinions of its authors. Christine Frison conducts a PhD research as junior affiliated researcher at the Université catholique de Louvain and at the Katholieke Universiteït Leuven (Belgium) on international law and governance of plant genetic resources for food and agriculture. Francisco López is Treaty Support Officer for the International Treaty on Plant Genetic Resources for Food and Agriculture and is based at the Food and Agriculture Organization (FAO) of the United Nations, Rome, Italy. José Esquinas-Alcázar is Director of the 'Catedra' of Studies on Hunger and Poverty at the University of Cordoba in Spain. Professor at the Politechnical University of Madrid, José Esquinas has worked as Secretary of the FAOs intergovernmental Commission on Genetic Resources for Food and Agriculture, and *interim* Secretary of the Treaty for 30 years. E-mail: jose.esquinas@upm.es

The editors are thankful to the Centre de Philosphie du Droit for its financial support out of the Interuniversity Attraction Poles programme IAP VI/06 project funded by the Belgian State – Belgian science policy (BELSPO) and to the Cátedra de Estudios sobre Hambre y Pobreza (CEHAP) of the Universidad de Córdoba (Spain) for its constant support.

1 It is argued that Article 9 of the Treaty contains weaker obligations than what the Treaty Preamble states in its §7 and 8 'Affirming that the past, present and future contributions of farmers in all regions of the world, particularly those in centres of origin and diversity, in conserving, improving and making available these resources, is the basis of Farmers' Rights;

'Affirming also that the rights recognized in this Treaty to save, use, exchange and sell farm-saved seed and other propagating material, and to participate in decision-making regarding, and in the fair and equitable sharing of the benefits arising from, the use of plant genetic resources for food and agriculture, are fundamental to the realization of Farmers' Rights, as well as the promotion of Farmers' Rights at national and international levels.'

2 It should be noted that during the publication process of this book, success stories in the implementation of Farmers' Rights have been reported at the 'Global Consultations on Farmers' Rights in 2010', which took place in Ethiopia in 2010, under the umbrella of the Fridtjof Nansen Institute of Norway. See the 'Note by the Secretary' document IT/GB-4/11/Circ.1.

3 See Second Session of the Governing Body, decision IT/GB-2/07/Report at § 66–68.

4 Genesys is a newly developed PGR portal that gives breeders and researchers a single access point to information of about a third of the world's gene bank accessions. It is an initiative by Bioversity International in partnership with the Secretariat of the Treaty on Plant Genetic Resources for Food and Agriculture and the Global Crop Diversity Trust.

5 It is an endowment fund, which provides funds in perpetuity to support long-term conservation of PGRFA and ensure the conservation and availability of PGRFA which are most relevant for food security and sustainable agriculture. See the Relationship Agreement between the Governing Body and the GCDT in particular Preamble §5, approved at the First Session of the Governing Body, in Madrid in June 2006, document IT/GB-1/06/Report §35–40.

6 See IT/AC-SMTA-MLS1/10/4.

7 See Convention on Biological Diversity, COP Decision X/1, § 6 and 11 of the preamble.

8 Besides, it should be noted that the newly reformed CFS provides a pioneer way to to facilitate the active participation of civil society organizations to the Committee's activities: the Civil Society Mechanism (CSM).

9 In addition to the few information systems mentioned by authors, we would like to point to the Global Biodiversity Information Facility (GBIF). GBIF is a multilateral initiative established by intergovernmental agreement (initially 17 countries) and based on a non-binding Memorandum of Understanding. It aims to make the world's biodiversity data freely and universally available via the internet.

10 Coordination between the different existing systems (such as WIEWS, and SINGER, for example) should focus on avoiding duplication. To overcome this duplication problem a database of databases could be created, to be eventually operated through the Genesys initiative. The further development of this Treaty obligation would contribute to partially mitigate many identified constraints, such as the lack of public awareness and policy coherence, or the limited implementation of capacity-building and non-monetary obligations.

References

ITPGRFA (2006) 'Report of the Governing Body of the International Treaty on Plant Genetic Resources for Food and Agriculture,' First Session Madrid, Spain, 12–16 June 2006, IT/GB-1/06/Report

ITPGRFA (2011) 'Vision paper on the development of the global information system in the context of Article 17 of the Treaty', Fourth Session of the Governing Body, Bali, Indonesia, 14–18 March 2011, IT/GB-4/11/19

Annex 1

History, Milestones and Calendar of Meetings of the FAO Commission on Genetic Resources for Food and Agriculture and the International Treaty on Plant Genetic Resources for Food and Agriculture

Appendix on the Global Plan of Action

This annex provides an overview on the major developments in FAO that led to the negotiation, adoption and implementation of the International Treaty on Genetic Resources for Food and Agriculture. It also presents the major milestones within the Commission on Genetic Resources for Food and Agriculture and the negotiations that led to the adoption of its multi-year programme of work for all genetic resources for food and agriculture.

Since its establishment in 1983, the FAO Commission on Genetic Resources for Food and Agriculture (CGRFA) has served as an intergovernmental forum for the development of international policies and for the negotiation of international agreements. This has led to the adoption in 2001 of the International Treaty on Plant Genetic Resources for Food and Agriculture (the Treaty) by the FAO Conference Resolution 3/2001. Table A1.1 summarizes the milestones in the history of the Commission on Genetic Resources for Food and Agriculture and the Treaty. In 2001, the FAO Conference decided that the Commission would act as the Interim Committee for the Treaty. Table A1.2 shows the list of the negotiation meetings until its Governing Body was set up. Finally, Table A1.3 lists the meetings which have taken place since the signature of the Treaty in 2001.

Table A1.1 *Milestones in the history of the Commission on Genetic Resources for Food and Agriculture and the Treaty*

Year	Adoption of resolutions and major policy recommendations
1983	The FAO Conference adopts the International Undertaking on Plant Genetic Resources (IU) (Resolution 8/83). At the time of its adoption, the IU and the Intergovernmental Commission for Plant Genetic Resources that was established to monitor this instrument, was the only United Nations intergovernmental forum dealing with biodiversity and genetic resources. The Commission is established in accordance with Article VI.1 of the FAO Constitution (Resolution 9/83). The development of the FAO Global System on Plant Genetic Resources begins with the establishment of the Commission (see the appendix to this annex for explanations on the Global System).
1989	The FAO Conference adopts an Agreed Interpretation of the IU (Resolution 4/89) and a resolution on Farmers' Rights (Resolution 5/89), that became Annexes I and II to the IU. By recognizing that plant breeders' rights are not inconsistent with the IU and simultaneously recognizing Farmers' Rights, the resolutions aim at achieving a balance between the rights of breeders (formal innovators) and farmers (informal innovators). Resolution 5/89 already mentions how Farmers' Rights should be understood and could be implemented, in order, inter alia, to 'allow farmers, their communities, and countries in all regions, to participate fully in the benefits derived, at present and in the future, from the improved use of plant genetic resources, through plant breeding and other scientific methods.' The Commission calls for the development of The International Network of Ex Situ Collections under the Auspices of FAO, in line with the IU, because of lack of clarity regarding the legal situation of the ex situ collections.
1991	The FAO Conference recognizes the sovereign rights of nations over their plant genetic resources in Resolution 3/91 that became Annex III to the IU. The Conference recognizes the important consensus reached on a number of other delicate issues such as access to breeders' and farmers' material and implementation of Farmers' Rights through an international fund. Recognizing the importance of plant genetic resources for food and agriculture, the Conference also agrees that a first State of the World's Plant Genetic Resources for Food and Agriculture should be developed in a country-driven process.
1992	In adopting the agreed text of the Convention on Biological Diversity (CBD),[1] countries adopt Resolution 3 of the Nairobi Final Act, which recognizes the need to seek solutions to outstanding matters concerning plant genetic resources in harmony with the CBD, in particular: (a) access to ex situ collections not addressed by the Convention, and (b) the question of Farmers' Rights. It was requested that these matters be addressed within FAO's forum. Chapter 14 of Agenda 21, on promoting sustainable agriculture and rural development, calls for the strengthening of the FAO Global System[2] on Plant Genetic Resources, and its adjustment in line with the outcome of negotiations on the CBD.

1993 The FAO Conference adopts Resolution 7/93 at it 27th session, requesting the FAO Director-General to provide a forum for the negotiation among governments, for (a) the Revision of the IU, in harmony with the CBD; (b) consideration of the issue of access on mutually agreed terms to plant genetic resources, including ex situ collections not addressed by the Convention; and (c) the issue of the realization of Farmers' Rights. The Conference *urged* 'that the process be carried out through regular and extraordinary sessions of the Commission on Plant Genetic Resources, convened, if necessary, with extra-budgetary financing, and with the help of its subsidiary body, in close collaboration with the Intergovernmental Committee on the Convention on Biological Diversity, and after the entry into force of the Convention, with its Governing Body'.

Conference adopts the International Code of Conduct for Plant Germplasm Collecting and Transfer, developed by FAO and negotiated through the Commission.

The Commission endorses Gene Bank Standards, developed by an expert consultation in 1992, and requests for the preparation of a rolling Global Plan of Action on Plant Genetic Resources for Food and Agriculture, in order to identify the technical and financial needs for ensuring conservation and promoting sustainable use of plant genetic resources.

1994 Twelve centres of the Consultative Group on International Agricultural Research (CGIAR), and subsequently other institutions sign agreements with FAO, placing most of their collections (some 500,000 accessions) in the realm of the IU under the auspices of FAO. Through these agreements, the Centres agree to hold the designated germplasm 'in trust for the benefit of the international community'. The agreements provide an interim solution, until the revision of the IU has been completed.

Following the mandate given by the FAO Conference at its 27th regular session in 1993, the negotiations for the revision of the IU start in the 1st extraordinary session of the Commission.

1995 The FAO Conference broadens the Commission's mandate to cover all components of biodiversity of relevance to food and agriculture. It renames the Commission the Commission on Genetic Resources for Food and Agriculture (Resolution 3/95).

1996 FAO launches The State of the World's Plant Genetic Resources for Food and Agriculture developed through a participatory, country-driven process under the guidance of the Commission.

The International Technical Conference on Plant Genetic Resources, held in Leipzig, Germany, welcomes the *Report on The State of the World's PGRFA* as the first comprehensive worldwide assessment of PGRFA. The Conference adopts a complementary Global Plan of Action for the Conservation and Sustainable Utilization of Plant Genetic Resources for Food and Agriculture, negotiated by the Commission. It also adopted the Leipzig Declaration.

1997 The Commission establishes, as 'sectoral working groups', the Intergovernmental Technical Working Group on Animal Genetic Resources for Food and Agriculture and the Intergovernmental Technical Working Group on Plant Genetic Resources for Food and Agriculture to deal with specific matters in their areas of expertise, thereby abolishing the previous single Working Group on Plant Genetic Resources of the Commission.

1999 After numerous negotiating sessions, the Commission, at its 8th regular session, decides to continue negotiations for the revision of the IU in a regionally balanced intergovernmental contact group. Between 1999 and 2001 the contact group holds six meetings.

The Commission also agrees that FAO should coordinate the preparation of a country-driven report on The State of the World's Animal Genetic Resources for Food and Agriculture.

2001 In June 2001, the Commission adopts the text of the IU with some brackets remaining, and 'requested the Director-General to transmit it, through the 72nd session of the Committee on Constitutional and Legal Matters (8–9 October 2001) and the 121st session of the Council (30 October–1 November 2001), to the 31st session of the Conference (2–13 November 2001), for its consideration and approval'.[3] At its 121st session, the FAO Council establishes an open-ended working group to obtain consensus on the bracketed text.

After seven years of negotiations in the Commission, the FAO Conference adopts the International Treaty on Plant Genetic Resources for Food and Agriculture (Resolution 3/2001). This legally binding treaty covers all PGRFA. The Treaty recognizes Farmers' Rights and establishes a multilateral system to facilitate access to PGRFA, and to share the benefits derived from their use in a fair and equitable way.

2002 Between 2002 and 2006, the Commission acts as the Interim Committee for the International Treaty on Plant Genetic Resources for Food and Agriculture. The Interim Committee initiates negotiations of the standard material transfer agreement (SMTA), the Treaty's funding strategy, financial rules, rules of procedure and procedures and mechanisms to promote compliance.

2004 The International Treaty on Plant Genetic Resources for Food and Agriculture enters into force on 29 June 2004.

The Commission requests the Secretariat to prepare an analysis of the status and needs of the different sectors of genetic resources for food and agriculture, including cross-sectoral matters, with the aim to adopt a multi-year programme of work, at its 11th regular session.

2006 The First Session of the Governing Body of the International Treaty on Plant Genetic Resources for Food and Agriculture is held in Madrid, Spain.

In accordance with Article 15 of the Treaty, 11 Centres of the Consultative Group on International Agricultural Research (CGIAR) and other international collections place their ex situ gene bank collections under the Treaty. The Article 15 agreements replace the former agreements concluded between the centres and FAO in 1994.

The agreement with International Crop Diversity Trust is adopted as an (integral) part of the funding strategy of the Treaty.

The SMTA of the Treaty is adopted.

2007 FAO launches *The State of the World's Animal Genetic Resources for Food and Agriculture* developed through a participatory, country-driven process under the guidance of the Commission. The International Technical Conference on Animal Genetic Resources for Food and Agriculture, held in Interlaken, Switzerland, welcomes the report and adopts the Global Plan of Action for Animal Genetic Resources, negotiated by the Commission, and the Interlaken Declaration.

The Commission adopts its Multi-Year Programme of Work, a rolling ten-year work plan covering the totality of biodiversity for food and agriculture.

The FAO Conference welcomes the Global Plan of Action and the Interlaken Declaration as milestones in international efforts to promote the sustainable use, development and conservation of animal genetic resources. The Conference also endorses the Commission's Multi-Year Programme of Work and requests the Commission to oversee and assess the implementation of the Global Plan of Action (Resolution 12/2007).

At the Second Session of the Governing Body of the Treaty, the multilateral system becomes operational thanks to the adoption of the SMTA in 2006.

Dr Shakeel Bhatti is appointed as Secretary of the Governing Body by the Director-General of FAO on 29 January 2007, following a recommendation addressed to him by the Chairman of the Second Session of the Governing Body of the International Treaty. The Treaty Secretariat is constituted.

2009 The FAO Conference adopts Resolution 18/2009 prepared by the Commission at its 12th regular session. The resolution stresses the special nature of genetic resources for food and agriculture in the context of the negotiations of the International Regime on Access and Benefit-sharing of the CBD.

The Conference also welcomes the outcomes of the Commission's 12th regular session, including the *Second Report on the State of the World's Plant Genetic Resources for Food and Agriculture*, the Strategic Plan 2010–2017 for the implementation of the Multi-Year Programme of Work, and the funding strategy for the implementation of the Global Plan of Action for Animal Genetic Resources.

In view of preparations of the *State of the World's Forest Genetic Resources*, the Commission establishes its Intergovernmental Technical Working Group on Forest Genetic Resources.

At the Third Session of the Governing Body of the Treaty, the funding strategy is finalized and the strategic plan for the implementation of the benefit-sharing fund establishes a target of US$116 million between July 2009 and December 2014; the third party beneficiary procedures are adopted. Other resolutions are adopted, inter alia on compliance issues to continue designing its procedures and operational mechanism with the help of an ad hoc working group, relationship with the GCDT, the CGIAR and other international organizations, the work programme and budget for the 2010–2011 Biennium and on Farmers' Rights.

Eleven projects are approved by the Bureau to be funded under the benefit-sharing fund.

2011 The Fourth Session of the Governing Body will consider the adoption of the financial rules; a revised draft business plan (expected to play multiple roles, including serving as a planning, fundraising and a communication tool); an instrument to promote sustainable use; procedures and operational mechanisms to promote compliance and address issues of non-compliance. The Governing Body will also review and assess the implementation of the multilateral system and the level of payments under the SMTA.

Table A1.2 *Dates and venues of the formal negotiation meetings leading to the adoption of the Treaty within the realm of the Commission (1994–2001)*

In order to carry out its negotiating mandate, the Commission organized its first extraordinary session in November 1994. Six extraordinary sessions and six inter-sessional meetings of the contact group took place between 1994 and 2001.

Date	*Name and venue of the meetings*
7–11 November 1994	First Extraordinary Session, Rome, Italy
22–27 April 1996	Second Extraordinary Session, Rome, Italy
9–13 December 1996	Third Extraordinary Session, Rome, Italy
1–5 December 1997	Fourth Extraordinary Session, Rome, Italy
8–12 June 1998	Fifth Extraordinary Session, Rome, Italy
20–24 September 1999	First Inter-sessional Meeting of the Contact Group, Rome, Italy
3–7 April 2000	Second Inter-sessional Meeting of the Contact Group, Rome, Italy
26–31 August 2000	Third Inter-sessional Meeting of the Contact Group, Tehran, Iran
12–17 November 2000	Fourth Inter-sessional Meeting of the Contact Group, Neuchâtel, Switzerland
5–10 February 2001	Fifth Inter-sessional Meeting of the Contact Group, Rome, Italy
22–28 April 2001	Sixth Inter-sessional Meeting of the Contact Group, Spoleto, Italy
25–30 June 2001	Sixth Extraordinary Session, Rome, Italy
30 October–1 November 2001	Open-ended Working Group established by the Council on the IU, Rome, Italy
8–9 October 2001	Seventy-second Session of the Committee on Constitutional and Legal Matters, Rome, Italy
30 October–1 November 2001	Hundred and Twenty-first Session of the Council, Rome, Italy
2–13 November 2001	Thirty-first Session of the FAO Conference adopting the International Treaty, Rome, Italy

Table A1.3 *Dates and venues of the meetings of the CGRFA acting as Interim Committee (2002–2006), and the Governing Body of the International Treaty on Genetic Resources for Food and Agriculture (2006–2011)*

Date	*Name and venue of the meetings*
9–11 October 2002	First Meeting of the CGRFA acting as the Interim Committee for the International Treaty on Plant Genetic Resources for Food and Agriculture, Rome, Italy
4–8 October 2004	First Meeting of the Expert Group on the Terms of the Standard Material Transfer Agreement, Brussels, Belgium
15–19 November 2004	Second Meeting of the CGRFA acting as Interim Committee for the International Treaty on Plant Genetic Resources for Food and Agriculture, Rome, Italy
18–22 July 2005	First Meeting of the Contact Group for the Drafting of the Standard Material Transfer Agreement, Hammamet, Tunisia
14–17 December 2005	First Meeting of the Open-ended Working Group on the Rules of Procedure and the Financial Rules of the Governing Body, Compliance, and the Funding Strategy, Rome, Italy
12–16 June 2006	First Session of the Governing Body of the International Treaty on Plant Genetic Resources for Food and Agriculture, Madrid, Spain
24–28 April 2006	Second Meeting of the Contact Group for the Drafting of the Standard Material Transfer Agreement, Alnarp, Sweden
27 October–2 November 2007	Second Session of the Governing Body of the International Treaty on Plant Genetic Resources for Food and Agriculture, Rome, Italy
16 May 2008	First Meeting of the Coordination Mechanism for Capacity Building for the Implementation of the International Treaty, Bonn, Germany 16–17 October 2008
	Third Meeting of the Ad Hoc Advisory Committee on the Funding Strategy, Rome, Italy
24–25 November 2008	First Meeting of the Ad Hoc Third Party Beneficiary Committee Rome, Italy
2–3 December 2008	Second Technical Consultation on Information Technology Support for the Implementation of the Multilateral System of Access and Benefit-sharing, Rome, Italy
12–13 March 2009	Fourth Meeting of the Ad Hoc Advisory Committee on the Funding Strategy, Geneva, Switzerland
26–27 March 2009	Second Meeting of the Ad Hoc Third Party Beneficiary Committee Rome, Italy
1–5 June 2009	Third Session of the Governing Body of the International Treaty on Plant Genetic Resources for Food and Agriculture, Tunis, Tunisia

27 October 2009	Meeting of the Legal Focus Group on information technology tools to support the implementation of the Multilateral System of Access and Benefit-sharing, Rome, Italy
18–19 January 2010	First Meeting of the Ad Hoc Technical Advisory Committee on the Standard Material Transfer Agreement and the Multilateral System, Rome, Italy
2–3 February 2010	First Meeting of Ad Hoc Working Group on Compliance Rome, Italy
26–27 May 2010	Fifth Meeting of the Ad Hoc Advisory Committee on the Funding Strategy, Geneva, Switzerland
31 August–2 September 2010	Second Meeting of the Ad Hoc Technical Advisory Committee on the Standard Material Transfer Agreement and the Multilateral System, Brasilia, Brazil
7–8 October 2010	Third Meeting of the Ad Hoc Third Party Beneficiary Committee, Rome, Italy
13–15 October 2010	Sixth Meeting of the Ad Hoc Advisory Committee on the Funding Strategy, Rome, Italy
17–18 January 2011	Inter-sessional Second Meeting of Ad Hoc Working Group on Compliance, Rome, Italy
14–18 March 2011	Fourth Session of the Governing Body of the International Treaty on Plant Genetic Resources for Food and Agriculture, Bali, Indonesia

Appendix: The Global System on Plant Genetic Resources for Food and Agriculture

The development of the Global System began in 1983 with the creation of the Commission on Genetic Resources for Food and Agriculture. Its philosophy and many of its components, including *The State of the World's Plant Genetic Resources for Food and Agriculture* and the Global Plan of Action for the Conservation and Sustainable Utilization of Plant Genetic Resources for Food and Agriculture, are now part of the ITPGRFA.

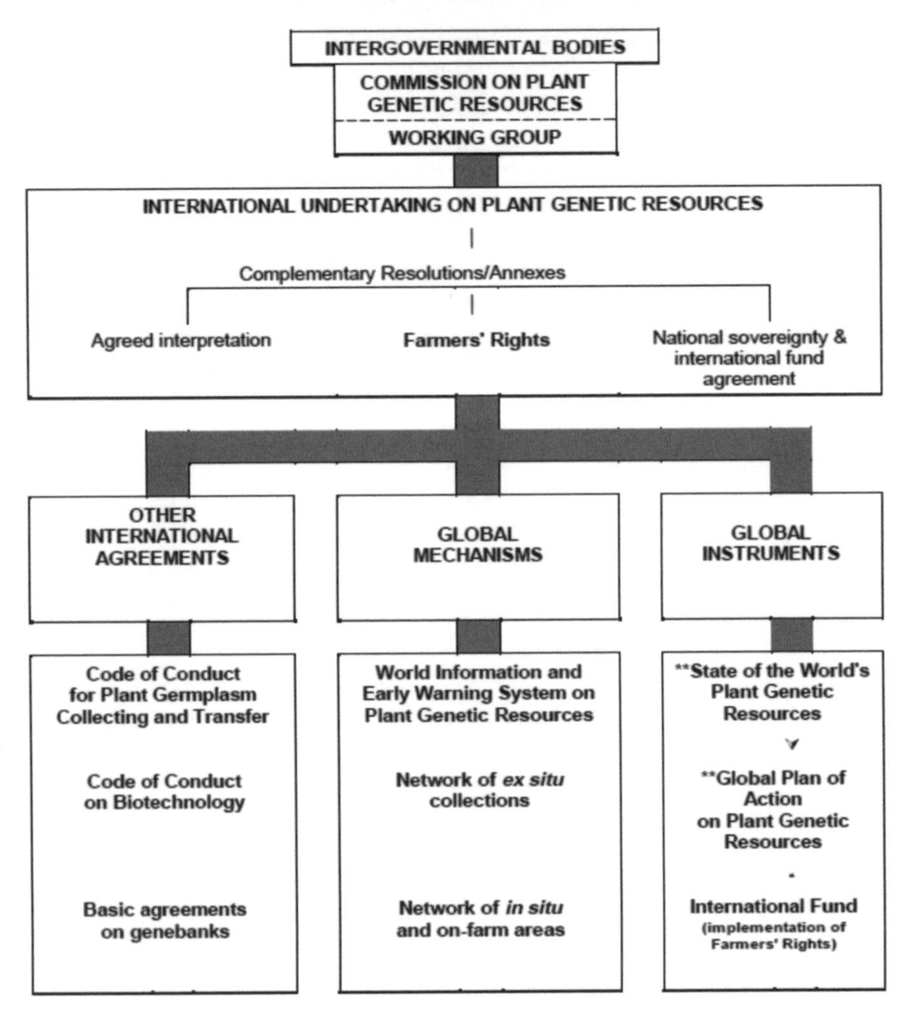

Source: Report of the Commission on Plant Genetic Resources, Sixth Session, Rome, 19–30 June 1995, CPGR-6/95/REP

Figure A1.1 *Constituting elements of the Global System*

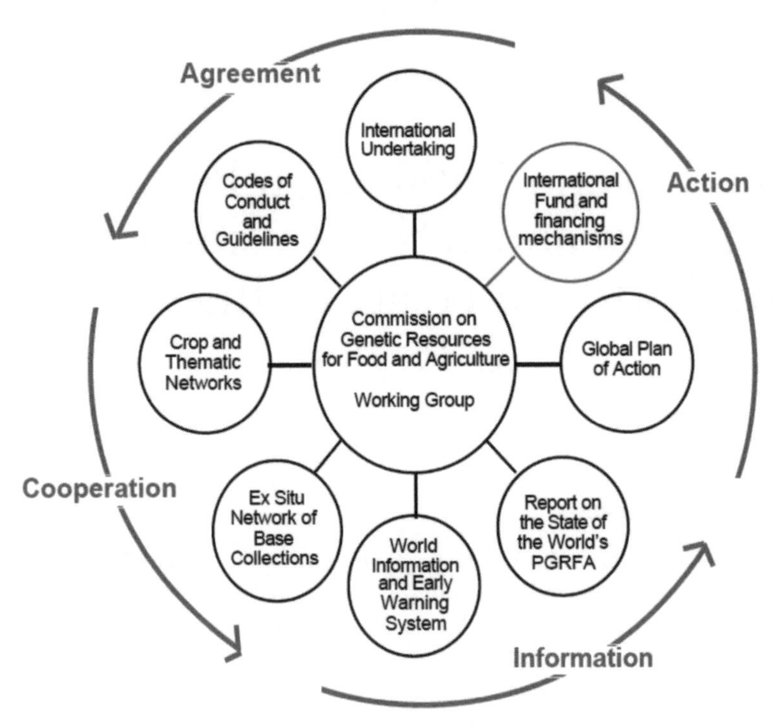

Source: Commission on Genetic Resources for Food and Agriculture, Seventh Session, Rome, 15–23 May 1997, Working Document 'Progress Report on the Global System on Plant Genetic Resources for Food and Agriculture', CGRFA-7/97/3

Figure A1.2 *Functioning scheme of the Global System*

Notes

1 The Nairobi Final Act was adopted on 22 May 1992. The Conference for the Adoption of the Agreed Text of the Convention on Biological Diversity was convened by the Executive Director of the United Nations Environment Programme (UNEP) pursuant to decision 15/34, adopted by the Governing Council of UNEP on 25 May 1989.

2 Nairobi Final Act, RESOLUTION 3 The Interrelationship Between the Convention on Biological Diversity and the Promotion of Sustainable Agriculture, §4.

3 FAO Conference, Thirty-first Session, Rome, 2–13 November 2001, document C 2001/16

Annex 2

Adoption and Ratification Process of the Treaty, and Table of Contracting Parties to the Treaty by the seven FAO Regions

Step 1: Negotiations

In 1993, the FAO Conference requests the Commission on Genetic Resources for Food and Agriculture to provide a forum for the negotiation among governments, for the Revision of the International Undertaking on Plant Genetic Resources. Negotiations take place until 2001. All members of the United Nations were allowed to participate in the negotiations. All international NGOs requesting to participate as observers were admitted in the forum. A text was designed and proposed to the FAO Conference for approval.

Step 2: Approval by the FAO Conference

The FAO Conference, at its 31st session (November 2001), through Resolution 3/2001, approved the International Treaty on Plant Genetic Resources for Food and Agriculture by consensus, with only two abstentions (the USA and Japan).

Step 3: Open for signature by States

In accordance with Article 25 of the Treaty, it was opened for signature at FAO Headquarters on 3 November 2001 and remained open for signature until 4 November 2002 by all Members of FAO and any States that are not Members of FAO but are Members of the United Nations, or any of its specialized agencies or of the International Atomic Energy Agency.

Step 4: Ratification process or equivalent

Under Article 26, the Treaty is subject to ratification, acceptance or approval[1] by the Members and non-Members of FAO referred to in Article 25. Instruments of ratification, acceptance or approval are deposited with the Director-General of

FAO. Under Article 27, the Treaty is open for accession by all Members of FAO and any States that are not Members of FAO but are Members of the United Nations, or any of its specialized agencies or of the International Atomic Energy Agency. Instruments of accession are deposited with the Director-General of FAO.

Step 5: Entry into force of the Treaty

In accordance with Article 28, the Treaty entered into force on the 90th day after the deposit of the 40th instrument of ratification, acceptance, approval or accession, provided that at least 20 instruments of ratification, acceptance, approval or accession have been deposited by Members of FAO. On 31 March 2004, 13 instruments (including the European Union) were deposited with the Director-General of FAO. Having reached the required number of instruments in order for the Treaty to enter into force, the date of entry into force is 29 June 2004.

Parties: 127

Up to 1 June 2011, 127 states were contracting parties to the Treaty. Below is a table providing the list of contracting parties to the Treaty and specifying the dates where the instruments were deposited. Another 12 countries, which have signed but not yet ratified the Treaty on 1 June 2010 are also listed.

Africa
39 Contracting Parties to the Treaty
3 Signatories but which have not yet ratified

Country	*Date of ratification or equivalent (see note)*
Algeria	13 Dec 2002
Angola	14 Mar 2006
Benin	24 Feb 2006
Burkina Faso	5 Dec 2006
Burundi	28 Apr 2006
Cameroon	19 Dec 2005
Cape Verde	Signed by the government (16 Oct 2002) but not yet ratified
Central African Republic	4 Aug 2003
Chad	14 Mar 2006
Congo	14 Sep 2004
Côte d'Ivoire	25 Jun 2003
Democratic Republic of the Congo	5 Jun 2003
Ethiopia	18 Jun 2003
Gabon	13 Nov 2006
Ghana	28 Oct 2002
Guinea	11 Jun 2002
Guinea-Bissau	1 Feb 2006

Kenya	27 May 2003
Lesotho	21 Nov 2005
Liberia	25 Nov 2005
Madagascar	13 Mar 2006
Malawi	4 Jul 2002
Mali	5 May 2005
Mauritania	11 Feb 2003
Mauritius	27 Mar 2003
Morocco	14 Jul 2006
Namibia	7 Oct 2004
Niger	27 Oct 2004
Nigeria	Signed by the government (10 Jun 2002) but not yet ratified
Rwanda	16 Oct 2006
Sao Tome and Principe	7 Apr 2006
Senegal	25 Oct 2006
Seychelles	30 May 2006
Sierra Leone	20 Nov 2002
Swaziland	Signed by the government (10 Jun 2002) but not yet ratified
Togo	23 Oct 2007
Tunisia	8 Jun 2004
Uganda	25 Mar 2003
United Republic of Tanzania	30 Apr 2004
Zambia	13 Mar 2006
Zimbabwe	5 Jul 2005

Asia

14 Contracting Parties

1 Signatory but which has not yet ratified

Country	*Date of ratification or equivalent*
Bangladesh	14 Nov 2003
Bhutan	2 Sep 2003
Cambodia	11 Jun 2002
Democratic People's Republic of Korea	16 Jul 2003
India	10 Jun 2002
Indonesia	10 Mar 2006
Lao People's Democratic Republic	14 Mar 2006
Malaysia	5 May 2003
Maldives	2 Mar 2006
Myanmar	4 Dec 2002
Nepal	19 Oct 2009
Pakistan	2 Sept 2003

Philippines	28 Sep 2006
Republic of Korea	20 Jan 2009
Thailand	Signed by the government (4 Nov 2002) but not yet ratified

Europe

35 Contracting Parties to the Treaty

3 Signatories but which have not yet ratified

Country	*Date of ratification or equivalent*
Albania	12 May 2010
Armenia	20 Mar 2007
Austria	4 Nov 2005
Belgium	2 Oct 2007
Bulgaria	29 Dec 2004
Croatia	8 May 2009
Cyprus	15 Sep 2003
Czech Republic	31 Mar 2004
Denmark	31 Mar 2004
Estonia	31 Mar 2004
European Union (Member organization)	31 Mar 2004
Finland	31 Mar 2004
France	11 Jul 2005
Germany	31 Mar 2004
Greece	31 Mar 2004
Hungary	4 Mar 2004
Iceland	7 Aug 2007
Ireland	31 Mar 2004
Italy	18 May 2004
Latvia	27 May 2004
Lithuania	21 Jun 2005
Luxembourg	31 Mar 2004
Malta	Signed by the government (10 Jun 2002) but not yet ratified
Montenegro	21 Jul 2010
Netherlands	18 Nov 2005
Norway	3 Aug 2004
Poland	7 Feb 2005
Portugal	7 Nov 2005
Romania	31 May 2005
Serbia	Signed by the government (1 Oct 2002) but not yet ratified
Slovakia	8 June 2010
Slovenia	11 Jan 2006
Spain	31 Mar 2004
Sweden	31 Mar 2004
Switzerland	22 Nov 2004

The Former Yugoslav Republic of Macedonia	Signed by the government (10 Jun 2002) but not yet ratified
Turkey	7 Jun 2007
United Kingdom	31 Mar 2004

Latin America and the Caribbean

16 Contracting Parties to the Treaty

3 Signatories but which have not yet ratified

Country	*Date of ratification or equivalent*
Brazil	22 May 2006
Colombia	Signed by the government (30 Oct 2002) but not yet ratified
Costa Rica	14 Nov 2006
Cuba	16 Sep 2004
Dominican Republic	Signed by the government (11 Jun 2002) but not yet ratified
Ecuador	7 May 2004
El Salvador	9 Jul 2003
Guatemala	1 Feb 2006
Haiti	Signed by the government (9 Nov 2001) but not yet ratified
Honduras	14 Jan 2004
Jamaica	14 Mar 2006
Nicaragua	22 Nov 2002
Panama	13 Mar 2006
Paraguay	3 Jan 2003
Peru	5 Jun 2003
Saint Lucia	16 Jul 2003
Trinidad and Tobago	27 Oct 2004
Uruguay	1 Mar 2006
Venezuela (Bolivarian Republic of)	17 May 2005

Near East

16 Contracting Parties to the Treaty

Country	*Date of ratification or equivalent*
Afghanistan	9 Nov 2006
Djibouti	8 May 2006
Egypt	31 Mar 2004
Iran (Islamic Republic of)	28 Apr 2006
Jordan	30 May 2002
Kuwait	2 Sep 2003
Kyrgyzstan	1 Jun 2009
Lebanon	6 May 2004
Libyan Arab Jamahiriya	12 Apr 2005

Oman	14 Jul 2004
Qatar	1 Jul 2008
Saudi Arabia	17 Oct 2005
Sudan	10 Jun 2002
Syrian Arab Republic	26 Aug 2003
United Arab Emirates	16 Feb 2004
Yemen	1 Mar 2006

North American Regional Group
1 Contracting Party to the Treaty
1 Signatory but which has not yet ratified

Country	*Date of ratification or equivalent*
Canada	10 Jun 2002
United States of America	Signed by the government (1 Nov 2002) but not yet ratified

South West Pacific
6 Contracting Parties to the Treaty
1 Signatory but which has not yet ratified

Country	*Date of ratification or equivalent*
Australia	12 Dec 2005
Cook Islands	2 Dec 2004
Fiji	9 Jul 2008
Kiribati	13 Dec 2005
Marshall Islands	Signed by the government (13 Jun 2002) but not yet ratified
Palau	5 Aug 2008
Samoa	9 Mar 2006

Note

1 Under international law being a contracting party to a treaty may result from the following different acts.

Where the 'signature' is subject to ratification, acceptance or approval, the signature does not establish the consent to be bound. However, it is a means of authentication and expresses the willingness of the signatory state to continue the treaty-making process. The signature qualifies the signatory state to proceed to ratification, acceptance or approval. It also creates an obligation to refrain, in good faith, from acts that would defeat the object and the purpose of the treaty [Arts.10 and 18, Vienna Convention on the Law of Treaties 1969].

'Ratification' defines the international act whereby a state indicates its consent to be bound to a treaty if the parties intended to show their consent by such an act. In the case of bilateral treaties, ratification is usually accomplished by exchanging the requisite instruments, while in the case of multilateral treaties the usual procedure is for the depositary to collect the ratifications of all states, keeping all parties informed of the situation. The institution of ratification grants states the necessary time-frame to seek the required approval for the treaty on the domestic level and to enact the necessary legislation to give domestic effect to that treaty [Arts.2 (1) (b), 14 (1) and 16, Vienna Convention on the Law of Treaties 1969].

'Adoption' is the formal act by which the form and content of a proposed treaty text are established. As a general rule, the adoption of the text of a treaty takes place through the expression of the consent of the states participating in the treaty-making process. Treaties that are negotiated within an international organization will usually be adopted by a resolution of a representative organ of the organization whose membership more or less corresponds to the potential participation in the treaty in question. A treaty can also be adopted by an international conference which has specifically been convened for setting up the treaty, by a vote of two thirds of the states present and voting, unless, by the same majority, they have decided to apply a different rule [Art.9, Vienna Convention on the Law of Treaties 1969].

The instruments of 'acceptance' or 'approval' of a treaty have the same legal effect as ratification and consequently express the consent of a state to be bound by a treaty. In the practice of certain states acceptance and approval have been used instead of ratification when, at a national level, constitutional law does not require the treaty to be ratified by the head of state [Arts.2 (1) (b) and 14 (2), Vienna Convention on the Law of Treaties 1969].

'Accession' is the act whereby a state accepts the offer or the opportunity to become a party to a treaty already negotiated and signed by other states. It has the same legal effect as ratification. Accession usually occurs after the treaty has entered into force. The Secretary-General of the United Nations, in his function as depositary, has also accepted accessions to some conventions before their entry into force. The conditions under which accession may occur and the procedure involved depend on the provisions of the treaty. A treaty might provide for the accession of all other states or for a limited and defined number of states. In the absence of such a provision, accession can only occur where the negotiating states were agreed or subsequently agree on it in the case of the state in question [Arts.2 (1) (b) and 15, Vienna Convention on the Law of Treaties 1969].

Annex 3

Overview of the Main Provisions of the International Treaty on Plant Genetic Resources for Food and Agriculture

Part	*Main provisions*
Part I – Introduction	• Article 1 sets out the objectives of the Treaty: the conservation and sustainable use of *Plant Genetic Resources for Food and Agriculture* (PGRFA) and the fair and equitable sharing of benefits arising from their use, in harmony with the Convention on Biological Diversity (CBD), for sustainable agriculture and food security. • Article 2 defines a number of key terms, such as 'genetic material', 'variety' and 'centre of origin'. • Article 3 expresses that the scope of the Treaty encom passes all PGRFA.
Part II – General Provisions on Conservation and Sustainable Use of PGRFA	• Article 4 requires Contracting Parties to ensure that national laws, regulations and procedures be in conformity with their obligations under the Treaty. • Article 5 calls for the promotion of an integrated approach to the exploration, conservation and sustainable use of PGRFA, and establishes a list of the main tasks related to the conservation, exploration, collection, characterization, evaluation and documentation of PGRFA to be complied with by Contracting Parties. • Article 6 requires Contracting Parties to develop appropriate policy and legal measures that promote the sustainable use of PGRFA, providing a non-exhaustive list of possible measures. • Articles 7 and 8 deal with national commitments, international cooperation and technical assistance. Contracting Parties are invited to provide assistance to each other for the conservation and sustainable use of PGRFA, especially to developing countries.

Part III – Farmers' Rights	• Article 9 recognizes the contribution that local and indigenous communities and farmers have made to the conservation and development of PGRFA, and encourages Contracting Parties to take measures to promote Farmers' Rights. These include the protection and promotion of (i) traditional knowledge relevant to PGRFA; (ii) the right to equitably participate in the sharing of benefits arising from the utilization of PGRFA; and (iii) the right to participate in making decisions at the national level with respect to the conservation and sustainable use of PGRFA. The responsibility for realizing Farmers' Rights rest with national governments. This article does not limit the right for farmers to save, use, exchange and sell farm-saved seed, subject to national law.
Part IV – The Multilateral System of Access and Benefit-sharing (MLS)	• Article 10 establishes the *Multilateral System of Access and Benefit-sharing* (the MLS). Contracting Parties, in the exercise of their sovereign rights over their own PGRFA, agree to grant each other facilitated access to the PGRFA they decide to include in the MLS, and to share, in a fair and equitable way, the benefits arising from the use of these resources. • Article 11 defines the coverage of the MLS. 64 food and forage crops, selected according to the criteria of interdependence among countries and their importance for food security, form part of the MLS. The list of crops is set out in *Annex I* to the Treaty. • The MLS also includes PGRFA listed in *Annex I* that are held by the CGIAR Centres or by other entities that have voluntarily included them in the MLS. • Under Article 12, the Contracting Parties agree to take the necessary measures to provide each other, as well as legal and natural persons under their jurisdiction, facilitated access to their PGRFA through the MLS. • Article 12 further states that recipients of material from the MLS must not claim Intellectual Property or other rights that limit the facilitated access to PGRFA in the form received from the MLS, including genetic parts or components thereof. Facilitated access is to be provided through the Standard Material Transfer Agreement (SMTA) of the Treaty. • Article 13 sets out the agreed terms for benefit-sharing under the MLS. Recognizing that facilitated access to PGRFA itself constitutes a major benefit of the MLS, it enumerates other mechanisms for benefit-sharing, including the information exchange, technology transfer, capacity building, and the sharing of commercial benefits.

Part V – Supporting Components	• These are activities outside the institutional structure of the Treaty that provide essential support to the achievement of its objectives. They include promoting the effective implementation of the rolling Global Plan of Action (Article 14), the encouragement of international plant genetic resources networks (Article 16), and the development of a global information system on PGRFA, including a periodic assessment of the state of the world's PGRFA (Article 17). • Article 15 deals with *ex-situ* collections of PGRFA held by the CGIAR Centres and other international institutions. The Treaty calls on the CGIAR Centres to sign agreements with the Governing Body to bring their collections under the Treaty. PGRFA listed in *Annex I* that are held by the CGIAR Centres are to be made available as part of the MLS.
Part VI – Financial Provisions	• In Article 18, Parties agree to implement a *Funding Strategy* (FS) to enhance the availability, transparency, efficiency and effectiveness of the provision of financial resources for the implementation of the Treaty. It includes the financial benefits arising from the commercialization of PGRFA under the MLS, as well as funds made available through other international mechanisms, funds and bodies, and voluntary contributions from Contracting Parties, the private sector, NGOs and others.
Part VII – Institutional Provisions	• Article 19 establishes a Governing Body composed of all Contracting Parties. This Governing Body acts as the supreme body for the Treaty and provides policy direction and guidance for the implementation of the Treaty and, in particular, the MLS. All decisions of the Governing Body are to be taken by consensus, although it is empowered to agree by consensus on another method of decision making for all matters other than amendments to the Treaty and to its annexes. The Governing Body is expected to maintain regular communication with other international organizations, especially the Convention on Biological Diversity, to reinforce institutional cooperation on issues related to genetic resources. • The Treaty also provides for the appointment of a Secretary of the Governing Body (Article 20). • Article 21 requires the Governing Body to approve procedures and mechanisms to promote compliance with the Treaty. • Settlement of disputes is covered by Article 22, which also provides for a third party to mediate, in the case Parties cannot reach an agreement by negotiation.

	• Articles 23–35 deal with amendments, annexes, signature, ratification, acceptance or approval, accession to and entry into force of the Treaty, relations with others, and provisions for withdrawals from or termination of the Treaty.
Annexes	• *Annex I* lists the crops covered under the MLS, while *Annex II* deals with arbitration and conciliation.

This table comes from the Secretariat of the Treaty. It has been elaborated for illustrative purposes and it is not meant as a substitute for professional legal advice. For more information contact:

International Treaty on Plant Genetic Resources for Food and Agriculture
Viale delle Terme di Caracalla – 00153 Rome – Italy
Tel: +390657053554 Fax: +390657056347
Email: pgrfa-treaty@fao.org
www.planttreaty.org

Annex 4

Country Case-study: Brazil – Actions and Reactions to the International Treaty on Plant Genetic Resources for Food and Agriculture

Lidio Coradin and Maria José Amstalden Sampaio

During the last millennia, farmers have domesticated plant wild varieties and through breeding and selection, made these plants viable for agriculture. The enormous development of global agriculture always relied on the work of farmers (see Chapter 13 on Farmers' Communities) and more recently on the breeding skills of modern breeders (see Chapter 15 on the plant breeders) and hence on the continuous supply of genetic variability found in germplasm samples. More and more, the genetic variability has proven fundamental to enable humankind to confront new threats such as climatic changes.

According to the Food and Agriculture Organization of the United Nations (FAO), since the beginning of the agricultural history humankind has already used more than 10,000 plant species for feeding. However, today's food is based on 150 species only, and only about 12 species provide more than 80 per cent of the food calories consumed by humans. In fact, only four species (corn, wheat, rice and potato) provide more than half of the required calories (FAO, 2008). Nevertheless, local crops add to the food consumed by millions every day and help to improve their nutrition.

As described in this annex, FAO member countries went to the extent of developing a specific Treaty – the International Treaty for Plant Genetic Resources for Food and Agriculture (ITPGRFA) to provide guidance and awareness about the need for conservation and permanent exchange and research with genetic resources for food and agriculture, not forgetting the need to share benefits and financial help with those farmers that have been developing and conserving these resources for generations.

The central pillar of the Treaty is the multilateral system of access and benefit-sharing (MLS), designed to provide facilitated access with pre-established, mutually agreed benefit-sharing provisions, in complementary bases and for mutual benefit. The reasoning behind this is that breeding programmes developed around the world need a constant flow of genetic material from different parts of the globe, as no country is entirely self-sufficient when looking at genetic resources for food and agriculture. Interdependency is a real fact and therefore parties to the Treaty recognize that the MLS is an enormous benefit for breeders and farmers. It is, however, worth clarifying that access to genetic resources, to be found in in situ conditions, must be acquired according to national legislation or, in its absence, according to rules to be established by the Governing Body of the Treaty. In the case of Brazil, who ratified the Treaty in 2006, the rules for in situ acquisition of genetic material are established by Law (Provisional Measure n° 2.186-16, 2001), in harmony with the CBD.

Taking note that the present Annex I of the Treaty includes only 64 crops representing 52 genera and 29 forage genera (www.planttreaty.org), which were defined basically in accordance with criteria related to (i) their importance for the production of food at global level and (ii) interdependency among nations regarding their utilization for food and agriculture, questions remain for those who were not so involved with the negotiations of the agreement as to why some other important crops for food and agriculture were not included and which are the rules for accession to those genetic materials?

The definition of the crops that were listed as the Annex I crops, required very skillful negotiations by countries' representatives. During the many meetings, countries had the opportunity to add or extract any species from the list (Moore and Tymowski, 2005). For Brazil, the final listing requires that the country provides genetic resources of cassava (*Manihot esculenta*), local varieties of rice (*Oryza sativa*), beans (*Phaseolus vulgaris*), maize (*Zea mays*) and sweet potato (*Ipomoea batatas*), as well as wild species of *Oryza* (*O. alta*, *O glumepatula*, *O. grandiglumis* and *O. latifolia*), *Solanum* (*S. calvescens*, *S. chacoense* and *S. commersonii*) and *Dioscorea* (*D. altissima*, *D. dodecaneura* and *D. trifida*).

Genetic material of peanuts (*Arachis* spp.), initially included in the list, was later removed, together with some other crops, such as soybean. Cassava is a crop that has an enormous social value as it is used in most countries as a staple component of the diet, mostly in poor regions of the globe. Brazil, as a supplier of this germplasm, can promote an important impact in Latin America, Asia and Africa. On the other side of the coin, thinking of the food security of the Brazilian people, it is important for the country to access genetic resources of rice, banana, potato, carrot, citrus, coconut, peas, beans, barley, cowpea, sunflower, apple, maize, sorghum, wheat, strawberry and some of the forage species.

The MLS is therefore a unique opportunity for Brazil to increase the genetic variability of its gene banks and use the material in breeding programmes, already well known for its excellent outputs in tropical agriculture. It is important that at the same time, Brazil makes a continuous effort to guarantee the equitable sharing of benefits derived from the use of those materials, to promote further their conservation, especially among farmers. According to de Jonge and Korthals (2006), the benefit sharing will not solve the world's hunger problem but it would be a mechanism to stimulate development and the distribution of basic needs that can contribute to social justice.

It is true that some other crops of global importance for food and agriculture were not included in the Annex I for lack of consensus. As a matter of fact, several other crops, even though agreed by many regions, were not included in the list. With the decision of not including soybean, some regions (Africa, Asia, Latin America and the Caribbean) decided to step back and removed some species already included in the list. These crops, some of major importance for Brazil, are: garlic and onion, peanut, oil palm, tomato, sugar-cane, minor millets, olive, pear, vine, fruit trees (*Prunus*), melon and cucumber, pumpkins and squashes and flax (see the first part of Chapter 6 by Modesto Fernandez). New solutions must be found for the exchange of these genetic materials, using the same collaborative spirit of the Treaty, in bilateral agreements which will have to consider national legislations, case by case. With time and hopefully with the success of the implementation of the MLS, the Annex I list could be increased, especially to incorporate the list of crops mentioned above which are also considered of primary importance. However, that can only be done by the consensus of all parties present at the meetings of the Governing Body of the Treaty. This was one of the contributions of Brazil to the ruling of the Treaty, because it worried that the Treaty should not have such an ample scope as to jeopardize the CBD. By ensuring that decisions to change the Treaty or its Annexes cannot be taken unless consensus is reached, Brazil wanted to guarantee equal opportunity for every country to have a voice, and therefore a better chance for total transparency in the decisions of the Governing Body. Also, because the parties of the Treaty are, in their great majority, parties of the CBD, an adequate balance should be present in the exercise of consensus.

Regarding the practical implementation at national level, the scope of the Treaty vis a vis that of the CBD still causes some discussions among policy makers. Questions refer mostly to non-Annex I plant genetic resources. Parties to the CBD (mostly the same as to the Treaty, as said above) have been discussing the text of a new binding protocol on access to genetic resources, associated traditional knowledge and benefit sharing. Due to its specific characteristics and problems, serious discussions are taking place on whether the genetic resources of primary importance used for food and agriculture, not only plants, but also domestic animals, microorganisms and aquatic species should receive a treatment similar to that, given to Annex I crops of the ITPGRFA.

There are legal constraints that must be resolved because under the CBD scope, there must be a guarantee of fair and equitable sharing of benefits arising from the use of all genetic resources. The Treaty only provides rules on how to deal with the issue of benefit sharing for the Annex I crops. Therefore, innovative solutions will have to be found during the implementation phase of the newly approved Nagoya–Cali Protocol on Access and Benefit Sharing (CBD, 2010).

Although relatively slow, national implementation of the Treaty is moving ahead. Regarding exchange of germplasm and the use of the MLS, Brazil has good collections of germplasm obtained from several sources and which were internalized during the 1980s, and therefore the entrance into force of the Treaty and the opportunities presented by the MLS have not yet raised much interest of breeders and research institutions. It could also be because Brazil still does not have a good and rapid quarantine service and delays in the introduction of material are often discouraging. The removal of this bottleneck in two years (new laboratories are being built) will probably boost the germplasm exchange, hence the impact of the Treaty and its MLS. Another boost will come from the perception that new genetic material, from regions that already face climate extremes, will be required by breeders devoting attention to these new challenges for the tropical agriculture.

Nonetheless, a major effort should be developed by countries, including Brazil, for the realization of Farmers' Rights, with the development of specific national policies. Informal discussions which took place in 2009/2010 have shown that it will not be a simple task to implement such policies because of the many stakeholders involved and the different views and concerns expressed by each group of participants. Brazil will continue to make its best efforts to discuss and implement these rights.

A more positive impact of the ratification of the Treaty by Brazil has been the need to provide better information about the accessions to be included in the MLS. This new responsibility has prompted the Brazilian Agriculture Research Corporation (Embrapa), holder of most of the public gene banks, to review its passport data and improve the characterization reports. The process has been relatively slow but with the necessary political will, the first results should be available in 2011. The new information system will be qualified to link with the new Germplasm Accession Portal Genesys (www.genesys-pgr.org), jointly funded by the CGIAR, the Global Crop Diversity Trust and the Treaty's Secretariat, to be also launched in 2011.

During recent years, several activities have been carried out in Brazil by the Ministry of the Environment in partnership with Embrapa and the National Institute for Amazonian Research (INPA) to make a complete inventory of landraces and wild relatives of some of the main crops cultivated in Brazil. These efforts encompass crops listed as Annex I crops, as the case of cassava, maize and rice, and non-Annex I, as it was the case of cotton, peanut, peach palm and pumpkin and squashes. These inventories will continue to cover other crops and their related gene pool, especially peppers, pineapple, passion fruit, beans, sweet potato and cashew. Activities include: (i) the definition of local landraces and wild relatives of each crop; (ii) mapping their geographical distribution; (iii) in situ, ex situ and on-farm conservation status; and (iv) major needs for the maintenance of landraces and wild relatives of each crop.

Another major effort to implement the Treaty is being launched in Brazil by the Ministry of the Environment in partnership with Embrapa, Federal Universities and non-governmental organizations for the identification of native plant species of actual or potential economic value used at local or regional level, also known as Plants for the Future. The main goal of this project was to prioritize potential species, including food species, and promote their sustainable use to: (i) identify new options for direct use by family farming; (ii) broaden the opportunities for industry investment on the development of new products; (iii) evaluate the degree of use of, and the existing gaps in, the scientific knowledge; (iv) increase food security and contribute to minimizing the vulnerability of the Brazilian food system; and (v) to develop partnerships towards the characterization of the nutritional value of these native plant species. Some of the activities developed on this initiative include: (i) inventory of native plant species, commercially sub-utilized, with emphasis on their potential for social, environmental and cultural benefits; (ii) survey of scientific literature to evaluate the state of technical and scientific knowledge regarding the species considered on the inventory; (iii) definition of priority species, taking into account the opening of markets for new products at local, regional, national and international levels; and (iv) integration of all different sectors as a challenge and as a way out for opening new markets to promote the utilization of local food species.

In parallel, with the support of Brazil's Foreign Affairs Ministry, Embrapa has been increasing its presence in Africa and other developing countries in Latin America in the last four to five years, through technology transfer and capacity-building projects in the agriculture sector. These include the transfer of improved genetic material of Annex I crops (mostly in the form of commercial varieties). Training includes field trials and the appropriate use of the necessary inputs for better production and yield. Therefore, Brazil has been implementing the Treaty with its actions abroad, as they relate to one of its major objectives: the sharing of benefits derived from research and the use of plant genetic resources.

A practical ongoing example is the programme called Africa-Brazil Agricultural Innovation Marketplace (www.africa-brazil.org). The Africa-Brazil Agricultural Innovation Marketplace aims to benefit smallholder producers, by enabling innovation through collaborative partnerships between Africa and Brazil and is supported by many partners such as the African national and sub-regional agricultural research and development organizations such as: the Forum for Agricultural Research in Africa (FARA), Embrapa, the United Kingdom Department for International Development (DfID), the International Fund for Agricultural Development (IFAD) and the the World Bank (WB). In 2010, 61 pre-proposals were found to be eligible and 20 of those were invited to be developed to full proposals which should be funded in 2011. All projects involve one Embrapa Research Center as national counterpart. African countries involved in this first call are: Madagascar, Uganda, Mozambique, South Africa, Kenya, Nigeria, Ethiopia, Togo, Tanzania, Burkina Faso and Ghana.

Conclusion

As seen in this short summary of the Treaty implications for Brazil, the easier part seems to have been its extensive negotiation. Now that some years have passed since its ratification, stakeholders are beginning to take stock of the need for action and implementation, as the issue of food security is receiving new attention due to food crises (high international prices), global availability versus demand, and mostly because of a better awareness about the impact of climate changes on the planet.

National policies are been reviewed to include more attention to genetic resources conservation, sustainable use and benefit sharing, as many new incentives are being discussed to advance Brazil's knowledge of its agricultural biodiversity, its implications for environmental services and the sustainability of the agricultural systems. The discussions regarding the Brazilian Government decision in 2009 to help with the reduction in greenhouse gases to mitigate global warming and the need to develop research to help farmers to adapt to the changing climate, have brought the issue of genetic resources for food and agriculture to the scientific and political screens. 2011–2012 should be special years to prepare and implement incentive policies, in preparation for the Rio plus 20 Conference on Sustainable Development, which will again take place in Rio de Janeiro, Brazil, in 2012.

References

CBD (2010) Tenth ordinary meeting of the Conference of the Parties at the Convention on Biological Diversity (UNEP/CBD.COP/10), available at www.cbd.int/decisions/cop/?m=cop-10

FAO (2008) *The International Treaty on Plant Genetic Resources for Food and Agriculture: Equity and Food for All*, Food and Agriculture Organization of the United Nations, Rome, Italy

De Jonge, B. and Korthals, M. (2006) 'Vicissitudes of benefit sharing of crop genetic resources: Downstream and upstream', *Developing World Bioethics*, vol 6, no 3, pp144–157

Moore, G. and Tymowsky, T. (2005) 'Explanatory guide to the International Treaty on Plant Genetic Resources for Food and Agriculture', Environmental Policy and Law Paper No. 57, International Union for Conservation of Nature, Gland, Switzerland and Cambridge, UK

Annex 5

Some Stories on the Inception of the International Treaty on Plant Genetic Resources for Food and Agriculture

José Ramón López Pontillo and Franciso Martínez Gómez

Available online at *ftp://ftp.fao.org/ag/agp/planttreaty/articles/pgrfs_annex_5.pdf* and at *www.bioversityinternational.org/publications.html*

Index

Note: page numbers followed by 'n' refer to end of chapter notes; 'Treaty' is used in index entries to refer to ITPGRFA